GAME THEORY IN BIOLOGY

Oxford Series in Ecology and Evolution

Edited by Paul H. Harvey, Robert M. May, H. Charles J. Godfray, and Jennifer A. Dunne

Game Theory in Biology

Concepts and Frontiers

JOHN M. McNAMARA
School of Mathematics, University of Bristol, UK

OLOF LEIMAR
Department of Zoology, Stockholm University, Sweden

OXFORD
UNIVERSITY PRESS

Great Clarendon Street, Oxford, OX2 6DP,
United Kingdom

Oxford University Press is a department of the University of Oxford.
It furthers the University's objective of excellence in research, scholarship,
and education by publishing worldwide. Oxford is a registered trade mark of
Oxford University Press in the UK and in certain other countries

© John M. McNamara and Olof Leimar 2020

The moral rights of the authors have been asserted

First Edition published in 2020
Impression: 1

Published in the United States of America by Oxford University Press
198 Madison Avenue, New York, NY 10016, United States of America

British Library Cataloguing in Publication Data
Data available

Library of Congress Control Number: 2020938654

ISBN 978-0-19-881577-8 (hbk.)
ISBN 978-0-19-881578-5 (pbk.)

DOI: 10.1093/oso/9780198815778.001.0001

Printed and bound by
CPI Group (UK) Ltd, Croydon, CR0 4YY

Links to third party websites are provided by Oxford in good faith and
for information only. Oxford disclaims any responsibility for the materials
contained in any third party website referenced in this work.

Contents

Acknowledgements

We started this book 3 years ago. During that time the book has evolved and improved, helped enormously by the feedback given to us by the following friends and colleagues: Redouan Bshary, Sean Collins, Sasha Dall, Tim Fawcett, Lutz Fromhage, Andy Higginson, Alasdair Houston, Patrick Kennedy, Eva Kisdi, Sean Rands, Brian Robathan, Susanne Schindler, Harry Suter, Pete Trimmer, and Chris Wallace. We are very grateful for their efforts. Thank you!

Part of the modelling work presented in this book was supported by the Swedish Research Council (grant 2018-03772).

1

Setting the Scene

1.1 Introduction

There is an extraordinarily diverse range of traits and behaviours in the natural world. Birds sing, flock, and migrate, male peacocks display extravagant tails, fish change sex, and ants 'farm' aphids. When faced with such phenomena it is natural to ask why. The term 'why' can have different meanings. In this book we are concerned with explanations in terms of the past action of natural selection. The action of natural selection tends to favour traits and behaviours that enhance the number of offspring left by an individual, i.e. it produces adaptations to the environment in which organisms live. For example, the aerodynamic shape of a bird's wing can be seen as an adaptation to increase flight efficiency. Many explanations in biology are similarly in terms of adaptation to the abiotic environment.

However, it is not just the abiotic environment that matters to organisms, but often equally or even more the characteristics of other organisms. The fitness of one individual, involving survival and reproductive success, often depends on how other population members behave, grow, and develop, i.e. depends on the strategies of others. Thus, the best strategy of one individual can depend on the strategies of others. When this is true, we say that the action of natural selection is frequency dependent. For example, oak trees in a wood shade each other. If surrounding trees are tall, then a tree must also grow tall to get more light. This growth requires the tree to divert resources that could otherwise be expended on seed production. The result is that in a wood it is best for an oak to be tall if other oaks are tall, but shorter if they are shorter. The shape of the crown is also predicted to be influenced by this competition (Iwasa et al., 1985).

Biological game theory attempts to model such frequency-dependent situations, and this theory is the subject of this book. The approach of the models is to predict the endpoints of the process of evolution by natural selection, rather than to describe the details of evolutionary trajectories. The overall objective is to understand the selective forces that have shaped observed behavioural and developmental strategies and make predictions that can be tested empirically.

The theory is concerned with the observable characteristics of organisms. These characteristics include both behaviour and morphological and physiological attributes, and are referred to as the phenotype of the organism. We do not try to model the details of the genetics that underlie the phenotype. Instead we usually

Game Theory in Biology: Concepts and Frontiers. John M. McNamara and Olof Leimar,
Oxford University Press (2020). © John M. McNamara and Olof Leimar (2020).
DOI: 10.1093/oso/9780198815778.003.0001

assume that the genetic system is capable of of producing strategies that do well in their natural environment; i.e. producing phenotypes that are adapted to the environment. This allows the theory to characterize evolutionary outcomes in phenotypic terms. The explanation of tree height in terms of trees shading each other can be understood in phenotypic terms, even though we might be ignorant of the details of the underlying genetics. This 'phenotypic gambit' is further explained in Section 2.2.

This book aims to present the central concepts and modelling approaches in biological game theory. It relates to applications in that it focuses on concepts that have been important for biologists in their attempts to explain observations. This connection between concepts and applications is a recurrent theme throughout the book. The book also aims to highlight the limitations of current models and to signpost directions that we believe are important to develop in the future.

Game theory is applied to a huge range of topics in biology. In Chapter 3 many of the standard games are described, with others being introduced in later chapters. We do not, however, attempt an encyclopedic account of the applications of game theory in biology; our focus is on game-theoretical concepts. Nevertheless, we illustrate these concepts in a range of situations in which the biology is important, incorporating central topics in life-history theory, ecology, and general evolutionary biology along the way.

1.2 Frequency Dependence

Examples of frequency dependence abound in biology. Here we list several to give the reader some sense of the diverse situations in which frequency dependence occurs.

Contesting a resource. When two individuals are in competition for the same food item or mate each might adopt a range of levels of aggression towards their competitor. The optimal level of aggression will depend on how aggressive an opponent is liable to be. If the opponent will fight to the death it is probably best not to be aggressive, while if the opponent is liable to run away if attacked it is best to be aggressive and attack.

Alternative mating tactics. In many species, different males in a population employ different mating tactics. In one widespread pattern some males attempt to defend territories and attract females to their territories, while other males attempt to sneak matings with females that enter these territories. This happens in Coho salmon, where the larger 'hook' males defend territories while the smaller 'jacks' attempt to sneak matings (Gross, 1996). But is it best to be a hook or a jack? It seems plausible that if almost all males are hooks and hold territories, then it is best to be a jack since there will be lots of opportunities to sneak. In contrast, if there are many jacks per hook then sneakers will be in competition with each other and so it is best to be a hook.

Search for resources. If members of a population are searching for suitable nest sites, individuals should be choosy and reject poor sites. However, they cannot afford to be too choosy, since if they delay too long the better sites will have been taken by others. But how quickly such sites disappear depends on the choice strategy of others. Thus the best choice strategy of one individual depends on the strategies employed by other population members.

Parental effort. Consider two parent birds that are caring for their common young. How much effort should each expend on provisioning the young with food? The more food the young receive the greater their survival prospects, although there are diminishing returns for additional food at high provisioning rates. However, parental effort is typically costly to the parent expending the effort. There are various reasons for this. Increased foraging effort might increase the time that the bird exposes itself to predation risk, and so increase its probability of dying. Foraging might tend to exhaust the bird, leading to an increased probability of disease and hence death overwinter. It might also reduce the time available for other activities such as seeking other mates. In short, each parent faces a trade-off: increased effort provisioning the current brood increases its reproductive success from this brood but reduces its reproductive success in the future. Typically the benefits of increased effort have diminishing returns but the costs (such as the probability of mortality) accelerate with effort. Faced with this trade-off, if a partner expends low effort on parental care it is best to expend high effort, as has been found for several species of birds (Sanz et al., 2000). In contrast, if the partner expends high effort, the young are already doing reasonably well and it is best to expend low effort to reduce costs. In other words, parental effort is frequency dependent.

Joint ventures. In the above example there is a common good (the young) but individual costs. Except for the case of lifelong monogamy, this means that there is a conflict of interest between the parents—each would prefer the other to expend the high effort so that they benefit themselves but do not pay the cost. This conflict of interest applies to many joint ventures. For instance, in a group of animals foraging under predation risk, vigilance benefits the whole group, but is costly to vigilant group members, since their feeding rate is reduced. Cooperative hunting is another example, where all group members benefit from prey capture, but the costs and risks of capturing and subduing prey go to the most active hunters.

Resource specialization. The beak size of a seed-eating bird is often best at exploiting seeds of a certain size. In an environment in which there is a range of seed sizes, if most individuals are specializing in one seed size the competition for these seeds will reduce their availability, so that it is better to have a different beak size and specialize in seeds of a different size. Under these circumstances there may be selection to evolve a range of beak sizes (cf. the evolutionary branching example of Section 4.3).

Timing of migration. For migratory birds, arriving early at the breeding grounds is advantageous in terms of getting the best nest sites, but it also involves risks. For American redstarts, for instance, male birds that arrived early at the breeding grounds in northern Michigan gained higher-quality territories, and early females produced earlier and heavier nestlings, with higher chances of survival to adulthood (Smith and Moore, 2005). A typical outcome of this competition for breeding sites for migratory birds is that individuals in good condition, who are better able to withstand the risks of early migration, tend to arrive earlier (Kokko, 1999).

Sex allocation. If a female can allocate limited resources between producing sons or producing daughters, her optimal allocation depends on the allocation strategies of other females in the population. At the heart of this dependence is the fact that every offspring produced has exactly one mother and one father. This implies that the total reproductive success of all males in a population must equal the total success of all females. Thus in a population in which other females are producing mostly sons, there will be more males than females and each son will, on average, have lower reproductive success than each daughter. It is then best for our focal female to produce daughters. Conversely, if the allocation strategy of other females results in a population that is female biased, it is best to produce sons (Section 3.8).

This example is of interest because it raises the question of what is fitness. Note that we are measuring the reproductive success of a female in terms of the numbers of matings of her offspring rather than her number of offspring (see Sections 3.8 and 10.4).

Producers versus scroungers. There are many gregarious bird species where flocks of birds search for food. In a flock, some individuals specialize in searching for and finding food, while others watch the searchers and move in to exploit any food source found. Thus a flock consists of 'producers' and 'scroungers'. Figure 1.1 shows

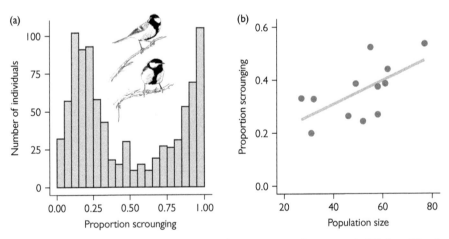

Fig. 1.1 Data from a producer–scrounger field experiment with great tits in Wytham Woods near Oxford (Aplin and Morand-Ferron, 2017a). Visits as producer or scrounger at puzzle-box feeders (Aplin et al., 2015) were recorded for individual birds. Because of frequency dependence, a mixture of producing and scrounging was seen, also for individual birds, but they tended to specialize as either producer or scrounger (a). There was more scrounging in bigger replicate sub-populations (b), which is consistent with the idea that scrounging is more rewarding in high-density populations. Illustration of great tits by Jos Zwarts. Published under the Creative Commons Attribution-Share Alike 4.0 International license (CC BY-SA 4.0).

data from a field experiment recording producer–scrounger behaviours in great tits. In many other situations certain individuals invest in something that becomes a common good of the group, while others exploit the fruits of these investments. For example, females in some bird species deposit their eggs in the nest of other females, so gaining the benefit of the investment of other females in nest building and care. As we outline in Section 3.1, bacteria can acquire iron by producing and releasing siderophores. Siderophores bind to environmental iron and can then be captured by any local bacteria that are present as well as the original releasing bacterium. Siderophore production is thus an investment in a common good that can be exploited by 'free riders' that just reap the benefits. There are strong analogies with many human activities in this context (Giraldeau et al., 2017). Exploiting the labour of others is liable to be very profitable when exploiters are rare, but will be less advantageous as a strategy when producers are rare (cf. the example of alternative mating tactics in salmon).

Cooperative hunting. Many social animals hunt in groups or packs, potentially being more efficient at capturing prey that are bigger and harder to subdue. Wolves, lions, and African wild dogs are among the species using this method. Cooperative hunting only works if several group members take part in it, and the rewards should go preferentially to active hunters for the behaviour to be favoured. If this is so it is referred to as a synergistic effect. In general, there is synergism if investing in an activity or a project is only worthwhile if others also invest and the benefits go primarily to those that invest. Synergism can be an important explanation for cooperation.

Warning colouration. Certain organisms are potential prey but are actually unsuitable as prey, for instance because they contain poisonous substances or they are otherwise dangerous or unpleasant to attack. Sometimes they have evolved a striking appearance that advertises their quality as unsuitable prey. The black and yellow stripes of stinging wasps and the striking colouration of monarch butterflies, which contain poisonous cardenolides, are examples of this. Predators learn through experience to avoid attacking such prey, which means that the warning signal is only helpful to prey if it is sufficiently common. There is thus synergism through positive frequency dependence for warning colouration.

Some of the above situations, such as the contest over a resource and parental effort, are concerned with individuals interacting pairwise. In analysing these situations one is dealing with a two-player game. At the other extreme, in the sex allocation example a female is not directly interacting with another female. Instead, her optimal allocation to daughters versus sons only depends on the ratio of males to females in the population, and this ratio is a product of the range of allocation strategies of all other females. The female is then 'playing the field', and the resultant game is referred to as a playing-the-field game. The situation of trees in a wood is somewhat intermediate, since a given tree is mostly affected by its local neighbours.

Despite the above seemingly simple classification, we will argue that pairwise competition over a resource is often not really a simple two-player game. Again consider two individuals contesting a food item. The optimal level of aggression of one individual not only depends on the aggressiveness of the opponent, but also depends on how easy it will be to access other sources of food. If other sources could also be contested, then the availability of alternative food depends on how other population members contest these items. One cannot then treat a single contest for food in isolation, and it is not really a two-player game (Section 9.4).

In most of the above cases that involve a continuous trait the best trait value of any given single individual decreases as the trait value in the rest of the population increases. We refer to this as negative frequency dependence. For a discrete trait such as in the jack versus hook example, a given strategy becomes less profitable as it becomes more common, and we also refer to this as negative frequency dependence. In contrast, there is positive frequency dependence in the example of trees in a wood, as the mean height of trees increases so does the best height for any individual tree. The example of warning colouration also has positive frequency dependence, as do some forms of group hunting.

1.3 The Modelling Approach

Evolutionary game theory aims to predict the evolutionary outcome in situations where frequency dependence exists. In particular it asks whether there is some final endpoint of the evolutionary process, and if so what that endpoint will be. For example, when two parents are caring for their common young, do we predict stable levels of care to eventually evolve? If so, at this stable endpoint is there equal care by both parents or mainly care by one sex? Will the conflict of interest between the parents over care be resolved in a way that results in high levels of care, or will there be low levels with the young suffering as each parent attempts to get its partner to do the bulk of the caring? In the case of sex allocation do we expect an equal number of males and females in populations that have reached an evolutionary endpoint? Also, can we predict how the sex of offspring might depend on local environmental conditions— for example can we explain why in some parasitic wasps females deposit sons in small hosts and daughters in large hosts? We deal with this last question in Section 3.11.

Our focus is on what evolves. By 'what evolves' we mean the strategy that evolves. We give a formal definition of strategies in Section 2.1, but we can think of a strategy as a rule that specifies how behaviour and other aspects of the phenotype depend on circumstances. For example, the growth strategy of a tree might be something as simple as to always grow to a given size, but could be more complex, specifying how the allocation of resources to growth depends on light levels and nutrient intake.

As we have explained, although the book is about evolutionary processes, we are mainly concerned with evolutionary endpoints. A formal specification of exactly what is meant by an endpoint is given in Section 2.4 and Chapter 4. Loosely, we can think of an endpoint as having the following property. A strategy will be referred to as the resident strategy when almost all population members follow this strategy. Then a necessary condition for a strategy x^* to be a stable endpoint of the process of evolution by natural selection is that no single individual with a different strategy can do better than x^*, when x^* is the resident strategy. We will need to specify exactly what we mean by 'do better'. In game theory in biology, invasion fitness provides a suitable performance measure. This concept is outlined in Section 2.3. For most of the situations analysed in this book, the invasion fitness of a strategy is just the mean number of surviving offspring left by an individual following the strategy. The simple measure of mean number of offspring is not adequate when not all offspring are the same, because the ability of these offspring to leave offspring themselves can then vary with their state. We defer the detailed presentation of this more complex case until Chapter 10.

If x^* is the resident strategy and no other strategy has greater invasion fitness than the resident strategy, we refer to x^* as a Nash equilibrium (Section 2.4). The Nash equilibrium condition is necessary but not sufficient for stability (Chapter 4). The typical approach to modelling a specific situation is first to specify the circumstances in detail, for example the actions that are possible, the time order of events, and so on. We then specify which strategies are possible. For the simplest models, for which an analytic approach is feasible, we can then evaluate how the invasion fitness of each strategy depends on the resident strategy in the population. This allows us to find all Nash equilibria. We then investigate further stability properties of each Nash equilibrium.

Often it is cumbersome to deal directly with invasion fitness and instead we work with a simpler performance measure (such as the rate of energy intake in some situations), which we know leads to the same Nash equilibria. The use of such fitness proxies is described in Section 2.5.

For many models an analytic approach is infeasible. One option is then to resort to evolutionary simulations. In a simulation a large population of virtual organisms interact, produce offspring, and die, and the offspring inherit the traits from their parent(s) with the occasional mutation producing a different trait. The simulation then follows the population forward over many generations until traits settle down to roughly equilibrium values, which are usually close to the values at a Nash equilibrium.

1.4 Scope of the Field and Challenges

To appreciate what game theory has achieved in biology, and to understand the challenges that still remain, some insights into the history of the field are helpful. Ideas with a game-theoretical flavour were developed already in the early 18th century by mathematicians interested in probability (see Bellhouse and Fillion, 2015), including by de Montmort and Nicolaus Bernoulli. They were interested in a card game called 'Le Her'. The crucial point was that for certain positions in the game it appeared that there was no single best thing for a player to do. Rather the player should randomize between different plays, in this way preventing the possibility that an opponent could predict the play and take effective countermeasures. We can recognize this as an instance of frequency dependence. R. A. Fisher, who was a prominent statistician as well as an evolutionary geneticist, took an interest in the matter, publishing a work concluding that indeed randomization can be the 'solution' to the problem of what is the best strategy for a player of Le Her (Fisher, 1934). Fisher was fascinated by the idea of possible important uses of randomization—most familiar is his insistence that treatments should be randomized in experiments—and when he much later wrote about polymorphism and natural selection (Fisher, 1958), he returned to the idea. By that time he was familiar with game theory as developed by mathematicians and economists, and he proposed that the outcome of natural selection could be that a population is polymorphic, effectively randomizing between phenotypes. The concrete example he had in mind was polymorphic Batesian mimicry in certain butterflies. The variation in appearances could prevent predators from quickly learning to attack a single prey appearance, which again is an instance of frequency dependence. However, Fisher did not develop the idea further.

It was Maynard Smith and Price (1973) who were the first to systematically use game theory in biology. They posed the question of why animal contests over resources, such as over a territory or a dominance position in a social group, are usually of the 'limited war' type, being settled without serious injury. They used the so-called Hawk–Dove game (Section 3.5) to conclude that, because of frequency dependence, the evolutionary endpoint could be a mixture or polymorphism of direct aggression and display behaviour; thus to some extent limiting direct aggression. This was a breakthrough partly because of the elegance and simplicity of the Hawk–Dove game, but even more because they presented a method to analyse important frequency-dependent situations in biology. Concerning animal contests, their work was just a starting point, with many further analyses of possible reasons for limited aggression appearing, for instance of the effects of assessment of relative fighting ability by contestants (Parker, 1974).

We describe parts of this work (Sections 3.5, 8.7, 9.4, and 9.5), which certainly has improved the understanding of animal contests. Nevertheless, the original question identified by Maynard Smith and Price (1973) is up to now only partially answered using worked-out models. An example where current models are lacking is aggressive interactions in social groups. This points to challenges for game theory in biology, and our aim is to address these in the book. One challenge lies in the sometimes intractable complexities of analysing representations of the information individuals gain about each other; for instance, information individuals acquire when they form dominance hierarchies in social groups. An equally important challenge is to make contact between models and the strategies and decision processes used by real animals. Our approach, as we outline in the next section, is to introduce behavioural mechanisms into models, including mechanisms inspired by animal psychology and neuroscience. Frequency dependence still plays a role in such models, but evolutionary outcomes are prescriptions of how to represent and respond to events, rather than a randomization between two or more actions.

The analysis of sex allocation is one of the major empirical successes of game theory in biology. The origin of ideas about why sex ratios should be close to 1:1

goes back to work by Düsing in the late 19th century, but it was Fisher (1930) who presented a highly influential formal model (Section 3.8). We have noted that frequency dependence is a crucial ingredient in such analyses (Section 1.2). Much of the theory and tests of sex allocation is surveyed by Charnov (1982), and we describe some of these models in the book (Sections 3.8 and 10.4).

As we have indicated, the roots of game theory as a formal tool to analyse behaviour lie in economics. There are many situations in economics where frequency dependence matters. For example, the best amount to produce or the best price to charge for a product depends on the amounts or prices chosen by other firms. These questions were analysed in early applications of game theory in economics and are referred to as Cournot and Bertrand competition. The ideas have also played a role in making biologists think in terms of markets (see Section 7.8). Other economic applications include hiring decisions in labour markets, the structuring of auctions, and the effects of social norms on behaviour. The definition of a Nash equilibrium in biology is mathematically the same as the standard Nash equilibrium condition of economic game theory. There are, however, differences in approach in the two areas (Box 2.1).

Perhaps the biggest influence from economics on biological game theory is an interest in cooperation, often presented in terms of one or several rounds of the so-called Prisoner's Dilemma (see Sections 3.2, 4.5, 7.3, 7.6, and 8.8). As information about other individuals is likely to be just as important in cooperative as in competitive interactions, these present another major challenge to game theory in biology. Our approach is to model such situations in ways that correspond in a reasonable way to biological reality (Sections 3.2 and 8.8), for instance using continuously varying investments for parents that jointly rear their young, and then to incorporate behavioural mechanisms into the model, for instance mechanisms that can be interpreted as negotiation between partners (Section 8.4). We discuss these matters in Section 8.8, pointing to successes and challenges for the current state of affairs in game theory.

1.5 Approach in This Book

Here we give an overview of major issues that we deal with in this book, indicating where in the book they occur. The central concepts of biological game theory, such as invasion fitness, the Nash equilibrium, and the phenotypic gambit are expanded on in Chapter 2. That chapter also deals with fitness proxies and the problem of embedding a game that occupies a small part of the life of an organism into a whole-life perspective, a topic we return to in Chapter 9.

A variety of game-theoretic models dealing with a range of phenomena have previously been developed. In Chapter 3 we present many of what have become the standard models. Some of these standard models deal with cooperation and the contribution to a common good, including parental care. We also introduce the simplest model of animal conflict over a resource: the Hawk–Dove game. Many animals signal to others, and we present a simple model showing that signals can evolve from cues, later returning to the question of why signals should be honest in Section 7.4. We also present the standard model of sex allocation, a topic we later return to in Section 10.4. Most of these simple models assume that all individuals are the same, so that if they take different actions this is because their choice has a random component. In reality it is likely that individuals differ in aspects of their state such as size or fighting ability, and different behaviours are a result of these difference. At the end of Chapter 3 we illustrate how such state-dependent decision making can be incorporated into models. The effects of state differences are harder to analyse when the state of offspring is affected by the state and action(s) of their parent(s). We defer the description of some standard models that have these features until Chapter 10, where we outline the theory needed to analyse inter-generational effects. We apply this theory to the problem of sex allocation when offspring tend to inherit the quality of their mother (Trivers and Willard theory) and to the case where female preference for a male ornament (such as tail length) and the ornament co-evolve (the Fisher process).

A population that has reached an evolutionary stable endpoint is necessarily at a Nash equilibrium. The converse is, however, not true. There are Nash equilibria that are not stable. Furthermore, there can be endpoints that would be stable if they could be reached, but cannot be reached by the evolutionary process. In Chapter 4 we return to general theory. There we set out the conditions needed for a strategy to be an Evolutionarily Stable Strategy (ESS); conditions that are stronger than those required for a Nash equilibrium, and which ensure stability. We also consider evolutionary trajectories using the concepts of adaptive dynamics and convergence stability (Section 4.2).

Evolution is concerned with the spread of genes. In most of the analyses in this book we can translate ideas about the spread of genes directly into the idea that at evolutionary stability each population member maximizes a suitable payoff. However, care must be taken when relatives affect each other, since relatives share genes. We describe the analysis of games between relatives in Section 4.5.

1.5.1 Challenges

The approach taken in the standard game-theoretic models often rests on idealized assumptions. This is important and helpful in providing easily understandable and clear predictions, but biologists might rely on models without careful examination of the consequences of the assumptions and limitations of the models. We believe the ideas used in the field need to be re-evaluated and updated. In particular, game theory needs to be richer, and much of the remainder of the book is concerned with ways in which we believe it should be enriched. Here we outline some of these ways.

Co-evolution. There is a tendency to consider the evolution of a single trait keeping other traits fixed. It is often the case, however, that another trait strongly interacts with the focal trait (Chapter 6). Co-evolution of the two traits can for instance bring about disruptive selection causing two morphs to coexist or giving rise to two evolutionarily stable outcomes. These insights might not be gained if the traits are considered singly.

Variation. There is usually considerable variation in natural populations. Many existing game-theoretical models ignore this and assume all individuals in a given role are the same. However, variation affects the degree to which individuals should be choosy over who they interact with and the value of expending effort observing others. Variation thus leads to co-evolution of the focal trait with either choosiness or social sensitivity. These issues are crucial for effects of reputation and the functioning of biological markets (Chapter 7). Variation is also crucial for phenomena such as signalling. We believe that models need to be explicit about the type and amount of variation present, and to explore the consequences.

Process. In many game-theoretical models of the interaction of two individuals each chooses its action without knowledge of the choice of their partner. Furthermore, neither alters its choice once the action of its partner has been revealed (a simultaneous or one-shot game). In reality most interactions involve individuals responding to each other. The final outcome of the interaction can then be thought of as resulting from some interaction process. The outcome can depend strongly on the nature of this process (Chapter 8). Since partners vary, the interaction often involves gaining information about the abilities and intentions of other individuals, emphasizing the importance of variation and learning (Chapters 5 and 8).

Timescales. Games can occur over different timescales. Many, such as a contest over a food item, are concerned with a brief part of the whole lifetime of the contestants. We may then ask to what extent it is possible to isolate the contest from the rest of the lifetime. We often isolate games by assuming suitable payoffs, but as we discuss in Chapter 9 this is not always possible and a more holistic view of an organism's lifetime is necessary if our model is to be consistent. Some games, such as those involving the inheritance by males of a trait that is attractive to females, occur over more than one generation (Chapter 10).

Behavioural mechanisms and large worlds. The number of subtly distinct circumstances encountered in the real world is vast. Evolution shapes behavioural mechanisms that must deal with this vast range. It is not realistic to suppose that this produces strategies that respond flexibly and appropriately to every possible circumstance. The behavioural mechanism needed to implement such a strategy could probably not evolve in the first place, and even if it did it would involve so much neuronal machinery that the maintenance costs of the machinery would outweigh the benefit. For this reason we expect the evolution of strategies that are implemented by psychological and physiological mechanisms of limited complexity, which perform well on average but may not be exactly optimal in any circumstance. Most game-theory models take what can be called a small-worlds perspective: they deal with optimal strategies in simple situations. The real world is large (Chapter 5) and models need to reflect this if we are to make more realistic predictions. To do so models need to explicitly consider psychological and physiological mechanisms, accepting that these are not exactly optimal.

To elaborate on this point, game-theory models in biology have traditionally not assumed particular behavioural mechanisms, but rather assumed that organisms can use any kind of information available to them in an optimal manner. Such models are certainly valuable when they can be achieved, because they allow a full accounting for the action of natural selection. However, limiting oneself to such models has two important drawbacks. One drawback is that these models can easily become too challenging to achieve, in practice preventing modellers from analysing many important situations where interacting individuals have partial information about each other. The difficulty is caused by the complexity of a complete representation of an individual's information about its world, including its social environment (technically, a complete representation would involve probability distributions in high-dimensional spaces). The other drawback is, as mentioned, that real organisms might not use information in an unconstrained manner, but rather rely on particular behavioural mechanisms. Models that do not allow for these mechanisms might then provide a weaker connection between model results and empirical observation.

It is difficult for theory to *a priori* predict the type of mechanism that will evolve, partly because many mechanisms can work well, but also because the effect of phylogeny is important. For most purposes we suggest that we should base models on observed mechanisms, such as known mechanisms for learning. We can then investigate how parameters of these mechanisms might be tuned in different environments. We can also ask how limitations of these mechanisms might change our predictions of observed behaviour, compared with the possibly Panglossian predictions of small-worlds models (e.g. Sections 8.5 and 11.2).

Because the world is complex and unpredictable we might expect learning to be important. Individuals engaging in a game will often enter the game not knowing the abilities of themselves or others, and may not know the game structure or the payoffs. Learning in games has hitherto received insufficient attention in biology. In Chapters 5 and 8 we explore the effect of a mechanism called actor–critic learning that arose in machine learning, but has links with known psychological mechanisms and neuroscience. For instance, we use this approach to model social dominance and individual recognition in groups (Section 8.6), which are phenomena that traditionally have been difficult for game theory to handle. In doing so we hope to promote more research on realistic learning mechanisms.

In general, we argue that game theory needs to be richer, by incorporating insights and approaches from neighbouring disciplines, including ideas from neuroscience, experimental psychology, and machine learning. We present more of our views on these issues in Sections 8.8 and 11.2.

2

Central Concepts

Game theory in biology is built on certain central concepts, which we introduce here. Among these are strategies (Section 2.1), which are regarded as being genetically determined and are hence a key concept in that they are passed on to future generations. However, we need some justification of why we can deal with strategies rather than having to deal with genes directly. In biological game theory this is usually discussed in terms of the phenotypic gambit (Section 2.2). We also need some measure of the performance of a strategy and this is provided by the concept of invasion fitness (Section 2.3). As we explain, this is a measure of the per-capita rate of increase in the number of individuals following the strategy when the strategy is rare. We are then in a position to formulate a necessary condition for the resident strategy in a population to be evolutionarily stable in terms of invasion fitness (Section 2.4). This stability condition is central to the book. It can, however, be helpful in many situations to formulate the stability condition in terms of a fitness proxy instead of working directly with invasion fitness (Section 2.5). Finally, many games of interest occur over a short part of the lifetime of an individual. In Section 2.6 we describe how one can relate the actions chosen in a part of the life of a single individual to invasion fitness, which concerns the growth of a cohort of individuals over many generations.

2.1 Actions, States, and Strategies

Strategies are simply rules that specify the action chosen as a function of state. In this section we unpack this statement, clarifying what we mean by an action and what we mean by a state.

2.1.1 Actions

As the examples in Section 1.2 illustrate, when we talk about strategies and actions this is meant in the widest sense, including behaviour, growth, and the allocation of the sex of offspring. The analysis in a specific context starts by specifying the range of possible actions each organism might take in that context. Example of actions include:

- For a reproducing female, two possible actions might be to produce a son or produce a daughter.

Game Theory in Biology: Concepts and Frontiers. John M. McNamara and Olof Leimar,
Oxford University Press (2020). © John M. McNamara and Olof Leimar (2020).
DOI: 10.1093/oso/9780198815778.003.0002

- In a contest over a resource such as a food item or potential mate, three possible actions might be to display to the opponent, to attack the opponent, or to run away.
- For a developing individual, two possible actions might be to disperse or remain at the birth site.
- For a foraging animal, the available actions might be to search for food or to rest.

The above examples all involve a finite set of actions. In many other examples possible actions lie on a continuum, so that there is a graded response available to the organism. For example:

- For a tree, an action might specify the allocation to growth of foliage as opposed to seed production.
- For a pathogen, an action might be its level of virulence.
- For a feeding animal, an action might be its level of vigilance while consuming food.

2.1.2 States

The action chosen by an individual is often contingent on the state of the individual. The idea of a state variable is very broad—basically it is a description of some aspect of the current circumstance of an organism. Examples of state variables include:

- *Energy reserves.* For a foraging animal, we might expect the decision to rest or search for food to depend on the animal's energy reserves. Similarly, the level of vigilance might depend on energy reserves.
- *Size.* In a contest over food an individual might base its decision on whether to attack an opponent on the opponent's size relative to its own size.
- *Opponent's last move.* The action in a contest may also depend on the previous behaviour of the opponent. In this sense we regard aspects of this previous behaviour as state variables.
- *Role.* In some interactions individuals have clearly defined roles that are known to the individuals. For example, in biparental care there are two roles, male and female. When an intruder challenges the owner of a territory there are two roles, owner and intruder. When there are clearly defined roles actions may depend on role, resulting in totally different evolutionary outcomes compared with the case in which there are no role asymmetries (Section 6.2).
- *Environmental temperature.* Environmental temperature is an important state variable for many plants, for example influencing vernalization and seed germination. In some reptiles environmental temperature affects the sexes differentially during development. Some temperatures favour male offspring over female, while the reverse is true at other temperatures. This probably accounts for the fact that in some species the sex of offspring is not genetically determined; rather whether males or females develop depends on the temperature (Section 3.11).
- *Social status.* This is an important state variable in many social species.

In general any aspect of physiology or any information that the organism possesses can act as a state variable. The action taken by an individual could potentially depend on the combination of values of its state variables. The action could also depend on time of day or year.

2.1.3 Strategies

Strategies are genetically determined rules that specify the action taken as a function of the state of the organism. For example, in a contest between two individuals over a resource, a strategy might specify whether to escalate a contest, display to the opponent, or retreat depending on perceived differences in fighting ability and the level of aggression already shown by the opponent. In temperature-dependent sex determination in reptiles, a strategy might be: if the temperature is greater than 30°C then develop as a female, if less than 30°C develop as a male. For a foraging animal a strategy might specify whether to forage or rest, depending on current energy reserves, the amount of food in the gut, and time of day. For a growing animal, a strategy might specify the relationship between environmental conditions and the time at which the animal switches from growth to reproduction. In this and other life-history contexts, such a reaction norm is just a strategy.

Genes code for strategies, but it should not be thought that there is a simple relationship between them. Models often assume that there are a few genetic loci that determine a strategy, and a simple relationship between the alleles at these loci and the strategy the individual employs. We should not regard such genotype–phenotype maps as being realistic. The phenotype is typically the product of developmental processes that are affected by many genes in a manner that may not be straightforward. However, this does not mean that the simple models should be dismissed (see the next section).

Strategies are the basic units through which selection acts. The reproductive success of an individual will typically depend on the strategy that the individual employs. The differential reproductive success of strategies affects the population frequency of alleles at genetic loci that code for the class of strategies that are possible.

In modelling a specific biological scenario it is important to keep the distinction between strategies and actions, being clear as to exactly what is regarded as a strategy. Strategies specify the action to take in all possible circumstances. Thus an organism following a given strategy may or may not take a specific action; it will depend on the state it finds itself in. The distinction needs to be kept clear because it concerns which characteristics are genetically determined and hence the level at which selection acts. To illustrate this, suppose that we are interested in the level of parental effort that evolves in a species in which both parents care for their common young. We might then regard the effort of a parent as her/his action. At the simplest level a strategy specifies a pair of efforts: the effort of an individual if female and the effort if male. Since strategies are genetically determined, this would be equivalent to assuming that male effort and female effort are genetically determined. In many cases this might be too simplistic, since there is evidence in many species that each parent adjusts her/his effort in response to the parental effort of the partner. We might instead regard a strategy as a sex-specific pair of negotiation rules where each rule specifies how own effort adjusts in response to the effort of the partner. From this perspective it is the negotiation rules that are genetically determined, not the efforts. Efforts are the result of the negotiation process. As we will see (Section 3.4 and 8.4), the predicted parental efforts differ between these two perspectives. When individuals learn which actions to take (Chapters 5 and 8), the behavioural mechanisms that guide learning correspond to strategies.

The behavioural response of an animal to stimuli is often influenced by the previous experience of the animal. So for example, the way in which an animal deals with a stressful situation as an adult may depend on its experiences when young. This might seem to contradict the notion that strategies are genetically determined. However, just as for the parental care case, it is important to analyse behaviour at the correct level. For the stress example, a strategy is not the response to stress as an adult but the rule that specifies how response to stress as an adult depends on early life experience. Similarly, there has recently been an interest in non-genetic inheritance, in particular, in epigenetic marks that are passed from mother to offspring and can affect the adult phenotype. In this context, a strategy is a genetically determined rule that specifies how the adult phenotype depends on the epigenetic marks and other influences.

2.2 The Phenotypic Gambit

To specify the evolutionary dynamics of strategies, one needs to make assumptions about the genetic determination. The approach in game theory is to first examine very simple dynamics, and to add specific details about the genetic determination of strategies only to the extent that this is important. A simple assumption is to think of each strategy as corresponding to a genotype. This could for instance be implemented as asexual inheritance of the strategy. Using the approach need not, however, imply a restriction to this single case, but rather that one grants that evolutionary endpoints can be the same for different kinds of genetic determination of strategies, including the very common case of diploid multilocus genetics. Alan Grafen coined the term 'phenotypic gambit' for this general idea (Grafen, 1984), and it is very widely used.

As we will describe, a fitness proxy is a function that assigns a measure of performance to each strategy (Section 2.5). If there is no frequency dependence one can think of a landscape of possible strategies, with the fitness proxy measuring the local height. The idea of the gambit is that evolution tends to increase this performance measure, so that the evolutionary dynamic is some form of hill climbing and that endpoints are strategies that maximize the fitness proxy. In other words, evolution takes the population onto a local peak. With frequency dependence, one needs to take into account that the fitness landscape depends on the strategies of population members, and so changes as these strategies change. One then has the idea that evolution takes the population to a hilltop in a landscape that is generated by the strategies in the population, when the population sits at this hilltop.

Even if some aspects of genetics are needed to describe a phenomenon, the basic idea of the gambit can still be used. For instance, for populations with two sexes and where each individual has one mother and one father, it is possible to work out fitness functions for sex allocation strategies (see Section 3.8), and typically one finds a sex ratio of 50% females as an endpoint. The phenotypic gambit then means that one assumes that a fairly wide range of evolutionary dynamics will lead to this endpoint.

Evolutionary change towards a predicted endpoint is illustrated by a selection experiment on Atlantic silverside fish (Fig. 2.1). In this species sex is determined by a combination of the temperature during a sensitive period, implemented as a genetically determined threshold above which an individual becomes male, and temperature-insensitive sex-determining genes. The sex-determining mechanisms present from the start in the laboratory population will influence the particular sex-ratio trajectory over the generations, but with a sufficient supply of different kinds of genetic variation, it is reasonable that the end point of a 1:1 sex ratio is approached. As seen in Fig. 2.1, this is also what happened in the experiment by Conover et al. (1992).

It is thus possible to adapt the phenotypic gambit slightly by taking major qualitative features of genetic inheritance into account, for instance by incorporating that each individual has one mother and one father into game theory analyses of sex ratios. Another example is to include effects of genetic correlations between co-evolving traits (see Section 6.1). Selection that changes one trait can then lead to a correlated

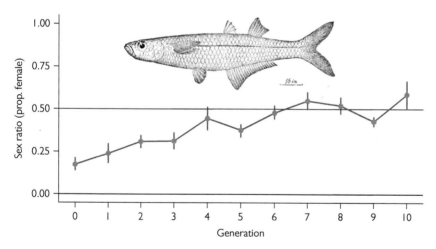

Fig. 2.1 Data from a sex-ratio evolution experiment with Atlantic silverside (*Menidia menidia*) fish (Conover and Van Voorhees, 1990; Conover et al., 1992). The species has combined temperature-dependent and genetic sex determination. The figure shows the mean and 95% confidence intervals of the proportion of females in a laboratory population (redrawn from fig. 1A in Conover et al., 1992). The population was collected as embryos from South Carolina. It was kept at a high temperature of 28°C, resulting in a skewed starting sex ratio. Over the generations, the population approached a predicted sex ratio of 0.5. Illustration of Atlantic silverside by William Converse Kendall, from the Freshwater and Marine Image Bank, University of Washington.

response in another trait. A widely used approach, which is in the spirit of the phenotypic gambit, is to assume some specific but simple genetic determination, such as haploid or single-locus diploid genetics, and to limit consideration to the invasion of a rare mutant allele into a genetically monomorphic population (see next section). Using this one can, for instance, extend the analysis to games between relatives (see Section 4.5).

The gambit can fail. Even though the phenotypic gambit is a helpful idealization, it should not be taken for granted. An interesting example where it fails is the alternative male mating types in the ruff, a species of wading bird (Fig. 2.2). The male type is determined by the genotype at a single Mendelian locus, but the alleles at this locus are actually supergenes, and were originally formed through an inversion of a block of 125 genes on a chromosome (Küpper et al., 2016). This event occurred around 3.8 million years ago and was followed by modification of the supergene content, including a rare recombination event between the original and inverted alleles 500,000 years ago (Lamichhaney et al., 2016), giving rise to three alleles. During the breeding season, male homozygotes for the non-inverted allele develop an elaborate head and neck plumage and strive to defend small territories. They are referred to as 'independents'. Heterozygotes between this allele and the oldest inverted allele, which

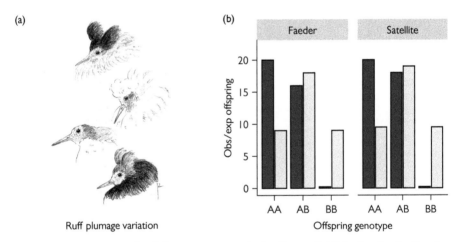

Fig. 2.2 (a) Illustration of ruff plumage variation, by Jos Zwarts. Published under the Creative Commons Attribution-Share Alike 4.0 International license (CC BY-SA 4.0). (b) Survival of different morph-determining genotypes in the ruff. Observed (dark grey) and Mendelian expected (light grey) number of surviving male plus female offspring when both father and mother have a heterozygous genotype AB, where A is the allele for the 'independent' male morph and B is either the allele for the 'faeder' male morph (left side) or the 'satellite' male morph (right side). Data were obtained from a pedigree for a captive genotyped population, and come from table 1 in Küpper et al. (2016). Note that both the faeder and satellite alleles are lethal when homozygous (genotype BB), and this is the case both for males and females.

are referred to as 'faeders', have a female-like appearance and sneak copulations. Heterozygotes with the recombined allele are called 'satellites', and develop a head and neck breeding plumage, but less colourful than those of the independents. They do not defend territories, but associate with independents, displaying on—and increasing the attractiveness of—territories, and attempt to sneak copulations. The ruff male breeding plumage is quite variable (Fig. 2.2a), but differs between the genetic types.

As an accidental consequence of the original inversion, a gene needed for successful development was disrupted, making homozygotes for inverted alleles, either faeder or satellite, lethal (Fig. 2.2b). The heterozygotes also suffer to some extent from the maladaptive consequences of the inversion, but male heterozygotes compensate by a high mating success, possibly because of high success in sperm competition (Küpper et al., 2016). The upshot is that the population frequencies of the ruff mating types are not at a Nash equilibrium (Section 2.4), and the genetic underpinnings have clearly maladaptive features. It might seem surprising that the long evolutionary history of the mating system has not removed these maladaptive features, but the empirical evidence for their origin and current existence is strong. It is worth noting that the ruff is a widespread and successful bird species. For the applicability of game theory in biology, examples such as this should be uncommon, which seems to be the case. Finally, even though inversions and supergenes may imply genetic constraints, they can still play important roles in evolution.

2.3 Invasion Fitness

Consider a large population in which almost all individuals employ a given strategy x. As a shorthand for this statement we will say that x is the resident strategy. We assume that for every resident strategy x, the population settles down to a unique demographic equilibrium (that can depend on x). Here, by the term 'demographic equilibrium' we mean that the population size has settled down and the proportion of population members in each possible state has also settled down. This equilibrium then sets the background in which we consider the fate of any mutants.

Now suppose that we introduce one or a few individuals following an alternative strategy x' into a population with resident strategy x, where the resident population has reached its demographic equilibrium. Since we have in mind that strategy x' might have arisen through mutation from the resident strategy, we refer to x' as a mutant strategy. We will define the invasion fitness of the mutant strategy $\lambda(x',x)$ as an appropriate measure of the per-capita growth in the number of individuals that follow the mutant strategy, while mutants are still rare compared with residents. For convenience, we measure this growth rate as the per-capita annual increase.

The precise definition of this 'appropriate measure' depends on circumstance. Before defining it we need to consider how stochastic events affect population members. At one extreme all population members might experience the same environmental factors such as weather, population size, and the size of prey populations.

We refer to random fluctuations in these factors as environmental stochasticity. Economists refer to this form of stochasticity as aggregate risk since all population members experience the same fluctuations. At the other extreme, different population members might find different numbers of prey items due to good and bad luck when foraging. Fluctuations that affect different population members independently (for given environmental conditions) are referred to as demographic stochasticity, which economists would call idiosyncratic risk.

To study invasion fitness for a mutant strategy we census the mutant population at time points with a suitable spacing, for instance annually at the start of each year, which we now use for concreteness. Let $N(t)$ denote the total number of x' mutants present at the start of year t. We are interested in how this number changes over time. In the very simplest case we assume the following conditions:

1. The population is asexual.
2. There are discrete, non-overlapping generations with a generation time of 1 year. Thus those individuals that are born at census time t reproduce and die over the course of the year, leaving surviving offspring at census time $t + 1$.
3. All mutants are in the same state at each annual census time.
4. There is no environmental stochasticity, so that all years have the same environmental conditions. This does not mean that there are no seasonal effects, but if there are, they are the same each year.

For this scenario consider a mutant individual following strategy x' that is present at the census time in year t. We can refer to the surviving offspring left at the next annual census time by this individual as its recruits. The number of recruits depends on the good and bad luck that the mutant experiences. Let $\lambda(x',x)$ be the mean number of recruits that also follow the strategy x', where in forming the mean we are averaging over demographic stochasticity. Then if there are $N(t)$ mutants following strategy x' in year t there will be approximately $N(t + 1) = \lambda(x',x)N(t)$ mutants in year $t + 1$. This approximation depends on $N(t)$ being sufficiently large so that we can use the law of large numbers to average over the (independent) good and bad luck experienced by different mutant individuals. Thus $\lambda(x',x)$ is the per-capita rate of increase in mutant numbers when mutants have become common enough to average over demographic stochasticity but are still rare compared with the resident population. We define $\lambda(x',x)$ to be the invasion fitness of the mutant. When considering whether a mutant strategy x' can invade into a resident x we should note that there is an initial phase where there are few mutant individuals. In this initial phase the mutant might go extinct because of demographic stochasticity, even if the number of mutants tends to increase on average. Our assumption is that there is some probability for mutant numbers to become large enough so that our approximation based on averaging over demographic stochasticity is valid. This initial randomness might thus require several mutants appearing before our invasion analysis can be applied. Assuming that this is so, if $\lambda(x',x) < 1$ mutant numbers will not increase further, while if $\lambda(x',x) > 1$ mutant numbers increase while rare. The background environment in which the cohort of mutant individuals live is determined by the resident strategy.

The stipulation that mutants should be rare compared with residents ensures that the background remains approximately constant when the growth rate is measured. If mutants become common, we would expect their per-capita rate of increase to change due to frequency- and density-dependent effects.

In the above motivation for invasion fitness it was important that all recruits were in the same state (condition 3); i.e. the population is unstructured. Suppose that this is not the case. For example, suppose that individuals differ in their energy reserves or parasite load at an annual census time. Then different x' mutants would have different potentials to leave surviving offspring. Furthermore, the state of any recruit might be correlated with the state of its parent. Under such circumstances it is not sufficient just to count numbers of recruits when defining invasion fitness. Instead, one needs to consider projection matrices that specify how parental state affects the number and state of descendants left the following year (Chapter 10). Invasion fitness is then taken to be the per-capita rate of increase when the distribution of states of the cohort of mutants has settled down to its limiting steady-state distribution, which is the leading eigenvalue of the projection matrix (Section 10.1). For most of our applications we will not need this machinery.

Our assumption also specifies that the population is asexual (condition 1). With two sexes the population is necessarily structured at the annual census time; there are males and females. Despite this, it is often sufficient just to take the number of female recruits left by a mutant female as invasion fitness. However, in the sex-allocation game of Section 3.8 this expedient fails. In that game a female has N recruits and must decide whether these recruits should be sons or daughters. The numbers of recruits left by a male depends on the sex ratio in the resident population, and can be different from the numbers of recruits left by a female. In this case, in principle one needs the full machinery of projection matrices. However, the analysis can be simplified as it can be shown (Section 10.4) that number of grandchildren left by a female acts as a fitness proxy (Section 2.5).

We further assumed that the population has discrete non-overlapping generations (condition 2). Suppose instead that an individual can live for several years. Then the population is typically structured at an annual census time as individuals of different ages, which can have different potentials to survive and reproduce. Nevertheless, one can simplify analyses when age is the only state variable, because all new recruits (age 0) are in the same state. To do so, one defines the lifetime reproductive success (LRS) of an individual as the total number of its recruits (discounted by relatedness if the population is sexually reproducing). As we explain in Section 2.5, mean LRS is what we refer to as a fitness proxy. We use this fitness proxy to analyse the evolutionarily stable scheduling of reproduction over the lifetime of an organism in Section 10.6.

In the cases we examine in this book condition 4 always holds. When condition 4 fails to hold, so that there is environmental stochasticity, then our argument that the actual change in mutant numbers approximately equals the mean change fails. For the definition of fitness in this more general scenario, see Metz et al. (1992) and chapter 10 in Houston and McNamara (1999).

2.4 Evolutionary Endpoints

As mentioned, rather than looking at evolutionary dynamics in detail, a large part of game theory in biology focuses on the endpoints of evolutionary change. By an endpoint we have in mind a strategy (or distribution of strategies) that is stable over time. In particular, no new mutant strategy that might arise can invade into the population and change its composition. In this section we introduce the concept of a Nash equilibrium, which provides an important necessary condition for evolutionary stability.

Let x be a resident strategy. We can consider a mutant strategy that is identical to this resident strategy. The invasion fitness of this mutant, $\lambda(x,x)$, is the per-capita annual growth of the resident population. For a population at its demographic equilibrium we must have $\lambda(x,x) = 1$. Now let x' be a mutant strategy that is different from the resident strategy. In the previous section we argued that if $\lambda(x',x) > 1$ then an individual mutant has a positive probability to invade, changing the composition of the population so that x would not be evolutionarily stable. Thus a necessary condition for stability is that $\lambda(x',x) \leq \lambda(x,x) = 1$. Motivated by this we define x^* to be a Nash equilibrium strategy if

$$\lambda(x',x^*) \leq \lambda(x^*,x^*) \quad (=1) \quad \text{for all} \quad x'. \tag{2.1}$$

We note that invasion fitness is often measured in terms of $r = \log \lambda$, the logarithm of the relative annual growth rate. We would then have $r(x',x^*) \leq r(x^*,x^*) \ (=0)$ for all x'.

As we have remarked, the Nash condition is exactly the same as that in economic game theory. However, there are some differences in the approach to game theory in biology and economics (Box 2.1).

Condition (2.1) is necessary for evolutionary stability, but not sufficient. In particular, in Section 4.1 we consider what happens when there are mutants with equal fitness to residents, introducing the idea of an Evolutionarily Stable Strategy (ESS). As we will see, the condition for a strategy to be an ESS is stronger than that required for a strategy to be a Nash equilibrium. Thus an ESS is necessarily a Nash equilibrium.

We can reformulate the Nash equilibrium condition as follows. A strategy $\hat{x}$ is called a best response to a resident strategy x if this strategy has fitness at least as great as any other mutant. That is

$$\lambda(\hat{x},x) = \max_{x'} \lambda(x',x). \tag{2.2}$$

Then condition (2.1) is equivalent to the condition that x^* is a best response to itself. There may be more than one best response to a given resident strategy. However, in many of the examples we analyse every resident strategy x has a unique best response, which we denote by $\hat{b}(x)$. When this is so, condition (2.1) is equivalent to the condition that $\hat{b}(x^*) = x^*$.

The above analysis is concerned with evolutionary stability for a monomorphic population; i.e. at stability resident population members all employ the same strategy.

Box 2.1 Economic versus biological game theory

In economic game theory the agents making decisions could be individuals, firms, or governments. In biology they are individuals, but these are only important in that they pass on the genes determining behaviour to future generations, so it is the genes rather than the individuals that are important in evolutionary terms. Approaches in the two areas differ in two major respects.

Payoffs. In biology the payoff of an action is an increment in invasion fitness (e.g. the change in mean lifetime number of offspring). We expect evolution to produce individuals that maximize invasion fitness by maximizing their payoffs.

In economics there are clear performance criteria, such as monetary return, in many applications. In others, where for example social standing is important, the performance criterion is not usually specified in advance. Instead each individual's utility is constructed from their observed behaviour. If behaviour satisfies certain consistency conditions, such as transitivity of choice, then one can construct a utility function such that observed behaviour maximizes expected utility. The utility of an action then acts as the payoff in a game.

In summary, in biology payoffs are normative—they specify what an organism should do to maximize invasion fitness. In economics, utilities (if they exist) are purely descriptive of what individuals actually do (Kacelnik, 2006; Houston et al., 2014).

Justification of the Nash equilibrium concept. The Nash equilibrium is also the equilibrium concept in economic games. Early justification of this centred around the rationality of players. For example, in order to argue that two players would each play their Nash equilibrium strategies it was necessary to assume not only that both were rational, but that each knew the other was rational, and each knew that the other knew each was rational. More recently, attention has switched to dynamic models of choice, where individuals experience payoffs and update their actions accordingly—the focus then is whether a particular learning rule will eventually lead to strategies being in Nash equilibrium.

In contrast, in biology the dynamic process is over generations and is dictated by natural selection. If this dynamic process converges we expect it to be to a Nash equilibrium.

It is also possible to have evolutionarily stable mixtures of strategies. The population is then said to be polymorphic.

Finally, when considering the invasion fitness of a mutant x' in a resident x^* population, it is helpful to think of the mutant as a modification of the resident strategy. Potentially, mutants at different genetic loci can modify the resident strategy, so it is not necessary that there is a single gene or allele coding for each strategy.

2.5 Fitness Proxies

It is often more convenient to work with a fitness proxy rather than working directly with invasion fitness $\lambda(x', x)$. We will refer to a function $W(x', x)$ of x' and x as a fitness proxy if

$$W(x', x) < W(x, x) \iff \lambda(x', x) < \lambda(x, x) \quad \text{for all} \quad x', x, \qquad (2.3)$$

with the analogous result holding when the two inequalities are reversed. If W is a fitness proxy, then λ satisfies condition (2.1) if and only if W satisfies

$$W(x', x^*) \leq W(x^*, x^*) \quad \text{for all} \quad x', \tag{2.4}$$

so that x^* is a Nash equilibrium strategy for the game with payoff W. Thus in attempting to find a strategy that is evolutionarily stable we can use a fitness proxy rather than invasion fitness. It is usual to refer to a fitness proxy as the payoff in the game being analysed. In Section 2.3 we noted that the mean number of grandchildren left by a female can act as a fitness proxy for a sex-allocation game.

To illustrate a very useful fitness proxy, consider a population whose members may live for more than 1 year and may reproduce at different times in their life. We suppose that all offspring that survive until their first annual census time at age 0 (i.e. recruits) are in the same state at this time. In this population, the LRS of a mutant individual is the number of its recruits that also follow the mutant strategy. As we show in Section 10.6, if the mean LRS of a mutant is < 1 then the mutant cannot invade the resident population, whereas if the mean LRS is > 1 the mutant can invade. Thus mean LRS is a fitness proxy.

LRS is concerned with the whole of the lifetime of population members. However, in many game-theoretical situations the focus is on just part of the lifetime of an individual. It is then crucial to what extent the game can be isolated from the rest of the individual's life. For example, suppose that all age 0 offspring are in the same state so that mean LRS is a fitness proxy. Let $K(x)$ be the mean reproductive success before the game. We allow this to depend on the resident strategy x as this strategy may affect factors, such as the resident population density, that affect reproductive success. Let $W(x', x)$ be the mean reproductive success of an x' mutant from the start of the game onwards. We can decompose the mutant's LRS as

$$\text{mean LRS} = K(x) + W(x', x), \tag{2.5}$$

so that W is also a fitness proxy.

We illustrate this decomposition for a simple game: the contest between two foraging animals over a food item in which a strategy specifies the level of aggression towards the opponent. There is no reproduction during this game, so that W specifies the mean reproductive success after the game. Suppose that the contest has three outcomes for a contestant: it (i) gains the item and lives, (ii) fails to gain the item and lives, and (iii) dies. Let $G(x', x)$, $L(x', x)$, and $D(x', x)$ be the probabilities of these outcomes when the contestant has strategy x' and opponent strategy x. These three probabilities sum to 1. Let V_G, V_L, and $V_D = 0$ denote the reproductive success after each of these three outcomes. Then

$$W(x', x) = G(x', x)V_G + L(x', x)V_L + D(x', x)V_D \tag{2.6}$$

is a fitness proxy.

Assuming that values for V_G and V_L are given independently of x may not be reasonable in certain circumstances. One reason is that the current resident behaviour may also be correlated with future behaviour. Thus in a contest over a territory, the value of gaining the territory may depend on how long the territory can be held, which may depend on the level of aggression shown by others in the future (Section 9.5). The resident strategy in a game may affect future resources and hence future reproductive success. For example, in Section 9.3 we consider a game between two parents over whether to care for their common young or desert and attempt to remate. In this game, the availability of future mates depends on who deserts and so depends on the resident strategy. Thus the payoff for desertion cannot be specified independently of the resident strategy. The topic of placing a game in an ecological setting in a consistent manner is explored in Chapter 9.

Despite these caveats, many fitness proxies that deal with a part of the lifetime of an organism work well. For example, in the producer–scrounger game members of a group of foraging animals each decide whether to search for food themselves or exploit the food sources found by others (Giraldeau and Caraco, 2000; Giraldeau et al., 2017). Since one expects the reproductive success of each group member to increase with the amount of energy it acquires, the mean net rate of energy gain is a valid fitness proxy. Whether it is reasonable to treat a part of an organism's life in isolation depends on the issues of interest and judgements based on knowledge of the biology of the system.

It is convenient to also use the concept of a strong fitness proxy. We make the following definition: a function $W(x', x)$ of x' and x is a strong fitness proxy if

$$W(x'_1, x) < W(x'_2, x) \iff \lambda(x'_1, x) < \lambda(x'_2, x) \quad \text{for all} \quad x'_1, x'_2, x. \tag{2.7}$$

A strong fitness proxy is necessarily a fitness proxy. Strong fitness proxies are useful when dealing with best responses. This is because for a strong fitness proxy W, $\hat{x}$ satisfies $\lambda(\hat{x}, x) = \max_{x'} \lambda(x', x)$ if and only if $W(\hat{x}, x) = \max_{x'} W(x', x)$. The mean rate of net energy gain is often a strong fitness proxy. Although mean LRS is a fitness proxy it is usually not a strong fitness proxy. This is because invasion fitness, which is a growth rate, depends on the age at which offspring are produced as well as the total number produced (see Exercise 10.7).

2.6 From Strategies to Individuals

Genes (and the strategies they code for) are central to evolution. Individuals are important in that they are the vehicles that pass on genes (Dawkins, 1976). They carry out the actions specified by their inherited strategy, and this influences their survival and reproduction and hence how many of their descendants follow the strategy. The combined actions of the cohort of population members that follow the strategy thus determine the per-capita growth rate of the cohort, i.e. the invasion fitness of the strategy. For this reason we do not talk about the fitness of individuals in this book. Fitness is assigned to strategies and is invasion fitness.

In contrast, game theory is largely pitched at the level of the individual. We can usually work on individuals since the maximization of the invasion fitness of a strategy is often equivalent to each member of the cohort maximizing some appropriate performance measure. For example, suppose that conditions 1–4 of Section 2.3 hold, so that there are discrete non-overlapping generations with a generation time of 1 year. Then the invasion fitness of a mutant strategy is the mean number of recruits left next year by a randomly selected mutant, and this is maximized if every member of the cohort of mutants maximizes its mean number of recruits. Similar remarks apply to fitness proxies such as mean LRS. There are, however, situations where the fitness of a strategy is not a simple linear sum of the performance of those individuals that follow the strategy. In particular, when environments fluctuate there is an implicit frequency dependence with the best action of one cohort member dependent on the actions of others (McNamara, 1995, 1998), although one can still find measures that are individually maximized (Grafen, 1999). We do not deal with fluctuating environments in this book.

3

Standard Examples

Various games have become part of the standard repertoire of behavioural ecologists and evolutionary biologists. In this chapter we present some of these games. In doing so we describe the value of each in understanding biological phenomena, but also highlight limitations and suggest modifications, many of which we follow up later in the book.

3.1 Contributing to the Common Benefit at a Cost

Consider a situation in which a pair of individuals engage in an activity, where each member of the pair decides how much effort to expend on some common good. The more effort that an individual expends, the greater the benefit each member of the pair derives from the common good, but the greater the cost paid by that individual. For example, when two birds forage together and must be wary of predators, we can equate effort with the proportion of time spent vigilant. The more time an individual devotes to vigilance, the more likely that the pair will be warned of the approach of a predator (the common benefit to the pair), but the less time that individual will be able to forage (the cost to the individual).

When two animals pair up to hunt down prey the probability that prey is captured increases with the effort of each, but the cost of hunting in terms of the energy expended by an individual and its risk of injury increases with its hunting effort. Predator inspection by a pair of fish might also involve a joint benefit at individual cost. In such scenarios, because there is a common good but individual costs there is a conflict of interest: each would prefer their partner to expend the majority of the effort so that they benefit from the common good but pay less cost. In order to model how this conflict of interest is resolved we first make a number of simplifying assumptions that are also used in many other games:

- The population is large.
- When individuals pair up they do so at random.
- When individuals choose their effort they do so without knowledge of the choice of effort of their partner. Furthermore, they are then committed to their chosen effort even when the effort of their partner becomes known. This assumption is often referred to by saying that moves are simultaneous, that the game is one-shot or that

Game Theory in Biology: Concepts and Frontiers. John M. McNamara and Olof Leimar,
Oxford University Press (2020). © John M. McNamara and Olof Leimar (2020).
DOI: 10.1093/oso/9780198815778.003.0003

there are sealed bids. In many cases the assumption is unrealistic, but it is made hoping that a simplified analysis can reveal essential features of a phenomenon.
- Individuals have the same abilities, benefits, and costs.
- The game has symmetric roles, so there is no labelling difference such as male and female or territory owner and intruder.

For this game a strategy just specifies the effort expended. Here we assume a very simple payoff structure for the game. This structure may not hold in detail for specific cases, but will suffice to illustrate key ideas. Suppose the payoff to an individual that expends effort x' is $W(x',x) = B(x' + x) - C(x')$ when the partner expends effort x. Here the common benefit to both individuals is some function $B(x' + x)$ of the sum of the efforts and the cost paid by the focal individual is a function $C(x')$ of its effort. Figure 3.1 illustrates possible benefit and cost functions. As is biologically reasonable, both the costs and the joint benefit increase with increasing effort. For the functions shown the marginal benefit of increasing effort decreases as total effort increases (diminishing returns), at least for large efforts, whereas the marginal cost of increased effort increases for all efforts (accelerating costs). Whether these properties are reasonable depends on the circumstances and the scale on which effort is measured.

Suppose that almost all population members expend effort x (i.e. the resident strategy is x). Then within this population the mutant with the highest payoff employs an effort x' that maximizes $W(x',x)$. We denote this best response by $\hat{b}(x)$. Formally we have

$$W(\hat{b}(x),x) = \max_{x'} W(x',x).$$

Figure 3.2 shows this best response function. As can be seen, the best response effort decreases as the resident effort increases. This property is true whenever the benefit

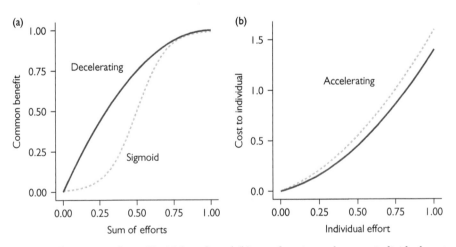

Fig. 3.1 Illustration of possible (a) benefit and (b) cost functions when two individuals put effort into a common good.

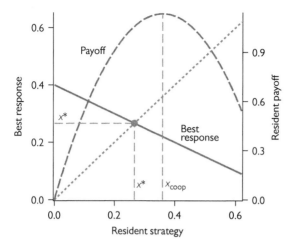

Fig. 3.2 The best response (solid curve) and Nash equilibrium level of effort x^* when two individuals put effort into a common good. The effort x^* is at the intersection of the best response curve and the 45° line (dotted), so x^* is a best response to itself. The dependence of the mean resident payoff on the resident strategy is also shown (dashed curve). This mean payoff is maximized at effort x_{coop}. Benefit and cost functions are given by the solid curves in Fig. 3.1.

function $B(x' + x)$ has diminishing returns and the cost function $C(x')$ is accelerating (Exercise 3.2).

If a resident strategy x^* is evolutionarily stable, then no mutant strategy can have greater fitness within a population following this strategy, so that x^* is a best response to itself (i.e. a Nash equilibrium). In the figure this corresponds to the strategy at which the best response curve intersects the dotted 45° line. The figure also shows how the payoff to population members following the resident strategy depends on this strategy. As can be seen, this payoff is not maximized at the Nash equilibrium— if both individuals were to expend the same effort then the common effort that would maximize the payoff to each is x_{coop}. This strategy would not, however, be evolutionarily stable, since the best response of a mutant within such a resident population would be to expend an effort that is below even x^*. Thus a resident strategy of x_{coop} is invadable by lazy mutants that exploit the fact that their partners are putting in high effort. The discrepancy between what is best for the individual and what is best for the population as a whole reflects the conflict of interest between pair members.

In the above model we have taken the payoff to be a benefit minus a cost. This captures the essence of the conflict of interest when individuals contribute to a common benefit, but payoffs in specific situations may have a different form. For example, when two birds forage together and must choose their level of vigilance, the benefit of increased vigilance is increased survival. In this case the payoff should really be the product of the survival probability and the expected benefit from food given a bird does survive (Exercise 3.3). Even when payoffs are of the form of benefit minus cost, the benefit may not be a function of the sum of the individual efforts. For example, when two individuals hunt together prey capture may require both individuals to coordinate their efforts and the probability of success may be closer to a function of the minimum of the two efforts rather than their sum. In such cases the best response effort may increase with the effort of the partner over a range of the partner's effort. These kinds of effects are sometimes described as synergistic selection (Maynard Smith, 1989; Leimar and Tuomi, 1998) and can be seen as promoting cooperation.

We have analysed a situation where two individuals contribute to their common good, but the idea can be extended to a group of more than two individuals. For example, the analysis extends directly to a group of foraging animals where each chooses their level of vigilance (Exercise 3.3, see also McNamara and Houston, 1992a) or a group hunting prey. At evolutionary stability each individual is maximizing its payoff given the behaviour of others in the group.

Siderophore production in bacteria provides another example of a game with a common benefit but individual costs (see e.g. West and Buckling, 2003). Iron can be a major limitation on bacterial growth. One way that iron can be acquired is through the production and release of siderophores, which bind to environmental iron and are then taken up by the bacteria (Ratledge and Dover, 2000). Siderophore production is costly to the producing bacteria but acts as a common good since they can be taken up by all bacteria in the local vicinity.

3.2 Helping Others: The Prisoner's Dilemma Game

The evolution of helping might well be the most studied problem in game theory in biology. The reason is in part that helping and cooperation are important for many organisms, but even more that the evolution of helping appears as a challenging problem for Darwinian theorizing (e.g. Sachs et al., 2004). So why would helping evolve—in particular if it is costly to the individual that is providing help? This conundrum is captured by the Prisoner's Dilemma model. Here we present a special case of that model. In this model population members meet in pairs. Each pair member takes one of two actions: one action, 'cooperate' (C), is to give help to the partner; the other, 'defect' (D), is to withhold help. An example of giving help might be sharing food with a hungry partner, as happens in vampire bats (Wilkinson et al., 2016). For the game the usual simplifying assumptions (above) such as random pairing apply. The payoff to a recipient of help is b. However, helping the partner reduces own payoff by c. It is assumed that $b > c$ so that the benefit to others of getting help exceeds the cost to the helping individual. Table 3.1 gives payoffs of combinations of actions. It can be seen that this game differs from that of contributing to a common benefit at a cost (Section 3.1), not just in having two rather than a continuum of actions, but also in that an individual's investment only benefits the partner and not itself.

Consider the case where there are just the two strategies: always cooperate and always defect. If the partner cooperates the best action of the focal individual is to

Table 3.1 Payoffs in a simplified version of the Prisoner's Dilemma game.

Payoff to focal		Action of partner	
		C	D
Action of focal	C	$b - c$	$-c$
	D	b	0

defect since $b > b - c$. If the partner defects the best action is also to defect since $0 > -c$. Thus the best response to any resident population strategy is always to defect, and this strategy is the unique Nash equilibrium. The dilemma in this example is that at the Nash equilibrium population members achieve a payoff of 0, whereas if they had always cooperated they would have achieved the higher payoff of $b - c$. Of course such a cooperative population would not be evolutionarily stable since mutant defectors would invade. The Prisoner's Dilemma is often presented with a more general payoff structure than here, although the dilemma remains the same: defection is the unique Nash equilibrium strategy, but population members would do better if they all cooperated.

The Prisoner's Dilemma model has become the focus of much research into conditions under which cooperative behaviour would evolve. As cooperation does not evolve in the basic model a variety of modifications and extensions have been considered, e.g. assortative pairing and repeated play of rounds of Prisoner's Dilemma with the same partner. But, as we will discuss (Section 8.8), the game structure of many real interactions with helping is not captured by the repeated Prisoner's Dilemma. In the Prisoner's Dilemma the best action of an individual does not depend on what its partner does; it is always best to defect. Most biological games where there is investment in others or in a joint project are not as extreme as this game. One might argue that if cooperation evolves for the Prisoner's Dilemma then it will certainly evolve for less extreme games. However, since the objective of the theory is to understand cooperation in real systems, which might be brought about by mechanisms that cannot operate for games as extreme as the Prisoner's Dilemma (see Section 7.6), it does not seem reasonable to focus so strongly on the Prisoner's Dilemma in biology.

3.3 The Tragedy of the Commons

When individuals share a resource and each behaves to maximize their own payoff the result is often the overexploitation of the resource to the detriment of all the users of the resource. This phenomenon has been referred to as 'the tragedy of the commons' (Hardin, 1968). It has important consequences for human issues such as overexploitation of the environment, but is also of relevance to many aspects of organismal behaviour where the resource could, for example, be food (for a herbivore) or light (for a plant). The tension between immediate self-interest and the destruction of the resource is central to the evolution of virulence in pathogens (West et al., 2006; Alizon et al., 2013). The fitness of a pathogen increases with its ability to transmit offspring to other hosts. The growth rate of a pathogen within a host, and hence the rate of transmission, increases with its virulence. However, too high a level of virulence will rapidly kill the host (the resource) so that no further transmission is possible. When many pathogens co-exist in the same host it may be in the self-interest of each to have high virulence so as to increase their share of the host resources, to the detriment of all pathogens (and the host!). For an analysis of the evolutionarily stable level of virulence, see, for example, Frank (1996).

3.4 Biparental Care: The Parental Effort Game

Biparental care of young (that is care of young by both the mother and father) is common in birds and is also found in other taxa. For example, in mammals it is found in several primate and rodent species, but is also found in some arthropod species such as burying beetles. When both parents care, how much effort do we predict each parent to invest in care? Survival of the young is a common benefit to both parents, and the more effort expended by a parent the greater the young's survival prospects. But care also incurs costs. Increased effort by a parent may reduce their opportunities for extra-pair matings. For each parent, in foraging to provision the young the parent may expose itself to increased predation risk and this risk is liable to increase at an accelerating rate with the amount of food delivered to the young. Increased provisioning effort may also reduce the probability that the parent survives the following winter (Daan et al., 1996).

The situation thus involves contributing to a common benefit at individual cost, but now there is a role asymmetry; one parent is in the female role and the other in the male role. To analyse evolutionarily stable levels of effort we regard a strategy as specifying a pair of efforts: the effort if female and the effort if male. Here we are envisaging the genes that control effort to be present in both males and females, but in determining effort they interact with the sex of the individual, so that different sexes can expend different efforts—in fact we will assume there is complete flexibility at no cost in this regard. Suppose the resident population strategy is (x, y), i.e. resident females expend effort x and resident males expend effort y. Consider a mutant strategy (x', y'). Then a mutant female expending effort x' is typically paired with a resident male who expends effort y. The payoff to the female might then be written as $W_f(x', y) = B(x' + y) - C_f(x')$, where $B(x' + y)$ is the female's benefit from the survival of the young and $C_f(x')$ is the cost paid by the female. Similarly the payoff to a mutant male might be written as $W_m(x, y') = B(x + y') - C_m(y')$. These are referred to as 'local' payoffs, because they apply to one of the roles in the game.

The overall fitness of the mutant strategy will be an increasing function of both of the local payoffs above. Thus, under the assumption of flexibility at no cost, a resident strategy (x^*, y^*) is at a Nash equilibrium if no individual of either sex can improve its payoff by changing its level of effort. That is

$$W_f(x^*, y^*) = \max_{x'} W_f(x', y^*)$$

and

$$W_m(x^*, y^*) = \max_{y'} W_m(x^*, y').$$

This is equivalent to efforts x^* and y^* being (local) best responses to one another. Specifically, if for a given male effort y the female local payoff $W_f(x', y)$ is maximized at $x' = \hat{b}_f(y)$, and for a given female effort x the male local payoff $W_m(x, y')$ is maximized at $y' = \hat{b}_m(x)$, then

$$x^* = \hat{b}_f(y^*) \quad \text{and} \quad y^* = \hat{b}_m(x^*).$$

(Houston and Davies, 1985).

Figure 3.3 illustrates the best response functions. As before, at equilibrium population fitness is not maximized. One can illustrate this in a diagram that shows all possible combinations of payoffs to the female and male as the combination of their efforts varies (Fig. 3.4). The region occupied by the combination of payoffs is bounded by a curve known as the Pareto front. If efforts are such that payoff combinations lie on this front, then any change in effort by the parents that increases the payoff to

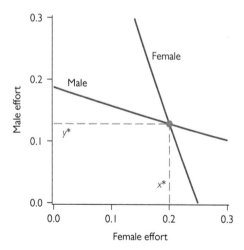

Fig. 3.3 The best response effort of a mutant male to the resident female effort and the best response effort of a mutant female to the resident male effort. The Nash equilibrium pair of efforts (x^*, y^*) are the coordinates of intersection of these two best response curves. The common benefit function is given by the solid curve in Fig. 3.1a and the cost functions for female and male effort are given by the solid and dotted curves, respectively, in Fig. 3.1b.

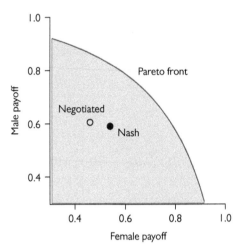

Fig. 3.4 The set of combinations of payoffs to male and female parents that are achievable through combinations of efforts, where benefit and cost functions are as in Figs 3.3 and 3.1 (only those combinations for which both parents have a payoff greater than 0.3 are shown). The solid point gives the payoffs at the Nash equilibrium. The open point gives the payoffs at the negotiated outcome predicted by a variant of the model of Section 8.4.

one will decrease the payoff to the other. We might therefore regard combinations of efforts that produce payoff combinations on the front as in some way efficient, in that it is not possible for both parents to increase their payoff. As can be seen, the pair of payoffs at the Nash equilibrium does not lie on this front. There are combinations of efforts for which both parents would do better than at the Nash equilibrium; however, such a combination of efforts would not be evolutionarily stable.

In the example illustrated the female has the greater ability to care (she pays less of a cost for a given level of effort than the male) and so expends more effort at the Nash equilibrium (Fig. 3.3). This results in her having a lower payoff than the male (Fig. 3.4). In pure optimization situations greater ability is always advantageous, but it may not be so in game-theoretic situations. Even in situations without sex roles, if individuals know the ability of themselves and their partner and adjust their efforts accordingly, individuals of greater ability may have a lower payoff than their partner in a game; if a low-ability individual is paired with a high-ability partner it is best for the low-ability individual to contribute little and the high-ability individual to contribute more. One consequence might be that ability does not always evolve to higher levels in such situations. Another is that there is selection for high-ability individuals to hide their ability so that they are not exploited by lower-ability partners.

Note that in the above analysis the expressions for the payoffs are approximate because they ignore some of the effects of the resident strategy on population processes and composition. For instance, a male's chance of obtaining extra-pair matings is likely to depend on whether other males are also seeking such matings. Thus, if reduced opportunity for extra-pair matings is a cost of male parental investment, self-consistency demands that this cost depends on the resident strategy. If a cost of increased parental effort is increased mortality, then this cost depends on the 'cost of death'. This latter cost is actually the loss of future reproductive success. Future reproductive success depends on the behaviour of the individual in future bouts of parental care. It also depends on the chances of finding future mates, which depends on the sex ratio in the population. This sex ratio depends on the differential mortality of males and females and hence depends on the resident strategy. The dependence on the resident strategy means that payoffs cannot really be specified in advance, as we have done here. As we emphasize in Chapter 9, a fully consistent solution to many games demands that we specify consequences in terms of survival and reproductive success rather than payoffs. Having identified a Nash equilibrium strategy for such a system, it is often possible to identify payoffs that are consistent with this equilibrium, and this analysis may give insight.

3.4.1 Parents Can Respond to Each Other

The above parental effort game is widely used and has an appealing simplicity: at evolutionary stability the effort of each parent is the best given the effort of the other. However, the model has a crucial assumption: parents do not observe one another and adjust their efforts accordingly; rather, their efforts are fixed, genetically determined quantities. This is biologically unrealistic as parents are known to respond to the behaviour of their partner (see, e.g. table 4 in Sanz et al., 2000), although this may be an indirect effect as they may be responding directly to the state of the young, which in turn is influenced by the provisioning rate of their partner. So do models that capture behavioural interactions make different predictions? At a conceptual level, once behavioural responses are allowed, so that parents negotiate effort by each adjusting their own effort in response to the behaviour of their partner, the picture completely changes. Now it is the rules for negotiation with a partner that are genetically determined rather than efforts. At evolutionary stability the negotiation rule of males is the best given that of females, and vice versa. The efforts that are negotiated by such rules will typically not be the best responses to one another (McNamara et al., 1999b). A basic reason for considering negotiation or similar processes where partners adjust their efforts is that individuals vary, for instance, in their ability and thus cost of providing care. It is an important task for game theory to throw light on such phenomena, but it is also a challenging task as the current behaviour of one individual during the negotiation could potentially depend on the individual's ability as well as any aspect of the previous behaviour of the two individuals.

Allowing negotiation may or may not produce more cooperative outcomes (McNamara et al., 1999b). This shows that the interaction process can significantly affect outcomes, a theme we return to in Chapter 8. However, in contrast to the approach taken by Roughgarden et al. (2006), cooperation should not be assumed (McNamara et al., 2006a; Binmore, 2010), rather it may or may not emerge as a result of individuals doing the best for themselves and their kin. Genes are selfish in the sense of Dawkins (1976), but this does not imply they code for behaviour that is 'selfish' in the sense that is usually used to describe human behaviour.

3.5 Contest Over a Resource: The Hawk–Dove Game

Contests between opponents over a resource do not always result in all-out fights. For instance, contests between male field crickets tend to follow a stereotyped escalating sequence of behaviours, where antennal fencing and mandible spreading, with no physical contact, appear early in aggressive interactions. Rillich et al. (2007) estimated that more than 35% of contests between size-matched males ended without any physical interaction, with the remainder progressing to mandible engagement and grappling. In some contests, however, the probability of serious or even fatal injury can be high. In male bowl and doily spiders, who fight by intense grappling, Austad (1983) recorded as many as 70% of staged fights between size-matched opponents that ended with serious injury.

In their classic paper, Maynard Smith and Price (1973) presented a model that aimed to show that even when population members have effective weaponry, limited fighting can be evolutionarily stable. In his book Maynard Smith (1982) gave a simplified version of this model. In the model two opponents pair up in order to contest a resource of value V (i.e. gaining the resource increments reproductive success by V). All the usual simplifying assumptions listed in Section 3.1, such as simultaneous choice and symmetry apply. In the game an individual takes one of two actions, labelled Hawk and Dove. If both take action Hawk the pair fight; each wins this fight with probability one half, with the winner gaining the resource and the loser paying a cost C due to injury incurred (i.e. future reproductive success is decreased by C due to injury). If one contestant takes action Hawk and the other takes Dove, the Hawk attacks and the Dove runs away, leaving the resource to the Hawk. If both take action Dove, they display to each other (at no cost), and each gains the resource with probability one half. These payoffs are summarized in Table 3.2. As we will see, model predictions depend on whether the value of the resource, V, exceeds the cost of injury, C. For some biological scenarios it seems reasonable that $V < C$, whereas for others (see below) it is likely that $V > C$.

Table 3.2 Payoffs in the Hawk–Dove game.

Payoff to focal		Action of partner	
		H	D
Action of focal	H	$\frac{1}{2}(V-C)$	V
	D	0	$\frac{1}{2}V$

For these payoffs, consider a resident population in which the probability that a randomly selected opponent plays Hawk is q. Within this population a mutant individual has average payoff (averaged over what the opponent does) of

$$W_H(q) = q\frac{1}{2}(V - C) + (1 - q)V \tag{3.1}$$

if it plays Hawk, and

$$W_D(q) = q \times 0 + (1 - q)\frac{1}{2}V \tag{3.2}$$

if it plays Dove. Figure 3.5a illustrates how these functions depend on q in a case for which $V < C$.

To analyse evolutionary stability we consider two different scenarios separately. In one scenario there are just two genotypes coding for the two strategies: always play Hawk and always play Dove. In the other, there are also strategies under which the action is chosen probabilistically.

Consider first the scenario in which individuals either always play Hawk or always play Dove. We begin by analysing the case in which $V < C$. When the population is composed almost entirely of Doves we have $q = 0$ and a Hawk does better than a Dove ($W_H(0) > W_D(0)$) because the Hawk always gets the reward. Thus in such a population the proportion of Hawks will tend to increase under the action of natural selection. When the population is composed almost entirely of Hawks, so that $q = 1$, a Dove does better than a Hawk ($W_D(1) > W_H(1)$). This is because a Hawk almost always fights another Hawk and so on average gets the negative payoff $\frac{1}{2}(V - C)$, whereas a Dove always runs away from a Hawk opponent and so gets a payoff of 0. Thus the proportion of Hawks will tend to decline in this population. Regardless of the initial proportion of Hawks in the population, we might expect the proportion of Hawks to approach the stable level q^* shown in Fig. 3.5a at which the payoff to a Hawk equals that to a Dove (i.e. $W_H(q^*) = W_D(q^*)$). It is easy to show that $q^* = V/C$. A population such as this, in which there are stable proportions of different genotypes, is referred to as a stable polymorphism.

In the case where the value of the resource exceeds the cost of injury ($V > C$) Hawks always do better whatever the population composition, so that the only evolutionarily stable proportion of Hawks is $q^* = 1$.

A strategy that specifies that an action is chosen with probability 1 is referred to as a pure strategy. In the above model the two strategies—always play Hawk and always play Dove—are pure strategies. More generally we might allow strategies that choose actions probabilistically; these are referred to as mixed strategies. To do so in the Hawk–Dove game, suppose that a strategy specifies the probability, p, that an individual plays Hawk in an interaction. We then seek a monomorphic Nash equilibrium, i.e. a Nash equilibrium at which all population members adopt the same strategy p^*.

Suppose that almost all population members follow strategy p, i.e. p is the resident strategy. Then since individuals choose their probability of playing Hawk independently and the population is large, the proportion of individuals in the population that play Hawk is $q = p$. Consider a single mutant individual that plays Hawk with probability p' in this population. Then, averaging over the actions chosen by the individual, its mean payoff is $W(p',p) = p'W_H(p) + (1-p')W_D(p)$, which we rewrite as

$$W(p',p) = W_D(p) + [W_H(p) - W_D(p)]p'. \tag{3.3}$$

Using eqs (3.1) and (3.2) we can also write this as

$$W(p',p) = \frac{1}{2}(V - pC)p' + \frac{1}{2}(1-p)V. \tag{3.4}$$

Alternatively, we could have derived this formula by averaging over the four combinations of Dove and Hawk for the mutant and resident to give

$$W(p',p) = (1-p')(1-p)\frac{1}{2}V + (1-p')p \times 0 + p'(1-p)V + p'p\frac{1}{2}(V-C), \tag{3.5}$$

which reduces to eq (3.4) on simplification.

Consider the case where $V < C$ and set $p^* = V/C$. If $p < p^*$, then since $W_H(p) - W_D(p) > 0$ (Fig. 3.5a) by eq (3.3) the payoff is maximized over p' at the value $p' = 1$. Conversely, when $p > p^*$, $W_H(p) - W_D(p) < 0$ and the payoff is maximized over p' at the value $p' = 0$. Finally, when $p = p^*$ we see that the payoff does not depend on p' so that any p' is a best response. These results also follow directly from eq (3.4) since

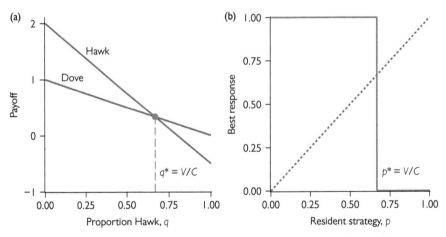

Fig. 3.5 The Hawk–Dove game for the case where $V = 2$ and $C = 3$. (a) The payoff to a mutant individual as a function of its action (Hawk or Dove) and the probability that a randomly selected opponent plays Hawk, q. When there are only two (pure) strategies, always play Hawk and always play Dove, the evolutionarily stable proportion of Hawks is $q^* = V/C$. (b) The best response to the resident population strategy when mixed strategies are allowed, so that a strategy is now specified by the probability p that the individual plays Hawk. The Nash equilibrium strategy is $p^* = V/C$. The 45° line (dotted) is also shown.

$V - pC > 0$ if and only if $p < p^*$. Figure 3.5b illustrates the best response function. For this game p^* is the only strategy that is a best response to itself, so that there is a unique monomorphic Nash equilibrium at $p^* = V/C$. Note that at this equilibrium the actions of playing Hawk and Dove both yield the same expected payoff. This is an illustration of a more general result: at a mixed strategy equilibrium every action chosen with positive probability under the equilibrium strategy has the same expected payoff—if this were not true then the strategy that chose the action with the maximum payoff all the time would have a greater payoff than the resident, so that the resident strategy would not be a best response to itself and hence would not be a Nash equilibrium strategy.

For $V < C$ the Nash equilibrium $p^* = V/C$ in the space of mixed strategies results in the same proportion of the population playing Hawk as the equilibrium we derived above for a polymorphism of pure strategies. The correspondence holds also for $V \geq C$, in which case the unique monomorphic Nash equilibrium is $p^* = 1$. There is this kind of correspondence between Nash and polymorphic equilibria for games in large populations where payoffs are linear functions of both mutant and resident action probabilities, as in eq (3.5).

3.5.1 Lethal Fighting

The Hawk–Dove game is a highly idealized model of animal contests, but there are situations it describes rather well. Thus, it can be a reasonable model of highly dangerous fighting, where the cost is serious injury or even death.

Suppose that the winner of a contest has an immediate gain V in reproductive success but the loser of a Hawk–Hawk fight dies. Then the expected total future reproductive success of a mutant playing p' when the resident plays p is $W(p',p) = A(p',p)V + S(p',p)W_0$, where $A(p',p) = \frac{1}{2}(1 - p')(1 - p) + p'(1 - p) + \frac{1}{2}p'p$ is the probability the mutant wins the contest, $S(p',p) = 1 - \frac{1}{2}p'p$ is the probability it survives the contest, and W_0 is the mutant's future reproductive success given that it survives. Thus

$$W(p',p) = W_0 + \frac{1}{2}(1 - p)V + \frac{1}{2}(V - pC)p', \qquad (3.6)$$

where $C = W_0$. This is in the form of eq (3.4) except there is an additive constant W_0 that does not affect the stability analysis. The cost of dying is $C = W_0$ because the cost of death is the loss of future reproductive success. It follows that if W_0 is smaller than or comparable with the contested value V, the Evolutionarily Stable Strategy (ESS) is either pure Hawk ($V \geq C$) or a high probability $p^* = V/C$ of playing Hawk.

This situation, with a small value W_0 of the future, occurs in nature. Among the best-studied examples are wingless males in some species of non-pollinating fig wasps (Fig. 3.6). These males develop in a fig fruit and, because they are wingless, can only mate with local females. If few females lay eggs in a fruit, there will be few females developing there, and few possibilities for future reproductive success for males that fight over access to females. From the simplified analysis of the Hawk–Dove game, we

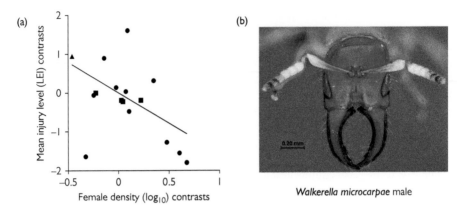

Fig. 3.6 (a) Data from a phylogenetic comparative study of fig wasp species (from fig. 2 of West et al., 2001), comparing an index of the level of injury in male–male contests with the number of females that develop in a fruit; each point represents a different species. Males are more severely injured in species with few females per fruit. Reprinted by permission from Springer Nature: S. A. West, M. G. Murray, C. A. Machado, A. S. Griffin, and E. A. Herre, Testing Hamilton's rule with competition between relatives, *Nature* 409, 510–513, Copyright (2001). (b) Anterior view of a wingless male fig wasp, with powerful mandibles. From fig. 6D of van Noort et al. (2013). Published under the Creative Commons Attribution 4.0 International license (CC BY 4.0).

would then expect more injuries in male–male contests in fig wasp species where few females develop in a fruit. This was found to be the case in a study by West et al. (2001), comparing many fig wasp species, as seen in panel (a) of Fig. 3.6. Panel (b) shows the powerful mandibles of a wingless fig wasp male, used as weapons in dangerous fights. This qualitative result, that a small value of the future promotes dangerous fighting, has also been found using more elaborated game theory models than the Hawk–Dove game (e.g. Enquist and Leimar, 1990; Leimar et al., 1991).

3.5.2 Information is Important in Real Fights

Returning to the original aim by Maynard Smith and Price (1973) to show that limited fighting can be evolutionarily stable, we conclude from the analysis of the Hawk–Dove game that if the value of the contested resource is sufficiently low in comparison with the cost of injury ($V < C$), there is 'limited fighting' in the sense that only a proportion of contests escalate to physical fighting. The question is then to what extent this explains limited fighting in nature.

The Hawk–Dove model brings out the frequency dependence in fighting; i.e. an individual should be less inclined to fight the more aggressive an opponent is likely to be. The model is less good at capturing details of the biology of an aggressive interaction. In particular it ignores the role of information. In real populations individuals can differ in fighting ability, energy reserves, motivation and commitment, and other variables. Even if individuals have no information on their opponent they might, for example, be able to estimate their own fighting ability, a phenomenon we model in Section 5.2. If they can do so then we expect the level of aggression shown by individuals to increase with their fighting ability. As we show in Section 3.11, this can substantially reduce the overall level of fights predicted.

If individuals can estimate the relative fighting ability of opponents by their relative size this is also likely to reduce the severity or number of fights since it might be best for the smaller individual not to attack an opponent or to quickly abandon a contest. This idea was developed by Parker (1974). As an example, for the contests mentioned above between males of the bowl and doily spider, fitting a game-theory model that allows for contestants to estimate their relative size Leimar et al. (1991) showed that the spiders do take relative size into account. The fitted model gives quantitative predictions for fight durations and probabilities of winning and injury as a function of relative size, in reasonable agreement with observation. For instance, observation and model both find that smaller males tend to give up quickly (Leimar et al., 1991), in this way reducing the risk of injury. We introduce a first step in modelling contests with assessment in Section 3.12.

In contrast to the Hawk–Dove game, most real contests are not one-shot inter-actions. For example, in the breeding season male red deer are in competition over access to females. In a contest males roar at one another and perform parallel walks. This display element is probably to gain information on aspects such as the size, fighting skill, and determination of the opponent. Display may be followed by actual physical contact by locking horns; again it is likely contestants are gaining further information during this phase since the level of further contact shown by pairs varies greatly (Clutton-Brock et al., 1979). Overall fighting is a sequential process in which information is gained and used, and this is likely to be an important explanation for the prevalence of limited fighting in nature. We consider this process and its consequences in Chapter 8.

3.6 The Evolution of Signalling: From Cue to Signal

A signal is a trait or behaviour that has been modified by selection for the purpose of influencing other organisms, the signal receivers, by giving them information. Traits can also act as cues, in the sense that they are used by others to gain information, without these traits having evolved for that purpose. An example of a cue in aggressive interactions would be an opponent's size, which can be directly assessed, and more readily for certain postures by the opponent. An important idea about the evolution of signalling is that signals can evolve from cues, through the selective effects of receivers on the cue/signal (Maynard Smith and Harper, 2003; Bradbury and Vehren-camp, 2011). Threat signals emphasizing or exaggerating information about size, for instance lateral display in fish, are possible results of this evolutionary process.

Male traits that are used by females in mate choice could have evolved as signals in this way. As an example, male fiddler crabs have an enlarged major claw that is used in territorial interactions with other males. The claw is metabolically and developmentally costly (Allen and Levinton, 2007), so that males in better condition have the capacity to develop a larger major claw. Fiddler crab males also attract females by performing a wave display with the claw, which is energetically costly (Murai et al., 2009). Fiddler crab females can thus mate with higher-quality males by preferring larger claws. Female choice of males with larger claws might then have resulted in evolutionary exaggeration of this trait, turning it into a signal of male quality.

We can make a simple model illustrating this kind of process. For males with varying quality q ($0 < q < 1$), let x be a trait that influences fitness outside the context of mate choice, in a way that depends on quality. Here we assume that q is uniformly distributed between 0 and 1 and that the trait influences reproductive success in a situation where females are not choosy. Let v denote this basic payoff to a male, which does not include the influence of responses by signal receivers. The basic payoff includes survival and, as in our example, can also include things like the degree of success in male–male competition over territories. For illustration we use the following expression:

$$v(x,q) = u(q)\big(1 - c(x - q)^2\big), \tag{3.7}$$

where $u(q)$ is the basic payoff for $x = q$, which should be an increasing function of q. For a male of quality q the trait x giving optimal reproductive success, apart from effects of female choice, is then $x = q$, and the parameter c sets the cost of deviating from this optimum. Females searching for high-quality males can then use x as a cue for male quality. The result is that male reproductive success is a product of the basic payoff and mating success:

$$w(x,q;\bar{x}) = (1 - b + b\frac{x}{\bar{x}})v(x,q). \tag{3.8}$$

Here b is a parameter ($0 \le b \le 1$) giving the importance of female selectivity. For $b = 0$, the trait does not influence male mating success, and for $b = 1$ the mating success is directly proportional to x. In eq (3.8), $\bar{x}$ is the mean trait in the population of males. The interpretation of the factor $x/\bar{x}$ is that the probability of being chosen is a weighted distribution over the population with the male trait values as weights (other assumptions about female choice are of course possible).

We can regard the male's trait x as a function of the quality q as a strategy $\hat{x}(q)$ in a game between males, with payoff given by eq (3.8). We can call this type of strategy a reaction norm, giving the trait as a function of the condition or state of the individual. For the case of $b = 0$, it is evident that $\hat{x}(q) = q$ is optimal, and thus evolutionarily stable. For $b > 0$, we can find the best response to a resident reaction norm by computing $\bar{x}$ for the resident population, and then finding the best x as a function of q from eq (3.8). For our simple model, we can use calculus, solving $\partial w/\partial x = 0$ analytically (see Exercise 3.5), but in general the optimal trait x can be found numerically. One way to find a Nash equilibrium is to compute the best response to a resident reaction norm and then let this best response be the resident strategy, and so on, until there is convergence, as discussed by McNamara et al. (1997). In the limit the reaction norm is a best response to itself. Figure 3.7 shows optimal

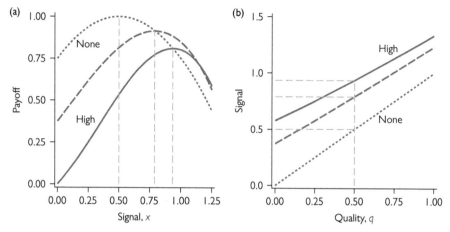

Fig. 3.7 Evolution from cue to signal as a result of receiver selectivity. Three cases that differ in the strength of receiver selectivity are shown: none (dotted), intermediate (dashed), and high (solid line). These correspond to, respectively, $b = 0$, $b = 0.5$, and $b = 1$ in eq (3.8). In all cases $c = 1$ in eq (3.7). (a) The payoff in each case to individuals with quality $q = 0.5$, as a function of their trait x, assuming $u(0.5) = 1$ in eq (3.7). For each case, the population of signallers follow the evolutionarily stable reaction norm for that case, and the vertical line indicates the optimal signal for quality $q = 0.5$. (b) Evolutionarily stable signalling reaction norms for the three cases. The vertical line indicates the quality $q = 0.5$ to which the payoff functions in (a) refer.

traits and evolutionarily stable reaction norms for three values of receiver (female) selectivity. It illustrates the idea of the evolutionary effect of receivers on a cue (the case $b = 0$), turning it into a signal ($b > 0$).

The above analysis does not account for why females should adopt a particular value of b. A complete model would also include details of the costs to females of being selective and the benefits to them of choosing a high-quality male. For instance, for our fiddler crab example, a female might prefer high-quality males because they are better at guarding and maintaining their territories. One would then seek a co-evolutionary outcome where the strategy of males is a Nash equilibrium given the female b value, and this b value maximizes female lifetime reproductive success given the male strategy.

If the quality of a male tends to be passed on to his male offspring the situation is more complex. In this situation if the resident female strategy is to prefer males with high quality ($b > 0$ in the example), then a further reason for a female to mate with a high-quality male is that her sons will be preferred by other females, so resulting in more grandchildren. There is then the potential for female preferences for high-quality males to strengthen, leading to further evolutionary changes in male traits that indicate quality. In order to analyse this phenomenon we need to consider effects that extend beyond one generation. Section 10.5 describes how this is done.

The stable reaction norms for $b > 0$, illustrated in Fig. 3.7, satisfy the so-called handicap principle (Grafen, 1990; Maynard Smith and Harper, 2003). The principle states that the marginal gain of increasing a signal x, defined as $\partial w / \partial x$, should be higher for higher-quality signallers (Exercise 3.6). This ensures that a low-quality signaller cannot gain by 'cheating' with a high-quality signal. Note that the handicap principle does not determine the costs for signallers of different quality at an evolutionary equilibrium. For instance, in the example in Fig. 3.7, low-quality signallers pay a higher cost than high-quality signallers, in the sense that their trait (signal) deviates more from the optimum when there is no female choice (dotted line in Fig. 3.7b). This applies to the particular model and is not a general result. Perhaps the terminology 'handicap' is not a very good guide to intuition for this kind of signal evolution.

The handicap principle has been and still is controversial. One recent argument is that signalling traits are determined (constrained) by general developmental principles (Emlen et al., 2012; Warren et al., 2013), according to which high-quality individuals develop larger traits. However, this is not fully convincing, because development can be tuned by evolution. Rather, developmental evolution is one way for signal evolution to occur, and if so it will occur in accordance with the handicap principle (Biernaskie et al., 2014). We return to the topic of the evolution of signalling in Section 7.4.

3.7 Coordination Games

In many situations two individuals must coordinate their actions. We highlight this scenario with a game that is very artificial but is nevertheless useful in illustrating concepts, both here and in Chapter 4. Suppose that on a particular day each member of a human population must decide whether to drive on the left or right hand side of the road. During the day they encounter precisely one other driver who is driving in the opposite direction. If both drivers have chosen the same side of the road they pass each other safely. If, however, they have chosen to drive on opposite sides of the road they crash head on. Hypothetical payoffs are given in Table 3.3.

A strategy for this game specifies the probability, p, that the right hand side is chosen. A formal analysis of this game is set as Exercise 4.1. Here we note the three Nash equilibria for the game. If everyone always drives on the left hand side (i.e. the

Table 3.3 Payoffs when individuals choose the side of the road on which to drive.

Payoff to focal		Action of other driver	
		Left	Right
Action of focal driver	Left	1	−1
	Right	−1	1

resident strategy is $p = 0$), then the unique best response is also to drive on the left hand side. Thus $p^* = 0$ is a Nash equilibrium strategy. Similarly, always driving on the right hand side is a Nash equilibrium ($p^* = 1$). Finally, if the resident strategy is to decide on which side to drive by the toss of a fair coin (i.e. $p = 0.5$), then any mutant strategy has zero expected payoff and so achieves the maximum possible payoff and is a best response to the resident strategy. In particular the resident strategy of tossing a coin is one of the best responses to itself and so is a Nash equilibrium! However, although the Nash equilibrium condition is necessary for evolutionary stability, as we will show in Chapter 4 it is not sufficient. In particular we show that a resident population with strategy $p^* = 0.5$ is neither stable nor would it have evolved in the first place.

There are various coordination games in the game-theoretic literature. In the Battle of the Sexes each of a couple decides whether to go to the cinema or a football match: they prefer to go together but the wife prefers the cinema and the husband prefers the football match. In the Stag Hunt game, each of two hunters decides whether to hunt a hare or hunt a stag: each can hunt a hare alone but a stag hunt will only be successful if both individuals choose this action. In both cases it is assumed that actions are simultaneous; i.e. each player chooses their action without knowledge of the choice of their partner and is then committed to their chosen action even when their partner's choice becomes known. This is not a reasonable assumption for many biological games in which individuals must coordinate their actions, since individuals will often signal intent or flexibly adjust behaviour in the light of new information.

The direction of coiling in Japanese land snails of the genus *Euhadra* provides a case where the simultaneous choice assumption is reasonable. In these snails the direction of coiling of offspring is determined by a single genetic locus of the mother. Individuals that have opposite handedness in coiling have difficulties mating, so that there is frequency-dependent selection to eliminate the rarer handedness in a population. This leads to the possibility of single-gene speciation (Ueshima and Asami, 2003).

3.8 Produce Sons or Daughters? The Sex-Allocation Game

Mammal and bird species have chromosomal sex determination, viz. the XY and the ZW system, and typically around one half of newborn offspring are sons and half are daughters. In other taxa, the sex ratio of offspring can deviate more strongly from a 1:1 sex allocation; see West (2009) for a review. To understand why a particular strategy for specifying the sex of offspring might be adaptive we first note that sex allocation is frequency dependent: if most individuals in a population produce daughters then it is best for a mutant female to produce sons since they will have the potential to mate with many females; conversely if most produce sons, each will on average produce few offspring because there are few females per male, so that it is best to produce daughters.

Here we analyse the evolutionarily stable sex-allocation strategy in the simplest case. For definiteness we assume discrete non-overlapping generations with a generation time of 1 year. New offspring are produced at the beginning of a year and their parents then die, the offspring mature by the breeding season, mate, produce offspring themselves at the beginning of the next year, and then die, and so on. During the annual breeding season each female chooses a mate at random from the males in the population. This mating results in the production of exactly N offspring. The female is assumed to decide the sex of these offspring.

Even in this simplest case the game payoff cannot be the number of offspring produced, as that is the same for all individuals regardless of their sex-allocation strategy. The sex-allocation strategy of a female does, however, affect the number of matings obtained by offspring. Although this latter quantity is not invasion fitness, as we will show (Section 10.4), it is a fitness proxy (Section 2.5). In our simple example this expected number of matings is proportional to the expected number of grand-offspring. Thus the mean number of grand-offspring also acts as a fitness proxy for the game. Note also that as we are using grand-offspring as the payoff, the genes coding for a strategy should have an autosomal location, because such genes are inherited to the same extent by all grand-offspring. A rare mutant strategy would then correspond to a heterozygote at an autosomal locus.

Consider the allocation decisions of those females that mate in year $t - 1$. These females produce the cohort of newborns present at the start of year t. The cohort of newborns produced at the beginning of year $t + 1$ are their grand-offspring. Each of these grand-offspring has exactly one mother and one father, and these two parents are present in the birth cohort from year t. It follows that the total number of offspring left by the males in the year t birth cohort equals the total number of offspring left by the females in this birth cohort. Thus for the females that mate in year $t - 1$, the total value (in terms of grand-offspring) of all of their sons must equal the total value of all of their daughters. So if the resident allocation strategy of these females results in more sons than daughters, then each son is worth less than each daughter and the best action for a mutant is to produce daughters. Conversely, if more daughters than sons are produced it is best to produce sons. Thus the population allocation strategy can only be evolutionarily stable when it results in the production of equal numbers of sons and daughters at birth.

The idea behind this argument goes back to Düsing (1884) (see Edwards, 2000) and it was further developed by Fisher (1930) and Shaw and Mohler (1953), all before the systematic development of game theory in biology. Note that the argument still holds if during development one sex has greater mortality than the other, so that the adult sex ratio in the breeding season is not 1:1. This can be understood by noting that increasing the mortality on one sex makes it rarer and hence more valuable, and this exactly compensates for the increased mortality. This compensation can be seen in the formal analysis of the case with mortality given in Box 3.1. In the scenario we have discussed a son and a daughter are equally costly to produce. When females have limited resources to allocate to offspring, and the production of an offspring of one sex requires more resources than the other, at evolutionary stability population members

---

Box 3.1 Sex allocation with mortality during maturation

As in the main text we consider a large population with discrete non-overlapping generations. At the start of a year mated females give birth to young and all adults then die. Each
newborn survives to maturity with probability f if it is female and probability m if it is male,
independently of the survival of others. Surviving individuals enter a mating pool in which
each female chooses a single male to father her offspring. She then gives birth to N offspring
at the start of the next year.

We take the sex-allocation strategy of a female to specify the probability, for each
offspring, of being a son. Suppose the resident strategy is to produce sons with probability p.
Then if there are K newborn individuals at the start of a year, $(1 - p)K$ will be female and pK
will be male, so that there will be $(1 - p)Kf$ females and pKm males in the mating pool. The
average number of matings achieved by each male in this pool is the ratio $s(p)$ of the
numbers of females to males, i.e. $s(p) = (1 - p)f/(pm)$.

Consider the allocation strategy of a mutant female within this resident population. Each
daughter she produces must first survive until maturity, in which case she will produce
N offspring. Thus by producing a daughter the mutant female gets an average of $W_f = Nf$ grand-offspring. Similarly, since each surviving son gets $s(p)$ matings on average, the
production of a son results in an average of $W_m = Nms(p)$ grand-offspring. From these
formulae and the expression for $s(p)$ we have

$$W_m - W_f = \frac{Nf}{p}(1 - 2p).$$

Thus if $p < 0.5$ we have $W_m > W_f$, so that the best response for a mutant female is to
produce all sons; i.e. $\hat{b}(p) = 1$. Similarly $\hat{b}(p) = 0$ if $p > 0.5$. Thus any strategy for which
$p \neq 0.5$ cannot be a best response to itself and cannot be a Nash equilibrium. When the
resident strategy is $p = 0.5$, sons and daughters are equally valuable and all mutant strategies
do equally well and are hence best responses. In particular since the resident strategy is a
best response to itself it is a Nash equilibrium.

allocate equal total resources to the production of each sex rather than producing
equal numbers. A proof of this assertion is set as Exercise 3.7.

There are many variants and extensions of this game. For example the equal sex
ratio argument applies if there are overlapping generations. The conclusion can,
however, be altered if there is parental care and differential mortality of sons and
daughters during this period. Competition between sibs can also change the predicted
sex ratio from 1:1. If there is limited dispersal of offspring the sons of a female will
compete with each other for access to matings. Under this local mate competition a
bias in favour of daughters is predicted (Hamilton, 1967). In contrast, if sons disperse
but daughters do not the daughters of a female will be in competition over resources.
Under this local resource competition a bias in favour of sons is predicted. For more
on these topics, see Charnov (1982), Godfrey (1994), and West (2009).

If the environmental circumstances (states) under which offspring are produced
vary within a population, then there will be selection for the sex-allocation strategy

to be state dependent (Section 3.11). If individuals can differ in quality (another aspect of state), and there is a tendency to inherit quality, then again there will be selection for state-dependent sex allocation, although the payoff currency may not be as simple as the number of grand-offspring (Section 10.4).

3.9 Playing the Field

In two-player games such as the Hawk–Dove game and Prisoner's Dilemma the payoff has a linear structure. To illustrate this suppose that the resident strategy in the Hawk–Dove game results in half the residents playing Hawk and half playing Dove. Then the payoff to a mutant individual in this population is the average of their payoff when the resident strategy is to always play Dove and their payoff when the resident strategy is to always play Hawk.

The sex-allocation game is a game in which individuals 'play the field' rather than being a two-player game, and does not have this linear structure. For example, consider the payoff to a mutant individual that produces all sons. This approaches zero when the resident strategy approaches production of only sons (since there are no females for the mutant's sons to mate with), approaches infinity when the resident strategy approaches production of only daughters (since during breeding every female mates with one of the mutant's sons), and is one when the resident strategy is to produce equal numbers of sons and daughters. The payoff of one is not the average of zero and infinity, so this game has a non-linear payoff structure.

Most playing-the-field games are non-linear. For example consider elephant seals, in which the largest males get almost all the matings. Suppose that a strategy specifies the size of a male (this is rather simplistic—it would be more realistic to have a strategy specify growth and foraging decisions). Then the payoff to being medium sized is large if all residents are small, is very small if all residents are large, but is also very small if half the residents are small and half are large (and hence larger than the medium-sized individual).

The signalling game in Section 3.6 is a non-linear playing-the-field game, and so is the following example.

3.10 Dispersal as a Means of Reducing Kin Competition

The seeds of plants and the offspring of many animals disperse before settling on a site, maturing, and reproducing themselves. Dispersal is clearly beneficial if the environment is heterogeneous with the possibility that there are empty sites to colonize. But what if the environment is spatially homogeneous with all possible sites essentially identical to the current site? Somewhat surprisingly, Hamilton and May (1977) showed that some dispersal of offspring is adaptive even when this holds. The idea behind their model is that dispersal reduces the competition among kin, so that a dispersing individual increases the prospects of kin left behind even though it may pay a cost itself. Here we present a simplified version of their model.

We consider a large asexual population with discrete, non-overlapping generations. The life cycle of population members is as follows. In the environment there is a large number of sites. During maturation each site is occupied by exactly one individual. Those individuals without a site die during this phase. Those with a site mature, produce m offspring, and then die. We assume that m is large. Each offspring disperses with some probability d independently of others. Dispersing individuals die with probability μ. Those that survive migrate to a site chosen at random. At each site, the offspring that did not disperse from the site and immigrants to the site all compete for possession of the site, with each competitor equally likely to be the lucky individual. This individual then matures, and so on. We assume that the dispersal probability d is genetically determined and seek its evolutionarily stable value.

Before analysing this model, we note that we might expect some dispersal to evolve. This is because a genotype that did not disperse would always be restricted to the same site and its line would go extinct if in some generation offspring were outcompeted for the site by an immigrant. As Hamilton and May say, at least one offspring must migrate whatever the odds of successful migration.

Let the resident dispersal probability be d. On a site previously occupied by a resident the $m(1 - d)$ offspring that remain compete with $md(1 - \mu)$ immigrants for possession of the site. The probability that each of these manages to secure the site is the inverse of the total number of competitors; i.e. is $p = 1/[m(1 - d\mu)]$. Consider a rare mutant strategy with dispersal probability d'. An individual following this strategy leaves $m(1 - d')$ offspring on its site and these compete with $md(1 - \mu)$ non-mutant immigrants for possession of the site, so that each secures the site with probability $p' = 1/[m(1 - d') + md(1 - \mu)]$. The expected number of mature offspring produced by a mature mutant is thus

$$W(d',d) = (1 - d')mp' + d'm(1 - \mu)p. \qquad (3.9)$$

In deriving this formula we have implicitly assumed that the dispersing offspring of a given parent go to different sites. Simple differentiation reveals that this function has a unique maximum when $d' = \hat{b}(d)$ where

$$\hat{b}(d) = 1 + d(1 - \mu) - \sqrt{d(1 - d\mu)}. \qquad (3.10)$$

Figure 3.8 illustrates this best response function. From eq (3.10) the dispersal probability d^* is a best response to itself if and only if

$$d^* = \frac{1}{1 + \mu}. \qquad (3.11)$$

The above shows that at evolutionary stability at least one half of offspring disperse, even when the mortality while dispersing approaches $\mu = 1$. For small mortalities the dispersal probability is close to 1.

We have derived the formulae under the assumption that the number of offspring, m, is large, so that the variation in the numbers of migrants and other measures is small compared to their mean, making it reasonable to take mean values as actual values. When m is not large, the actual distribution of immigrants and those who do not migrate must be taken into account (Comins et al., 1980). Furthermore, if there is

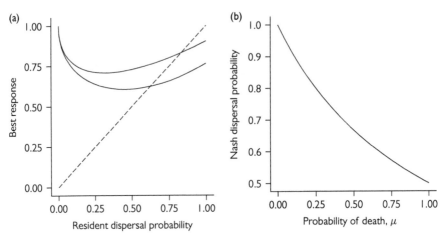

Fig. 3.8 Dispersal to avoid kin competition. (a) The best response dispersal probability as a function of the resident dispersal probability (upper curve $\mu = 0.2$, lower curve $\mu = 0.6$, dashed line is the 45° line). (b) The Nash equilibrium dispersal probability as a function of the probability of death during dispersal.

variation in offspring number, then we might expect kin competition, and hence the evolutionarily stable dispersal probability from a site, to be plastic and depend on the actual number of offspring at the site (Kisdi, 2004).

In the model each site supported exactly one individual. When a site can support a number of maturing individuals the situation becomes more complex, since measuring the mean number of descendants left by a mutant is no longer an appropriate fitness measure; it matters where these descendants are. Dispersal reduces local numbers at the natal site in the next generation and hence relaxes competition for those relatives left behind. One approach to this situation is to use the 'direct fitness' method of Taylor and Frank (1996). We describe this method in Section 4.5. However, there are complications even with this method, since the best dispersal rate depends on the relatedness to others on a site, but relatedness depends on the dispersal strategy. Thus whatever method is used it is necessary to implicitly or explicitly consider the distribution of the numbers of mutant individuals within sites over successive generations. For an example of the analysis of a model with more than one mature individual per site, see Cadet et al. (2003).

3.11 Threshold Decisions

There is always phenotypic variation in populations. Some of this is due to genetic variation, but even within genetically identical individuals such as identical twins, there is phenotypic variation that has arisen because of developmental noise and differences in past experience. Past experiences can affect physiological states, e.g.

through chance while foraging or whether a disease is caught, and can affect what an individual learns about the environment or the social group. We can describe phenotypic variation in term of state variables: energy reserves, body size, muscle mass, skills, social standing including position in the dominance hierarchy, etc. Some variables might be relatively fixed, others vary over shorter timescales. Whatever their origin, it is often adaptive for the actions chosen by an organism to depend on these state variables. For example, whether it is best for a male to challenge another over access to a female may depend on the size of the male and his rival. The choice of action may also depend on knowledge of factors external to the organism: an estimate of current food availability, perceived threat from conspecifics or predators, current weather conditions, and so on. Such information can also be considered as part of the state of the organism. As discussed in Section 2.1 a strategy is a rule that chooses actions contingent on state. We regard strategies as genetically determined and seek strategies that are evolutionarily stable.

In many circumstances an organism must make several decisions during a time interval over which its state varies. Here, however, we restrict attention to state-dependent decision making when each individual is characterized by a single state variable and makes a single choice between two actions based on the value of this variable. Three examples are considered: pairwise contests over a resource when fighting abilities differ, alternative male mating strategies when the size as a juvenile varies, and environmental sex determination when environmental temperature varies. For the payoff structure assumed in each of these examples, the evolutionarily stable decision rule is to choose one of the actions when state is below a critical threshold and choose the other action above the threshold.

3.11.1 Pairwise Contests Where Individuals Know Their Own Fighting Ability

Consider the Hawk–Dove game, modified so that population members differ in their fighting ability. Ability lies between 0 and 1 and is set during development; irrespective of genotype the ability of each individual is drawn from a uniform distribution between 0 and 1, independently of other individuals. In a fight between an individual of ability q' and an opponent of ability q, the probability that the q' individual wins the fight is a function $\alpha(q', q)$ of these two abilities. We assume that the probability an individual wins a fight increases as the individual's ability q' increases or the opponent's ability q decreases. In a contest each individual knows their own ability but not that of their opponent, so that the choice of action can depend on own ability. We seek a strategy (i.e. a rule for choosing whether to play Hawk or Dove as a function of own ability) that is evolutionarily stable. Note that since individuals following a strategy vary in ability, the payoff to a mutant strategy (for a given resident strategy) is an average over the payoffs obtained by these various mutant individuals (Box 3.2). Thus, if under a mutant strategy each individual maximizes their own expected payoff, then the payoff to the mutant strategy is maximized, and is hence a best response to the resident strategy—a principle of individual optimization.

Box 3.2 Payoffs when individuals know their own fighting ability

Suppose that the resident strategy has threshold x. An individual of ability q' within this resident population can meet residents of different abilities. Those residents with ability $q < x$ play Dove, while those with ability $q > x$ play Hawk. Averaging over these abilities, if the individual plays Dove its mean payoff is

$$W_D(q';x) = \int_0^x \frac{V}{2} dq. \tag{3.12}$$

Similarly

$$W_H(q';x) = \int_0^x Vdq + \int_x^1 \left[\alpha(q',q)V - (1 - \alpha(q',q))C\right]dq \tag{3.13}$$

is the mean payoff if the individual plays Hawk. The payoff $W(x',x)$ to a mutant strategy with threshold x' when the resident threshold is x is an average over the payoffs of mutants of different abilities; i.e.

$$W(x',x) = \int_0^{x'} W_D(q';x)dq' + \int_{x'}^1 W_H(q';x)dq'. \tag{3.14}$$

Differentiating W in (3.14) with respect to x' gives

$$\frac{\partial W}{\partial x'}(x',x) = W_D(x';x) - W_H(x';x). \tag{3.15}$$

From this equation it can be shown that the best response, $\hat{b}(x)$, is unique and satisfies

$$W_D(\hat{b}(x);x) = W_H(\hat{b}(x);x) \tag{3.16}$$

when $V < C$. Thus a Nash equilibrium threshold x^* satisfies

$$W_D(x^*;x^*) = W_H(x^*;x^*). \tag{3.17}$$

Exercise 3.9 gives an analytic expression for x^* when the function α is given by eq (3.18).

For this scenario it can be shown that, whatever the resident strategy, the best response has a threshold form; i.e. there exists x such that the best response strategy is:

if ability $q < x$ play Dove

if ability $q > x$ play Hawk.

Since a Nash equilibrium strategy is necessarily a best response to itself, it will also have this form. Thus in seeking a Nash equilibrium strategy we restrict attention to threshold strategies. In particular, we henceforth regard a strategy as specified by a threshold. We illustrate ideas using the specific function

$$\alpha(q',q) = \frac{1}{2} + \frac{1}{2}(q' - q) \tag{3.18}$$

for the probability of winning a fight. For this function, two opponents with the same ability each win the fight with probability 0.5. As the difference in ability increases

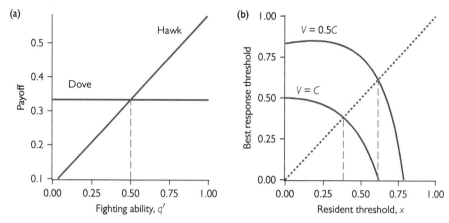

Fig. 3.9 The Hawk–Dove game with differences in fighting ability. (a) The payoff to a mutant individual as a function of its ability (q') and its action (Hawk or Dove) when the resident strategy is to play Hawk if ability is above the threshold $x = 2/3$ ($V = 1, C = 2$). The best response against this resident population is to use threshold $\hat{b}(2/3) = 0.5$, where the payoffs from playing Hawk and Dove are equal. (b) The best response threshold as a function of the resident threshold. Two cases are shown: $V/C = 0.5$, which has a Nash equilibrium at $x^* = 0.613$, and $V = C$, which has a Nash equilibrium at $x^* = 0.382$.

so does the probability that the more able opponent wins. At the extreme when one individual has ability 1 and the other ability 0, the ability 1 individual always wins.

Consider the expected payoff to a mutant individual of ability q' that plays Dove. This is an average over that obtained as the possible qualities of the opponent varies, taking into account the behaviour of the opponent; and this latter is determined by the resident strategy (Box 3.2). Similarly, the expected payoff from action Hawk is an average. Figure 3.9a shows these expected payoffs as a function of q' for a specific resident strategy. It can be seen that it is best to choose action Dove if q' is less than some critical ability $\hat{x}$ and it is best to choose Hawk if ability is above $\hat{x}$. By the above principle of individual optimization the (unique) best response to the resident strategy is specified by the critical threshold $\hat{b}(x) = \hat{x}$. Figure 3.9b illustrates the best response function for two values of the resource V. As can be seen, for each value of V there is a unique Nash equilibrium.

One of the main conclusions from this specific model is that the level of fighting at evolutionary stability is reduced in comparison with the standard Hawk–Dove game (McNamara and Houston, 2005). This can be illustrated by the case where $V = C$ in Fig. 3.9b. In this case the Nash equilibrium threshold is $x^* = 0.382$, so that a proportion $1 - 0.382 = 0.618$ of the population plays Hawk, resulting in $0.618^2 \approx$ 38% of contests resulting in fights. In comparison the Nash equilibrium proportion of Hawks is $p^* = V/C = 1$ in the standard Hawk–Dove game, so that every contest is settled by a fight. Fighting is reduced in the state-dependent formulation because only the individuals with high fighting ability play Hawk.

This model also illustrates a number of very general conclusions that also apply to the cases of alternative mating tactics and environmental sex determination that we present below:

- Selection acts on strategies, i.e. rules for choosing actions as a function of state. For this and many other scenarios the payoff to a strategy is an average of the payoffs of those individuals following the strategy. Individuals following a strategy are only important insomuch as they contribute to this average (a selfish gene perspective). The payoff to a strategy is maximized when each individual following the strategy maximizes their own payoff; a principle of individual optimization.
- Unlike the corresponding game without state differences, when individuals differ in state there are no mixed strategies at equilibrium; the choice of action is deterministic, but depends on state. Since state differences abound in nature this will very often be true. In this sense modelling mixed strategies is mathematically convenient but biologically unrealistic in most scenarios.
- In the corresponding game without state differences, at the mixed or polymorphic Nash equilibrium all population members have equal payoffs. This is no longer true once state differences are included. In the Hawk–Dove game with differences in ability, at the Nash equilibrium those individuals that play Hawk are all doing better than those that play Dove.

3.11.2 Alternative Male Mating Tactics For Pacific Salmon

In many species some males attempt to defend territories and attract females to their territories, while other males attempt to sneak matings with females that enter these territories. For example, in Coho salmon the larger 'hook' males defend territories while the smaller 'jacks' attempt to sneak matings.

At one extreme it might be that the reproductive tactics adopted are both genetically determined, for example they might be the result of alternative alleles at a single locus. If this is the case the two alternative reproductive strategies will have equal fitness at equilibrium.

At the other extreme, the tactic adopted might be conditional on some state variable such as body size. For example, in Coho salmon it appears that the tactic adopted may depend on size as a juvenile; larger individuals mature early as jacks while smaller individuals grow for longer and mature as the larger hooks. If the tactic is a result of such a conditional strategy, then as in the Hawk–Dove game where individuals know their own fighting ability, we would not expect each tactic to do equally well on average. In Coho salmon the suggestion is that the jacks, who have an early growth advantage, do better (Gross, 1996). For more on these ideas, see Gross (1996) and for a game theoretical analysis, see chapter 3 of Charnov (1993).

3.11.3 Environmental Sex Determination

In many species the sex of offspring is not determined by the offspring genotype but can depend on other circumstances of the mother or the developing offspring. In other words, sex determination is state dependent.

In many reptile species the sex of offspring is determined by the temperature at which the eggs develop. For example, in their study of sex determination in alligators, Ferguson and Joanen (1982) found that at hatching all offspring were female if the eggs developed at less than 30°C and all were male if eggs developed at greater than 34°C. Ferguson and Joanen suggest possible adaptive reasons for this phenomenon, but it is, of course, not easy to perform controlled experiments to find out what would have happened if the offspring had developed as the other sex. However, through hormonal manipulation of eggs the differential effects of temperature on the sexes can be experimentally explored. Using this technique on Jacky dragon lizards, where normally females are produced at low and at high temperatures and a mixture of males and females at intermediate temperatures, Warner and Shine (2008) found that males from intermediate incubation temperatures had greater lifetime reproduction than those from either low or high temperatures, and there was a tendency towards an inverted pattern for females. This kind of explanation for temperature-dependent sex determination—that the sexes differ in how their growth, survival, or reproduction are influenced by temperature during development, was put forward by Charnov and Bull (1977). After much work, it has proven difficult to establish general empirical support for the hypothesis, and this research is ongoing (e.g., Pen et al., 2010; Schwanz et al., 2016).

Motivated by the above we present a model of temperature-dependent sex determination in a hypothetical reptile. As in Section 3.8 we assume non-overlapping generations and a yearly life cycle in which each female mates once, choosing her mate at random. She then lays N eggs. The population is large and the temperature experienced by different clutches is drawn independently from some temperature distribution that is the same each year. These are simplifying assumptions; reptiles are often long-lived, have overlapping generations where the sexes differ in age at maturity, and the distribution of temperatures typically varies from year to year and between local habitats and populations, which all can matter for the analysis. In our model, if an egg experiences temperature q, then the developing individual survives to maturity with probability $f(q)$ if it develops as female and with probability $m(q)$ if it develops as male. We assume that the ratio $f(q)/m(q)$ is an increasing function of q.

For this scenario, in finding the Nash equilibrium strategy we can again restrict attention to strategies of threshold form; i.e. there is a threshold x such that males are produced if the temperature lies below x and females are produced if the temperature exceeds x.

Suppose that the resident population strategy has threshold x. Let $F(x)$ denote the average number of surviving female offspring left by a randomly selected female member of this population. Note that this is an average over the different temperatures experienced by the developing offspring of different population members. Similarly, let $M(x)$ be the average number of surviving male offspring left by a randomly selected female. Thus the ratio of females to males in the breeding population is $s(x) = F(x)/M(x)$. Since females choose mates at random, each male in the breeding population (whether it is a resident male or a mutant male) gets an average of $s(x)$ matings.

Now consider the cohort of individuals following the rare mutant strategy with threshold x' in this resident population. For this cohort the numbers of surviving daughters and sons left per cohort female are $F(x')$ and $M(x')$, respectively. Each surviving daughter mates once. Each surviving son mates an average of $s(x)$ times. Thus the offspring of a randomly selected female in this cohort mate an average of

$$W(x',x) = F(x') + M(x')s(x) \qquad (3.19)$$

times. As we show in Section 10.4 $W(x',x)$ is a fitness proxy. Thus in seeking a Nash equilibrium strategy we can take $W(x',x)$ to be the payoff function. $F(x')$ and $M(x')$ can be written as integrals (cf. the analogous integrals for the Hawk–Dove game in Box 3.2). Then setting the partial derivative of $W(x',x)$ with respect to x' equal to zero at $x' = \hat{b}(x)$, it can be seen that the best response to x satisfies $f(\hat{b}(x)) = m(\hat{b}(x))s(x)$. Not surprisingly, $\hat{b}(x)$ is the temperature at which male and female offspring do equally well. At the Nash equilibrium the critical threshold x^* satisfies

$$\frac{f(x^*)}{F(x^*)} = \frac{m(x^*)}{M(x^*)}. \qquad (3.20)$$

Note that at the Nash equilibrium the sex ratio of developing females to males is no longer necessarily 1:1 (cf. Exercise 3.10).

Since $W(x',x)$ is a fitness proxy, so is $W(x',x)/F(x)$; i.e.

$$F(x')/F(x) + M(x')/M(x) \qquad (3.21)$$

is a fitness proxy. This proxy is what Charnov (1982) calls the Shaw–Mohler equation. We also note that in some special cases the product $F(x)M(x)$ is maximized at the Nash equilibrium (see e.g. Leimar, 2001).

The above model was motivated by temperature-dependent sex determination in reptiles, but applies with minor modification to host size in parasitic wasps. For some wasp species, females that develop from large hosts are bigger and more fecund than those from small hosts, while there is no comparable large-host advantage for developing males. Consequently the Nash equilibrium laying strategy for the parasitoid is to deposit sons in small hosts and daughters in large hosts, and this pattern is observed (Godfrey, 1994).

3.12 Assessment and Bayesian Updating

In all of the two-player games that we have dealt with so far, when a pair is formed partners have no specific information about each other. In real interactions this is often not the case. For example, when two individuals contest a resource, each may recognize their relative size. There may also be threats and displays that have the potential to give information on abilities or intentions to the opponent. In this section we develop a model in which individuals differ in quality and each individual uses an observation in order to assess the difference in quality between itself and the opponent. They then base the decision of whether to be aggressive or submissive in the contest on this observation. For the analysis we use Bayes' theorem to take information into account. In this approach, the distribution of qualities in the population gives what is referred to as a prior distribution of quality differences. Once the observation is made, Bayes' theorem combines the prior distribution and the observation to form a posterior distribution, which incorporates all current information. The action of each contestant then depends on their own posterior distribution.

3.12.1 A Model of Animal Conflict with Assessment

Individuals vary in some quality q, which we assume is normally distributed in the population with mean μ_q and standard deviation σ_q. We might, for instance, have the logarithm of body size as q, or some other aspect of fighting ability.

When a pair meet there are two stages. In the first stage, the contestants observe some measure ξ of relative quality. For the interaction between contestants with qualities q' and q, the observation by the q' individual is

$$\xi' = q' - q + \epsilon', \tag{3.22}$$

where ϵ' is an error of observation, assumed to be normally distributed with mean 0 and standard deviation σ. Similarly, the observation by the quality q individual is $\xi = q - q' + \epsilon$. We assume that the errors ϵ' and ϵ are independent.

In the second stage each performs one of two actions, A and S, where A is an aggressive display or actual fighting behaviour and S is a subordinate or submissive display. If both contestants use action S, each has a 50% chance of obtaining a resource of value V (or, alternatively, they share it). If one uses A and the other S, the aggressive individual gets the resource. If both use A, we assume that they are preoccupied with aggressive behaviour, and neither gets the resource (for instance, some other, unspecified agent gets the resource, or, if it is a potential mate, that the resource moves on). An AA interaction incurs a cost from exhaustion or injuries. The cost to the q' contestant is

$$C\exp(-q' + q). \tag{3.23}$$

The observation in eq (3.22) is statistically related to this cost.

Let x be a resident strategy, specifying a critical observation threshold: given the observation ξ, be submissive if $\xi \le x$ and aggressive if $\xi > x$. Consider a rare mutant in this resident population. Before the mutant observes its partner, the distribution of qualities in the population specifies the probability distribution of possible values of the difference $h = q' - q$ (Fig. 3.10a). Once the mutant observes ξ' this prior information can be updated to give a revised distribution of h, known as a posterior distribution (Box 3.3). Figure 3.10a illustrates the posterior distribution for two values of ξ'. Using the posterior distribution, the payoffs for using A vs. S for a mutant that observes ξ' can be worked out (Box 3.3). These payoffs determine which is the best action given the observation. As Fig. 3.10b illustrates, this determines the best response, $\hat{b}(x)$, to the resident strategy. As usual a Nash equilibrium strategy x^* satisfies $\hat{b}(x^*) = x^*$ (Fig. 3.10c). For this model there is always a unique Nash

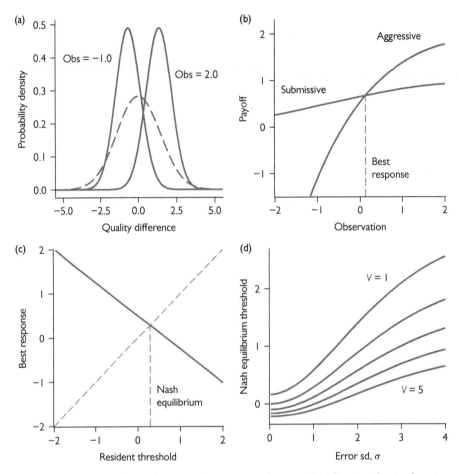

Fig. 3.10 Animal conflict with an initial assessment of partner. (a) The prior density function for quality differences (dashed line) and the posterior density functions for observations $\xi' = -1.0$ and $\xi' = 2.0$. (b) The payoffs to the two actions as a function of the observation ξ' when the resident strategy is to use observation threshold $x = 0.5$. The best response to this resident strategy is to use observation threshold $\hat{b}(x) = 0.1255$. (c) The best response threshold $\hat{b}(x)$ as a function of the resident threshold x. The Nash equilibrium threshold is at $x^* = 0.2857$. (d) The Nash equilibrium threshold as a function of the error standard deviation, σ, for five values of the resource $V = 1, 2, 3, 4, 5$. $\sigma = 1$ and $V = 2$ in (a)–(c). $\sigma_q = 1$ and $C = 1$ throughout.

equilibrium strategy for given parameter values. This strategy is also the unique best response to itself. Figure 3.10d shows the dependence of the Nash equilibrium strategy on the error standard deviation σ for a range of V values. This kind of Nash equilibrium, where players use Bayesian updating of information about each other, is sometimes called a Bayesian Nash equilibrium.

Box 3.3 Bayesian updating with relative quality

The qualities, q' and q, of two randomly selected contestants are independent and normally distributed with mean μ_q and variance σ_q^2. The difference in quality, $h = q' - q$ is normally distributed with mean 0 and variance $2\sigma_q^2$. This is the prior distribution of quality difference before any observation. On obtaining the observation ξ' the q' contestant updates this prior to a posterior distribution using Bayes' theorem. This theorem shows that the posterior density function for h is proportional to the product

$$\text{prior density of } h \times \text{ conditional density of } \xi' \text{ given } h. \tag{3.24}$$

From eq (3.22), the observation ξ' given h is normally distributed with mean h and variance σ^2. After some manipulation, using eq (3.24), one finds that the posterior distribution of h given ξ' is normal with mean $\kappa\xi'$ and variance $\kappa\sigma^2$, where $\kappa = 2\sigma_q^2/(2\sigma_q^2 + \sigma^2)$.

Now let x be the resident threshold strategy of using action A if $\xi > x$. For a mutant–resident pair with difference h, the probability that the resident individual uses A is then $p_A(h;x) = P(\epsilon > x + h)$. The payoffs for choosing action A and S, given a true quality difference h is then

$$w_A(h;x) = (1 - p_A(h;x))V - p_A(h;x)Ce^{-h} \tag{3.25}$$

$$w_S(h;x) = (1 - p_A(h;x))\frac{V}{2},$$

where we used eq (3.23) for the cost. The mutant individual of course does not know h. Instead, using the posterior distribution, the mutant payoffs for A and S given the observation ξ' are

$$W_A(\xi';x) = \int_{-\infty}^{\infty} w_A(h;x)f(h|\xi')dh \tag{3.26}$$

$$W_S(\xi';x) = \int_{-\infty}^{\infty} w_S(h;x)f(h|\xi')dh,$$

where $f(h|\xi')$ is the posterior probability density function.

3.12.2 How the Model Can Be Used

This model was originally used as a building block for the sequential assessment game (Enquist and Leimar, 1983), which we deal with in Section 8.7. We also use the model to study social dominance, in Sections 5.4 and 8.6. In social dominance interactions, individuals stay together in a social group where they recognize each other and show aggressive and submissive behaviours. It turns out that the kind of Bayesian updating we use here (Box 3.3) in practice is unworkable for that situation. This is because of the complexity of a complete representation of an individual's information about other group members, involving probability distributions in high-dimensional spaces, as mentioned in Section 1.5. Instead, in Chapter 5 we will assume that individuals have special behavioural mechanisms for learning the actions to use during their stay in a group.

3.13 Exercises

Ex. 3.1. Consider a two-player game in which there is a common benefit but individual costs. In this game the payoff to an individual that contributes effort x' when its partner contributes x is $W(x',x) = B(x'+x) - C(x')$, where $B(u) = 2u - u^2$ for $0 \leq u \leq 1$ and $B(u) = 1$ for $u > 1$, and $C(x') = x'^2$. Find the best response function $\hat{b}(x)$ for this game and show that the unique Nash equilibrium is $x^* = \frac{1}{3}$.

Ex. 3.2. Consider a two-player game in which the payoff to an individual that contributes effort x' when its partner contributes x is $W(x',x) = B(x'+x) - C(x')$. Assume that B is increasing with diminishing returns; i.e. $B'(u) > 0$ and $B''(u) < 0$ for all $u \geq 0$. Assume that $C'(0) = 0$ and that costs are increasing and accelerating; i.e. $C'(x') > 0$ and $C''(x') > 0$ for all $x' \geq 0$. Show the best response function $\hat{b}$ satisfies $B'(\hat{b}(x) + x) = C'(\hat{b}(x))$. By differentiating both sides of this expression with respect to x show that

$$\hat{b}'(x) = \frac{B''(\hat{b}(x) + x)}{C''(\hat{b}(x)) - B''(\hat{b}(x) + x)}.$$

Deduce that $-1 < \hat{b}'(x) < 0$, so that it is best for an individual to compensate incompletely for the reduced effort of its partner.

Ex. 3.3. A group of n animals are foraging together. Each must decide the proportion of the day spent feeding as opposed to being vigilant for predators. During the day there will be one predator attack. If at least one of the group detects the predator as it attacks, all survive. If none detects the predator, the predator kills one randomly selected group member.

Suppose that if an individual spends a proportion of time x feeding it detects the predator attack with probability $1 - x^4$. Consider a mutant individual that feeds for a proportion of time x' when all others feed for x. Then assuming all group members detect the predator attack independently, the probability the attack is not detected is $x'^4 x^{4(n-1)}$, so that the probability the focal individual dies is $x'^4 x^{4(n-1)}/n$. Suppose that the payoff to the mutant is x' if it survives the day. Thus its mean payoff is $W(x',x) = (1 - x'^4 x^{4(n-1)}/n)x'$. Show that the Nash equilibrium proportion of the day feeding, $x^*(n)$, is given by $x^*(1) = 0.669$, $x^*(2) = 0.892$, $x^*(3) = 0.958$, $x^*(4) = 0.986$ with $x^*(n) = 1$ for $n \geq 5$.

In this example vigilance decreases with group size for two reasons. The 'many eyes effect' means that each relies on others to detect the predator attack. The 'predator dilution effect' means that individuals are relying on the predator taking another group member if the attack is successful. When $n \geq 5$ individuals are not vigilant at all and are relying on this latter effect. For more on the influence of group size on vigilance levels, see McNamara and Houston (1992a).

Ex. 3.4. Two parents are caring for their common young. If the female contributes effort x and the male contributes y the payoffs to the female and male are $W_f(x,y) = B(x+y) - C_f(x)$ and $W_m(x,y) = B(x+y) - C_m(x)$, respectively. Here the probability the young survive is given by $B(u) = 2u - u^2$ for $0 \le u \le 1$ and $B(u) = 1$ for $u > 1$. The costs are $C_f(x) = x^2$ and $C_m(y) = 2y^2$, so that the female pays a lower cost than the male for a given level of effort. Find the pair of best response functions $\hat{b}_f(y)$ and $\hat{b}_m(x)$. Hence show that the Nash equilibrium efforts are $x^* = \frac{2}{5}$ and $y^* = \frac{1}{5}$. Note that at this equilibrium the cost paid by the female is greater than that paid by the male.

Ex. 3.5. Using eq (3.8), show that the optimal value of the trait x for a male of quality q is given by

$$\hat{x}(q) = \frac{2}{3}q - \frac{(1-b)\bar{x}}{3b} + \sqrt{\frac{1}{3c} + \frac{q^2}{9} + \frac{2(1-b)q\bar{x}}{9b} + \frac{(1-b)^2\bar{x}^2}{9b^2}},$$

where $\bar{x}$ is the mean trait in the resident population.

Ex. 3.6. Show that the payoff function in eq (3.8) satisfies the handicap principle, i.e. that $\partial/\partial q(\partial w/\partial x) > 0$.

Ex. 3.7. In a population with discrete non-overlapping generations each breeding female has total resource R to invest in offspring. Each daughter produced requires resource r_f, each son requires r_m. Thus if a female invests a proportion x of total resources in sons, ignoring mortality and the fact that numbers of offspring must be integers, she contributes $\frac{(1-x)R}{r_f}$ daughters and $\frac{xR}{r_m}$ sons to the breeding population in the next generation. Take the payoff to a female to be the expected number of matings obtained by her offspring (under random mating). Show that if a mutant female invest a proportion x' of her resources into sons when the resident proportion is x her payoff is

$$W(x',x) = \frac{R}{r_f}\frac{1}{x}(x+x'-2xx').$$

Deduce that the Nash equilibrium proportion of resources allocated to sons is $x^* = \frac{1}{2}$.

Ex. 3.8. Suppose that when two population members contest a resource, they do not fight, rather each waits for the other to give up and go away. The other then immediately gets the resource. A contest of this sort is known as a War of Attrition (Maynard Smith, 1974). Let the value of the resource be v and suppose that there is a cost c per unit time spent waiting. This latter cost is paid by both contestants, but only the more persistent individual gets the reward. The payoff to each contestant is the expected reward minus the expected waiting cost.

If the resident strategy were to wait for a fixed time t before giving up, then a mutant that waited a tiny amount longer would do better than residents. Thus no fixed time can be evolutionarily stable. To analyse the game, restrict attention to waiting times that have an exponential distribution. Let $W(x',x)$ be the payoff to a mutant with a waiting time that is exponentially distributed with parameter x' when the resident strategy is to wait for an exponential time with parameter x. Show that

$$W(x',x) = \frac{xv - c}{x + x'}.$$

Deduce that the evolutionarily stable waiting time is exponential with parameter $x^* = \frac{c}{v}$ (so that the mean waiting time is $\frac{v}{c}$). Note that at evolutionary stability the payoff to residents and any mutant is zero.

Ex. 3.9. Use the formulae in Box 3.2 to find the functions $W_D(q';x)$ and $W_H(q';x)$ when α is given by eq (3.18) and $V < C$. Hence show that

$$\hat{b}(x) = \frac{3C - V - 2Cx - (V+C)x^2}{2(1-x)(V+C)}. \tag{3.27}$$

Show that the Nash equilibrium is given by

$$x^* = \frac{(r+2) - \sqrt{2r^2 + 2r + 1}}{(r+1)}, \tag{3.28}$$

where $r = V/C$.

Ex. 3.10. Each year each mature female in a population lays N eggs. The population is large and the temperature experienced by different clutches is drawn independently from a distribution that is uniform between 20°C and 30°C. If an egg experiences temperature q, then the developing individual survives to maturity and breeds with probability $f(q) = 0.1(q - 20)$ if it develops as female and with probability $m(q) = 0.5$ if it develops as male. Restrict attention to strategies of threshold form: thus for threshold x males are produced if $q < x$ and females if $q > x$. Let $20 < x < 30$. In the notation of the main text show that $F(x) = (1 - \beta^2)/2$ and $M(x) = \beta/2$ where $\beta = (x - 20)/10$. Hence show that the Nash equilibrium threshold value is $x^* = 20 + 10/\sqrt{3} = 25.77$.

Note that at the Nash equilibrium the proportion of eggs that develop as males is $1/\sqrt{3} = 0.577$, which is not equal to $1/2$. The ratio of breeding females to males is $2/\sqrt{3} = 1.155$.

4

Stability Concepts: Beyond Nash Equilibria

Gene frequencies in a population change over time as result of natural selection and random processes. So far we have not specified the details of these evolutionary dynamics. Nevertheless, in Chapter 1 we argued that if the dynamics have a stable endpoint, then at this endpoint no mutant strategy should outperform the resident strategy so that the resident strategy is a Nash equilibrium; i.e. condition (2.1) holds for invasion fitness λ, or equivalently condition (2.4) holds for a fitness proxy W. However, we have yet to deal with two central issues, which form the main focus of this chapter:

- *Stability against invasion by mutants.* The Nash condition is necessary for stability but is it sufficient? For example, in the Hawk–Dove game with $V < C$, at the Nash equilibrium every mutant strategy does equally well as the resident. So can mutant numbers increase by random drift, changing the population composition? In considering conditions that are sufficient to ensure stability we will assume that mutants arise one at a time and their fate is determined before any other mutant arises. As we describe, this leads to the concept of an Evolutionarily Stable Strategy (ESS).
- *Dynamic stability and attainability.* Even if a Nash equilibrium cannot be invaded by new mutants, if the initial population is not at the equilibrium, will the evolutionary process take the population to it? To consider this question we introduce the idea of convergence stability. As we describe, a strategy x^* is convergence stable if the resident strategy evolves to this strategy provided that the initial resident strategy is sufficiently close to x^*. In other words, if the whole population is perturbed away from x^* then it will evolve back to x^*.

As we will illustrate, some strategies are both an ESS and convergence stable. However, there are ESSs that are not convergence stable, and so are not attainable. Furthermore there are strategies that are convergence stable but not an ESS, so that evolution can home in on a strategy that is not evolutionarily stable, and indeed may be a fitness minimum. Adaptive dynamics (Sections 4.2) is an approach to study the evolution of continuous traits, and it can also be used to analyse evolution to a fitness minimum, followed by evolutionary branching (Section 4.3). Replicator dynamics (Section 4.4) is another approach that is frequently used in game theory. It describes the change

Game Theory in Biology: Concepts and Frontiers. John M. McNamara and Olof Leimar,
Oxford University Press (2020). © John M. McNamara and Olof Leimar (2020).
DOI: 10.1093/oso/9780198815778.003.0004

in the frequencies of a fixed number of strategies. We end the chapter by examining evolutionary stability in populations that are structured through genetic relatedness (Section 4.5).

4.1 Evolutionarily Stable Strategies

Consider a large population in which individuals are paired at random for a two-player game. An individual following strategy x' in this game gets payoff $W(x', x)$ when the partner follows strategy x. Maynard Smith (1982) defines a strategy x^* to be an ESS if, for each strategy $x' \neq x^*$, one of the two conditions (ES1) and (ES2) holds:

(ES1) $W(x', x^*) < W(x^*, x^*)$

(ES2)   (i) $W(x', x^*) = W(x^*, x^*)$ and (ii) $W(x', x') < (W(x^*, x')$

The motivation behind this definition is as follows. Suppose the resident strategy is x^* and x' is a mutant strategy that is different from x^*. Initially the mutant will be very rare. It will thus almost always partner residents (as do residents). If condition (ES1) holds it does worse than residents, and so will be selected against and disappear. Suppose that condition (ES1) fails to hold, but condition (ES2)(i) holds. Then the mutant will have equal fitness to residents when very rare. If mutant numbers increase by random drift, both mutants and residents will occasionally partner other mutants. Condition (ES2)(ii) says that when this happens mutants do worse than residents in such contests. This will tend to reduce mutant numbers so that the mutant again becomes very rare. Overall, the mutant cannot invade, in the sense that the proportion of mutants will always remain very small. These ideas are formalized in Box 4.1 where we derive an equation for the rate of change in the proportion of mutants over time.

Note that if x^* is an ESS it is also a Nash equilibrium strategy since the Nash condition in eq (2.4) is that all mutants must either satisfy (ES1) or (ES2)(i). The converse is not true. As we will illustrate, there are many examples of Nash equilibrium strategies x^* that are not ESSs. However, if x^* is the unique best response to itself, then (ES1) holds for all mutants, so that x^* is an ESS. At least one Nash equilibrium exists for (almost) any game, but because the ESS criterion is stronger than the Nash equilibrium criterion, there are many games for which there is no ESS. For example, the standard form of the Rock–Paper–Scissors game (Exercise 4.9) has no ESS. In games in which the players make a sequence of decisions there is often no ESS, although evolutionary stability of some Nash equilibria can be achieved by introducing errors in decision making (Section 7.2).

4.1.1 Stability in the Hawk–Dove Game

We analyse stability in the version of the Hawk–Dove game in which a strategy specifies the probability of playing Hawk (Section 3.5). For this game eqs (3.1) and (3.2) give the payoffs $W_H(q)$ and $W_D(q)$ to playing Hawk and Dove, respectively, when the probability that a resident population member plays Hawk is q. Consider

Box 4.1 Rate of change of the proportion of mutants

Suppose that a proportion ϵ of a large population uses strategy x and a proportion $1-\epsilon$ uses strategy x^* in a two-player game with payoff function W. Then the expected payoff to each strategy can be written as

$$w(\epsilon) = (1-\epsilon)W(x,x^*) + \epsilon W(x,x)$$
$$w^*(\epsilon) = (1-\epsilon)W(x^*,x^*) + \epsilon W(x^*,x).$$

The difference $\Delta = w(\epsilon) - w^*(\epsilon)$ in payoff is

$$\Delta = (1-\epsilon)[W(x,x^*) - W(x^*,x^*)] + \epsilon[W(x,x) - W(x^*,x)]. \tag{4.1}$$

If condition (ES1) holds, the first square bracket on the right hand side of eq (4.1) is negative, so that $\Delta < 0$ for sufficiently small ϵ. If condition (ES2) holds, the first square bracket is zero and the second is negative, so that $\Delta < 0$ for all ϵ. Thus the Maynard Smith conditions ensure that $\Delta < 0$ if the proportion of mutants is sufficiently small; in fact these conditions are equivalent. However, the conditions say nothing about what will happen if the proportion of mutants is allowed to become large.

We now consider the rate of change in ϵ over time. Let $n(t)$ and $n^*(t)$ denote the numbers of individuals following strategies x and x^*, respectively at time t, so that $\epsilon(t) = n(t)/(n(t) + n^*(t))$. Taking the time derivative, we get $d\epsilon/dt = \epsilon(1-\epsilon)\left[\frac{1}{n}dn/dt - \frac{1}{n^*}dn^*/dt\right]$. Let the per-capita rate of increase of numbers following a strategy be equal to the payoff to the strategy in the game plus (or minus) a quantity that is the same for each strategy. That is $dn/dt = [w(\epsilon(t)) + w_0(t)]n(t)$ and $dn^*/dt = [w^*(\epsilon(t)) + w_0(t)]n^*(t)$. Here $w_0(t)$ might depend on $n(t)$ and $n^*(t)$, and so incorporate density-dependent effects. Using our definition of Δ we get that

$$\frac{d\epsilon}{dt} = \epsilon(1-\epsilon)\Delta. \tag{4.2}$$

Thus the condition for a strategy x^* to be an ESS is equivalent to the condition that in a large population comprising a mixture of x and x^* strategists, the proportion of x strategists will tend to zero if the initial proportion is sufficiently small. Equation (4.2) is a special case of the continuous-time replicator equation, when there are two types (Section 4.4). Derivation of the discrete time analogue is set as an exercise (Exercise 4.2).

the case in which $V < C$. In this case the unique Nash equilibrium strategy is $p^* = V/C$. Let $p \neq p^*$ be a mutant strategy. Since $W_H(p^*) = W_D(p^*)$ we have $W(p,p^*) = W(p^*,p^*)$ by eq (3.3). Thus condition (ES1) fails but (ES2)(i) holds. By eq (3.3) we also have

$$W(p^*,p) - W(p,p) = (p^* - p)[W_H(p) - W_D(p)]. \tag{4.3}$$

But from Fig. 3.5a it can be seen that $W_H(p) > W_D(p)$ if $p < p^*$ and $W_H(p) < W_D(p)$ if $p > p^*$, so that $W(p^*,p) > W(p,p)$ whatever the value of p. Thus condition (ES2)(ii) holds. Since this is true for all $p \neq p^*$, p^* is an ESS.

Intuitively stability can be understood as follows. Suppose that a mutant is less aggressive than residents ($p < p^*$). Then if the number of mutants increases, the probability that a random opponent plays Hawk is reduced below p^*. It is then best to be as aggressive as possible, and the residents, who are more aggressive than mutants, do better. An analogous argument applies to mutants that are more aggressive than residents ($p > p^*$): an increase in mutants increases the probability an opponent plays Hawk above p^* and it is best to be as Dove-like as possible, so that residents, who are less aggressive than mutants, do better.

In a Hawk–Dove game with $V > C$ the Nash equilibrium probability of playing Hawk is $p^* = 1$. This value is the unique best response to itself, and so is an ESS.

As described in Section 3.5, the Hawk–Dove game also has an equilibrium in pure strategies. For $V < C$ this equilibrium is a polymorphic population, with a proportion V/C of Hawks. As explained, this equilibrium is a stable polymorphism. However, if we also allow mixed strategies, it is easy to see that any polymorphism that results in the population probability $p^* = V/C$ of playing Hawk is an equilibrium. These various polymorphisms are neutral with respect to each other, and this is a kind of degeneracy arising from linear payoff functions.

One can contrast the Hawk–Dove game with the coordination game with payoffs given by Table 3.3. In that game each player decided on which side of the road to drive. We noted that one Nash equilibrium strategy was to toss a fair coin in making this decision. Suppose that was the resident strategy, and consider a mutant that, say, always drove on the right. When present singly this mutant would do equally well (badly!) as residents. If the proportion of mutants were to increase by random drift, then a randomly selected population member would drive on the right with a probability that was greater than 0.5. Under these circumstances it is best to also drive on the right and the mutants would do strictly better than residents. Thus the Nash equilibrium is not an ESS. A formal analysis of this case is set as Exercise 4.1. The strategy of always driving on the left is an ESS since it is the unique best response to itself. Similarly, the strategy of always driving on the right is an ESS.

These cases illustrate a general principle: in a two-player game with two actions, negative frequency dependence means that a mixed Nash equilibrium strategy is also an ESS, whereas positive frequency dependence means that a mixed Nash equilibrium strategy is not an ESS (Exercise 4.3).

4.1.2 Playing the Field

Condition (ES2) for evolutionary stability is applicable to a two-player game. When the game is against the field, payoffs are non-linear functions of the second argument (Section 2.9) and condition (ES2)(ii) must be modified to

(ES2′) (ii) $W(x', x_\epsilon) < W(x^*, x_\epsilon)$ for all sufficiently small positive ϵ.

Here, in an abuse of notation, the term $W(x, x_\epsilon)$ denotes the payoff to an individual with strategy x when the resident population comprises a proportion ϵ of x' strategists and a proportion $1 - \epsilon$ of x^* strategists. One can then show that the original (ES2) condition is a special case of this more general condition when payoffs are linear in the second argument, as is the case in a two-player game (Exercise 4.4).

In the sex-ratio game of Section 3.8 and Box 3.1 the unique Nash equilibrium strategy is to produce daughters and sons with equal probability. When this is the resident strategy, condition (ES2′) holds for all mutants, so the strategy is an ESS (Exercise 4.5). The intuition behind this result is similar to that for the Hawk–Dove game. For example, if the resident strategy is to produce sons and daughters with equal probability and a mutant arises that produces mainly daughters, then the breeding population becomes more female-biased if this mutant starts to become common, and the mutant strategy then does worse than the resident.

4.1.3 Illustration of Stability for a Continuous Trait: A Predator Inspection Game

It is common for fish to approach a predator in order to gain valuable information on the threat posed (Pitcher et al., 1986). Approach is dangerous, so a fish often approaches together with another fish in order to dilute the risk to themselves, and this is a form of cooperation. The phenomenon has been the focus of much empirical study. Figure 4.1 illustrates a typical laboratory set-up.

Under such circumstances we might expect the behaviour of each of the two fish to depend on the other. To model the game between the fish we loosely follow the approach of Parker and Milinski (1997). We are critical of the biological realism of our model (Section 8.4), but this simple game serves to illustrate the ESS analysis for a continuous trait.

Two fish are initially unit distance away from a predator. Each fish chooses how far to travel towards the predator. Thus an action is a distance x in the range $0 \le x \le 1$, where $x = 0$ is the action of not approaching the predator at all and $x = 1$ the action of travelling right up to the predator. If a fish moves distance x' while its partner moves distance x, the focal fish survives with probability $S(x',x)$ and gains information of value $V(x')$ if it survives. Its payoff from the game is $W(x',x) = S(x',x)V(x')$.

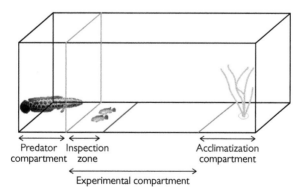

Fig. 4.1 The experimental set up of Hesse et al. (2015) in which two fish approach a predator. The predator is actually separated from the cichlids by a transparent perforated plastic sheet that permitted visual and olfactory cues. From fig. 1 of Hesse et al. (2015). Published under the Creative Commons Attribution 4.0 Generic license (CC BY 4.0).

We consider two choices of the pair of functions S and V. Both formulations of the game capture the idea that information increases with distance travelled but so does risk, although this reduces as the distance the other fish travels increases. Figure 4.2 illustrates how the fitness of a mutant depends on its distance travelled for fixed distance travelled by the resident in both formulations. In each formulation there is a unique best response to the resident approach distance x for every value of x (Fig. 4.3). In formulation I there is a single ESS at $x^* = 0.5$. In formulation II there are three ESSs and the middle one of these is $x^* = 0.5$. We contrast the two formulations in Section 4.2.

4.1.4 The Strength of Selection

Consider a situation in which the trait of interest is one-dimensional, and can take any value on some interval of the real numbers. Let the resident trait value be x. Let $W(x',x)$ denote the fitness of a mutant individual with trait value x' in this population. We can analyse how the fitness of this mutant depends on its trait value by plotting

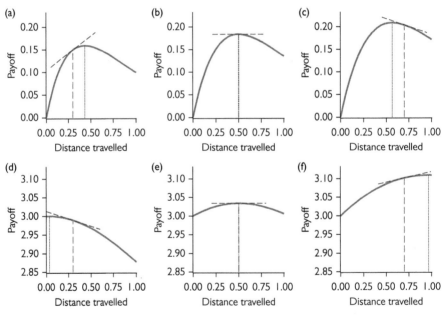

Fig. 4.2 Mutant fitness for two formulations of the predator inspection game. In formulation I (panels (a), (b), and (c)) the probability of survival is $S(x',x) = \exp\{-3x'/(1+x)\}$ and the value of information is $V(x') = x'$. In formulation II (panels (d), (e), and (f)) $S(x',x) = \exp\{-3x'/7(1+x)\}$ and $V(x') = 3+x'$. The curves show $W(x',x) = S(x',x)V(x')$ as a function of the mutant distance travelled x' for fixed resident distance x. The vertical dashed lines indicate the x values; $x = 0.3$ in (a) and (d), $x = 0.5$ in (b) and (e), $x = 0.7$ in (c) and (f). The distances maximizing mutant fitness are indicated by the vertical dotted lines. The tangent line to the curve at the resident distance is also shown. The slope of this tangent is the strength of selection $D(x) = \frac{\partial W}{\partial x'}(x,x)$.

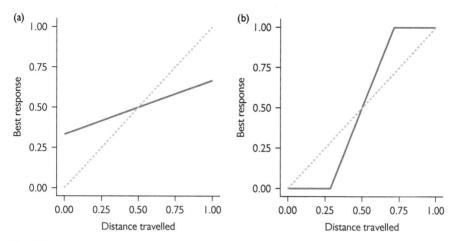

Fig. 4.3 Best mutant responses (solid lines) for the two formulations of the predator inspection game of Fig. 4.2. The 45° lines (dotted) are also shown. (a) Formulation I; the unique Nash equilibrium is at $x^* = 0.5$. (b) Formulation II; there are Nash equilibria at $x_1^* = 0$ $x_2^* = 0.5$ and $x_3^* = 1$. Since each Nash equilibrium is the unique best response to itself, all Nash equilibria are ESSs.

$W(x',x)$ as a function of x' for fixed x. The slope of the function, $\frac{\partial W}{\partial x'}(x',x)$, gives the rate of change of mutant fitness as its trait value increases when this value is close to x'. In particular, if we evaluate this derivative at $x' = x$, we obtain the rate of change of mutant fitness as its trait value increases when this value is close to the resident trait value. We denote this rate of increase by $D(x)$; i.e.

$$D(x) = \frac{\partial W}{\partial x'}(x,x). \qquad (4.4)$$

We refer to $D(x)$ as the strength of selection when the resident trait value is x. Figure 4.2 illustrates the slopes given by $D(x)$ for the two formulations of the predator inspection game.

Strictly speaking, the strength of selection should refer to the rate of change of invasion fitness rather than the rate of change in the game payoff, which is a fitness proxy; i.e. it should refer to $\frac{\partial \lambda}{\partial x'}(x,x)$ rather than $\frac{\partial W}{\partial x'}(x,x)$ (Section 2.5). However, for strong fitness proxies, which satisfy condition (2.7), the two derivatives have the same sign. This means that conclusions from reasoning about the strength of selection using a strong fitness proxy often also hold for invasion fitness, including the conclusions about convergence stability in the next section.

4.2 Adaptive Dynamics

In Box 4.1 we examined how the mutant proportion of a resident population changes with time. There are in principle three possibilities for the fate of a mutant. It can either fail to invade by approaching zero representation (which was examined in

Box 4.1), take over the population by instead approaching fixation, or remain in the population over the longer term by forming a polymorphism with the former resident. A simple conceptualization of evolutionary change is as a sequence of successful mutant invasions and fixations, which we refer to as adaptive dynamics. In reality evolutionary change need not follow such a simple scenario, but can instead proceed with substantial ongoing polymorphism, as well as with different types of genetic determination of strategies (e.g. diploid, multi-locus genetics), but the study of simple adaptive dynamics is useful for game theory in biology. In fact, the study of the invasion of an initially rare mutant into a resident population has proven to be a very productive link between game theory and evolutionary dynamics.

For games where strategies or traits vary along a continuum, it is of particular interest to examine if a mutant strategy x' in the vicinity of a resident x can invade and replace the resident. The sign of the strength of selection $D(x)$ is key to this issue. For x' in the vicinity of x we have the approximation

$$W(x',x) - W(x,x) \approx (x' - x)D(x). \tag{4.5}$$

Suppose now that $D(x)$ is non-zero for a resident strategy x, so that $W(x',x) > W(x,x)$ for an x' close to x. Then the mutant x' can invade the resident x (i.e. when the mutant is rare it will increase in frequency). It can also be shown that the mutant x' keeps increasing in frequency and will eventually replace the resident x. Part of the reasoning behind this is that $W(x,x') < W(x',x')$, because

$$W(x,x') - W(x',x') \approx (x - x')D(x') \approx -(x' - x)D(x) \approx W(x,x) - W(x',x) < 0,$$

so a mutant x cannot invade a resident x'. This means that the two strategies cannot form a protected polymorphism (where x' can invade x and x can invade x'). So when $D(x) \neq 0$, invasion implies fixation for some mutants near x (Geritz, 2005).

We now consider how the resident trait value x changes over evolutionary time under the action of natural selection. Let us assume that this change in the trait over time follows the adaptive dynamics described above, in the form of a sequence of mutant invasions and fixations, where each invading mutant is near the current resident. The trait will then increase over time if fitness increases with increasing trait value in the neighbourhood of x. Similarly, the trait value will decrease over time if fitness increases with decreasing trait value. We can then summarize the change of x under adaptive dynamics as follows:

$$D(x) > 0 \implies x \text{ increases}, \quad D(x) < 0 \implies x \text{ decreases}. \tag{4.6}$$

Our analysis below is entirely concerned with the direction of evolutionary change under adaptive dynamics, but one can also write the speed of change as

$$\frac{dx}{dt} = K\frac{\partial \lambda}{\partial x'}(x,x), \tag{4.7}$$

where the constant K can be written as the product of terms for the rate of mutation, the variance in mutation size, and the population size (Dieckmann and Law, 1996). Equation (4.7) is known as the canonical equation of adaptive dynamics.

4.2.1 Convergence Stability

Any resident trait value x^* that is an equilibrium point under adaptive dynamics must satisfy $D(x^*) = 0$. Such a trait value is referred to as an evolutionarily singular strategy (Metz et al., 1996; Geritz et al., 1998). But what would happen if an initial resident trait is close to such an x^*? Would it then evolve towards x^* or evolve away from this value? To analyse this we consider the derivative of $D(x)$ at $x = x^*$. Suppose that $D'(x^*) < 0$. Then for x close to x^* we have $D(x) > 0$ for $x < x^*$ and $D(x) < 0$ for $x > x^*$. Thus by the conditions in (4.6) x will increase for $x < x^*$ and decrease for $x > x^*$. Thus, providing the resident trait x is close to x^*, it will move closer to x^*, and will converge on x^*. We therefore refer to the trait x^* as being convergence stable if

$$D'(x^*) < 0. \tag{4.8}$$

We might also refer to such an x^* as an evolutionary attractor. Conversely, if $D'(x^*) > 0$, then any resident trait value close to (but not exactly equal to) x^* will evolve further away from x^*. In this case we refer to x^* as an evolutionary repeller. Such a trait value cannot be reached by evolution. Figure 4.4 illustrates these results.

If x^* is a Nash equilibrium value of the trait, then $W(x', x^*)$ has a maximum at $x' = x^*$. Thus if x^* lies in the interior of the range of possible trait values we must have $\frac{\partial W}{\partial x'}(x^*, x^*) = 0$; i.e. $D(x^*) = 0$. Thus x^* is an equilibrium point under adaptive dynamics; i.e. it is an evolutionarily singular strategy. Since every ESS is also a Nash equilibrium, every internal ESS is also an evolutionarily singular strategy. The above analysis of the convergence stability of an ESS was originally developed by Eshel and Motro (1981) and Eshel (1983). Eshel and Motro (1981) refer to an ESS that is also

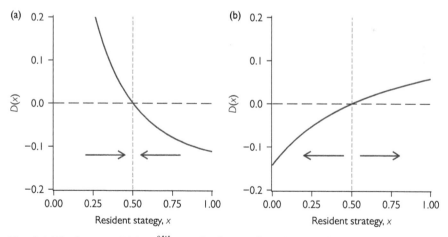

Fig. 4.4 The function $D(x) = \frac{\partial W}{\partial x'}(x, x)$ for the two formulations of the predator inspection example in Fig. 4.2. (a) Formulation I, (b) Formulation II. When $D(x) > 0$, x increases with time; when $D(x) < 0$, x decreases with time. Note that the slope of D is negative in (a), so that the trait value 0.5 is an evolutionary attractor, and is positive in (b), so that 0.5 is an evolutionary repeller.

convergence stable as a Continuously Stable Strategy (CSS). Some ESSs are also CSSs, others are not and are unattainable.

Figure 4.4 shows the fitness derivative $D(x)$ at $x = 0.5$ for each of the two formulations of the predator inspection game. In formulation I, $D(x) > 0$ for $x < 0.5$ and $D(x) < 0$ for $x > 0.5$, so that the trait $x = 0.5$ is convergence stable and is hence a CSS. In contrast, in formulation II, $D(x) < 0$ for $x < 0.5$ and $D(x) > 0$ for $x > 0.5$, so that the trait $x = 0.5$ is an evolutionary repeller and is unattainable. Figure 4.2 illustrates why the two formulations result in the sign of $D(x)$ behaving differently. The two formulations differ in the dependence of the survival of the mutant on the mutant and resident distances travelled when these distances are similar. In formulation I an increase above 0.5 in the distance travelled by the resident does not reduce the danger sufficiently for the mutant to also increase its distance by this much.

Eshel and Motro (1981) were the first to demonstrate that an ESS may not be convergence stable and so might not be attainable. In their model the trait under selection is the maximum risk that an individual should take in order to help kin. Since then unattainable ESSs have been demonstrated in a very wide variety of contexts. For example, Nowak (1990) shows that there can be an unattainable ESS in a variant of the repeated Prisoner's Dilemma model.

4.2.2 The Slope of the Best Response Function and Convergence Stability

The condition $D'(x^*) < 0$ for convergence stability in eq (4.8) can be translated into a condition on the second derivatives of the payoff function W (Box 4.2). This condition can then be used to show that if the slope of the best response function is less than 1 at an ESS, then the ESS is also convergence stable and is hence a CSS. Conversely if the slope exceeds 1 the ESS is unattainable (Box 4.2). In the predator inspection game the slope of the best response function at the ESS is less than 1 in Formulation I and greater than 1 for Formulation II (Fig. 4.3), in agreement with our previous findings for these two cases. We can also apply this result to other examples in this book. For example in the model of pairwise contests where individuals know their own fighting ability (Section 3.11), we can infer that the two ESSs shown in Fig. 3.9b are also convergence stable and are hence CSSs.

4.2.3 Pairwise Invasibility Plots

A pairwise invasibility plot (PIP) (van Tienderen and de Jong, 1986; Matsuda, 1995) shows, for each possible resident strategy, the range of mutant strategies that can invade. PIPs are useful graphical representations from which one can often ascertain whether a Nash equilibrium x^* is also an ESS at a glance. Figure 4.5 illustrates these plots for the two cases of the predator inspection game that we have analysed. In both cases it can be seen from the figure that when the resident strategy is $x^* = 0.5$ any mutant different from x^* has lower fitness. Thus $x^* = 0.5$ is the unique best response

Box 4.2 Conditions for convergence stability

Let $D(x) = \frac{\partial W}{\partial x'}(x, x)$. Then by differentiation we have

$$D'(x) = \frac{\partial^2 W}{\partial x'^2}(x, x) + \frac{\partial^2 W}{\partial x' \partial x}(x, x). \tag{4.9}$$

Now let x^* be a singular point satisfying $\frac{\partial W}{\partial x'}(x^*, x^*) = 0$. Then

$$D'(x^*) = \frac{\partial^2 W}{\partial x'^2}(x^*, x^*) + \frac{\partial^2 W}{\partial x' \partial x}(x^*, x^*). \tag{4.10}$$

Since the condition $D'(x^*) < 0$ is sufficient for convergence stability of x^*, we see that if

$$\frac{\partial^2 W}{\partial x'^2}(x^*, x^*) + \frac{\partial^2 W}{\partial x' \partial x}(x^*, x^*) < 0 \tag{4.11}$$

then x^* is convergence stable (Eshel, 1983).

Now suppose that for each x there is a unique best response $\hat{b}(x)$ satisfying $\frac{\partial W}{\partial x'}$ $(\hat{b}(x), x) = 0$ and $\frac{\partial^2 W}{\partial x'^2}(\hat{b}(x), x) < 0$. Then differentiating the first of these conditions we have

$$\frac{\partial^2 W}{\partial x'^2}(\hat{b}(x), x)\hat{b}'(x) + \frac{\partial^2 W}{\partial x' \partial x}(\hat{b}(x), x) = 0. \tag{4.12}$$

Let x^* be an ESS satisfying $\hat{b}(x^*) = x^*$, so that

$$\frac{\partial^2 W}{\partial x'^2}(x^*, x^*)\hat{b}'(x^*) + \frac{\partial^2 W}{\partial x' \partial x}(x^*, x^*) = 0. \tag{4.13}$$

Then by eqs (4.10) and (4.13)

$$D'(x^*) = \frac{\partial^2 W}{\partial x'^2}(x^*, x^*)[1 - \hat{b}'(x^*)]. \tag{4.14}$$

By assumption we have $\frac{\partial^2 W}{\partial x'^2}(x^*, x^*) < 0$. Thus if $\hat{b}'(x^*) < 1$ then $D'(x^*) < 0$, so that x^* is convergence stable and is hence a CSS. Conversely, if $\hat{b}'(x^*) > 1$ then x^* is an evolutionary repeller and cannot be reached by evolution.

to itself and is hence an ESS. PIPs are, however, not so useful in cases where some mutants have equal fitness to residents when the resident strategy is x^*. This is because the ESS criterion then relies on the second-order condition (ES2)(ii).

PIPs also can be used to infer whether x^* is convergence stable. In Fig. 4.5a if the resident trait is initially less than 0.5 then those mutants that can invade have trait values greater than the resident trait, so that the resident trait will tend to increase over time. Conversely if the resident trait is initially greater than 0.5 it will tend to decrease over time. The trait value $x^* = 0.5$ is therefore convergence stable. In contrast, in Fig. 4.5b if the resident trait is initially less than 0.5 those mutants that can invade have trait values less than the resident trait, so that the resident trait value will tend to decrease over time. Similarly, if the resident trait is initially greater than 0.5 it will tend to increase over time. The trait value $x^* = 0.5$ is therefore an evolutionary repeller in this case.

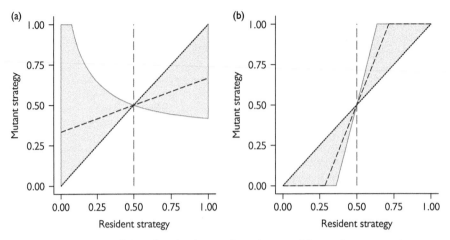

Fig. 4.5 Pairwise invasibility plots for the two formulations of the predator inspection game in Fig. 4.2. In the shaded regions the mutant strategy can invade the resident strategy. The 45° line (dark dotted line) borders one side of each region. The best response to the resident strategy is also shown (dark dashed line). (a) Formulation I (where $x^* = 0.5$ is an ESS and is also convergence stable), (b) Formulation II (where $x^* = 0.5$ is an ESS but is not convergence stable).

As we will illustrate in the next section, PIPs are also able to easily identify cases in which traits are convergence stable, but evolution leads to a local fitness minimum. For more on these plots, including a more nuanced classification of the plots than is presented here, see Geritz et al. (1998).

4.3 Evolution to a Fitness Minimum

At a singular point x^* of the adaptive dynamics the strength of selection $D(x^*) = \frac{\partial W}{\partial x'}(x^*, x^*)$ vanishes. If the second derivative of mutant fitness is negative at the singular point (i.e. $\frac{\partial^2 W}{\partial x'^2}(x^*, x^*) < 0$), mutant fitness has a strict local maximum at the resident trait value x^*, so that x^* is a (local) ESS. As we saw in the previous section, such an ESS may be convergence stable or may be an evolutionary repeller and so not be accessible under adaptive dynamics. If $\frac{\partial^2 W}{\partial x'^2}(x^*, x^*) > 0$, mutant fitness has a strict local minimum at the resident trait value x^*. Again this singular point may be convergence stable or may be an evolutionary repeller. Thus singular points may be any combination of fitness maximum or minimum and convergence stable or otherwise. A classification of these four cases in terms of the partial derivatives of the payoff function is given by Geritz et al. (1998). In this section we focus on the, perhaps surprising, case in which a fitness minimum is convergence stable. In other words, evolution leads a population towards a state at which residents have lower fitness than all nearby mutants. There is then disruptive selection on the resident trait, which can result in evolutionary branching, giving rise to a polymorphic population. We illustrate this with an example based on the evolution of generalism and specialism in a patchy environment that is roughly based on similar models by, for example, Brown and Pavlovic (1992) and Meszéna et al. (1997).

4.3.1 Two Local Habitats Linked by Dispersal

Many environments can be thought of as a collection of local habitats linked by dispersal. Habitats may vary in conditions such as the type of food or temperature, and the best strategy in a habitat may depend on these conditions. So, for example, for a seed-eating bird, seed size may vary between habitats, with optimal beak size varying accordingly. If habitats vary in their temperature it may be best to have high insulation in cold habitats and less insulation in warmer habitats. We might then ask whether we expect evolution to produce a species that is a generalist and does reasonably well in all habitats or produce specialist phenotypes that are mainly found in those habitats to which they are adapted. We consider this question in the following model.

4.3.1.1 Two-Habitat Model

An asexual species occupies an environment that consists of two local habitats linked by dispersal. This species has discrete, non-overlapping generations with a generation time of 1 year. Individuals are characterized by a genetically determined trait x where $0 \leq x \leq 1$. An individual with trait x in habitat i leaves $Ng_i(x)$ offspring where N is large. The functions g_1 and g_2 are illustrated in Fig. 4.6a. As can be seen, it is best to have a low trait value in habitat 1 and a high value in habitat 2. Each newborn individual either remains in its birth habitat (with probability $1 - d$) or disperses to the other habitat (with probability d). Here $0 < d < 0.5$. Each habitat has K territories, where K is large. After the dispersal phase each of these territories is occupied by one of the individuals present, chosen at random from all individuals currently in the habitat. Those individuals left without territories die, while the K that possess territories grow to maturity.

Note that for this example the per-capita growth rate of any resident population (i.e. the mean number of offspring obtaining a territory per resident parent) is 1 since the population is at constant size. Frequency dependence acts through competition for territories. For example, suppose the resident trait value is $x = 0$ and consider the growth rate of a subpopulation comprising individuals with mutant trait value $x' = 1$ while the size of this subpopulation is small compared with the resident population. Since residents will leave more offspring in habitat 1 than in habitat 2 after the dispersal phase, there will be greater competition for territories in habitat 1. However, since mutant individuals do better in habitat 2 mutants will tend to accumulate in habitat 2. They will then suffer less competition than the average competition experienced by residents and so will have a growth rate that is greater than 1. The mutant strategy will thus invade the resident population.

The derivation of the precise formula for invasion fitness $\lambda(x',x) \equiv W(x',x)$ of a mutant x' in a resident x population is deferred to Section 10.7. From this payoff

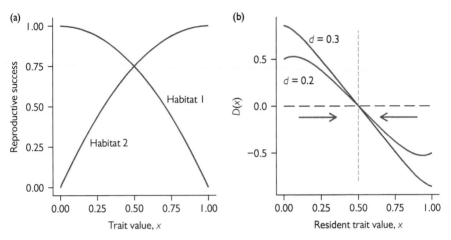

Fig. 4.6 Two habitats linked by dispersal. (a) The relative number of offspring produced on each habitat as a function of trait value. (b) The strength of selection as given by the directional derivative $D(x) = \frac{\partial W}{\partial x'}(x, x)$, illustrated for dispersal probabilities $d = 0.3$ and $d = 0.2$.

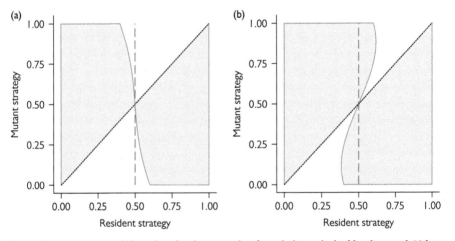

Fig. 4.7 Pairwise invasibility plots for the example of two habitats linked by dispersal. Values of the mutant trait that can invade the resident population are shaded. The 45° line (dark dotted line) borders one side of each shaded region. The two cases illustrated are (a) dispersal probability $d = 0.3$ and (b) $d = 0.2$.

function one can obtain the strength of selection. Figure 4.6b illustrates this function for two values of the dispersal probability d. As can be seen, for each of these dispersal probabilities the trait $x^* = 0.5$ is the unique convergence stable point under adaptive dynamics.

Figure 4.7 shows the pairwise invasibility plots for each of the dispersal probabilities considered in Fig. 4.6b. We can also infer the convergence stability of $x^* = 0.5$ from this figure. To see this, note that when the resident trait x is less than 0.5, mutants with trait values just greater than x can invade, so that x will increase under adaptive dynamics. Similarly when $x > 0.5$ mutants with trait values just less than x can invade, so that the resident trait value will decrease. This reasoning applies to both of the cases $d = 0.3$ and $d = 0.2$. However, the two cases differ fundamentally. For dispersal probability $d = 0.3$ (Fig. 4.7a), if the resident trait value is $x^* = 0.5$ then any mutant with a different trait value has lower fitness. Thus when the resident trait is $x^* = 0.5$, mutant fitness has a strict maximum at the resident trait value, so that $x^* = 0.5$ is an ESS as well as being convergence stable (and is hence a CSS). The evolutionary simulation in Fig. 4.8a illustrates convergence to this trait value. In contrast when $d = 0.2$ (Fig. 4.7b), if the resident trait value is $x^* = 0.5$ then any mutant with a different trait value has higher fitness. In other words, mutant fitness is minimized at the resident trait value, so that evolution leads the population to a fitness minimum.

As the next section describes, once evolution takes a population to a fitness minimum there is then further evolution.

4.3.2 Evolutionary Branching

Consider a population that evolves to a fitness minimum under adaptive dynamics. When the resident population is at or close to this minimum any mutant that arises has greater fitness than residents, so that there will be disruptive selection, leading to trait values having a bimodal distribution with peaks either side of the fitness minimum. This phenomenon is referred to as evolutionary branching.

Figure 4.8b illustrates evolutionary branching for the case of two habitats linked by dispersal. In this model the population always evolves towards the trait value $x^* = 0.5$. This represents a generalist strategy because individuals with this trait value do reasonably well in both habitats. If the dispersal probability is sufficiently high (Fig. 4.8a) and this generalist strategy is the resident strategy, then all other strategies have lower invasion fitness, so that the strategy is an ESS and hence a CSS. In contrast, for low dispersal probability (Fig. 4.8b), once the population approaches $x^* = 0.5$ there is disruptive selection and the population evolves to be polymorphic. In this polymorphic population half the population members do very well in habitat 1 but badly in habitat 2. These individuals are also mostly found in habitat 1 (not shown). The other population members are specialists for habitat 2 and are mostly found there. When there are more than two habitats, with different optima on each, further branching into habitat specialists can occur. Not surprisingly, the number of stable polymorphisms can never exceed the number of different habitat types (Geritz et al., 1998). In the model the population size on each habitat is regulated by limiting resources on that habitat (soft selection). Similar results on generalism versus specialism can be found in a number of models in which there is soft selection and limited dispersal (e.g. Brown and Pavlovic, 1992; Meszéna et al., 1997; Debarre and Gandon, 2010).

The standard adaptive dynamics framework assumes clonal inheritance; the phenotype is inherited. This is a major limitation when investigating evolutionary

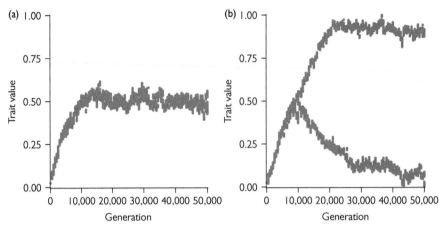

Fig. 4.8 Evolutionary simulation for the model of two local habitats linked by dispersal. (a) When the dispersal probability is $d = 0.3$ evolution is to a stable state. (b) In contrast, when $d = 0.2$ there is evolutionary branching. There are $K = 256$ territories on each habitat. In generation 0 all individuals have trait value $x = 0.025$. The mutation probability is 0.02 with mutant traits differing from parental traits by a random amount with a uniform distribution on $[-0.01, 0.01]$ Since the mutation rate is not small, in contrast to the assumptions underlying the theory of adaptive dynamics, new mutations arise before the population can evolve to be monomorphic after a previous mutation. Thus the population remains polymorphic throughout the simulation.

branching in a diploid sexually reproducing population. However, the framework can be extended to such populations. Kisdi and Geritz (1999) considered a two-habitat model with soft selection in which the phenotypic trait was determined by the two alleles at a single diploid locus. They again concluded that evolutionary branching could occur resulting in specialist phenotypes.

When the analysis of Kisdi and Geritz (1999) predicts a polymorphic outcome with specialist phenotypes for the two habitats, the distribution of the alleles present differs between habitats. This raises the possibility that the alleles that are present in an individual can act as cues to the individual as to its location (Leimar et al., 2006). There is then the potential for developmental plasticity, with a genetic modifier making use of this information in determining the phenotype. When this occurs the phenotype is expected to be more of a generalist than the two specialists that evolve without modification. This is because the specialists tend to find themselves in their appropriate habitat, whereas the modifier system is spread across both. There is then a genetic conflict between the modifier system and the alleles at the cue locus since their evolutionary interests are different (Leimar et al., 2019).

Evolutionary branching has now been predicted in a very wide range of settings (Doebeli and Dieckmann, 2000). Here we briefly describe three of these settings: seed size polymorphism, levels of cooperation, and virulence levels.

Geritz et al. (1999) present a model of seed size where due to resource limitation the more seeds that a plant produces, the smaller these seeds have to be. Large seeds are better competitors than small seeds. Using an adaptive dynamics framework they show that there can be a sequence of branchings in which a lineage splits into two daughter lines, each of which may then itself split. This process can result in a range of seed sizes coexisting.

In Section 3.1 we presented a model in which two players contribute to a common good at a cost to themselves. In the examples used to illustrate this game the common benefit increased with the the the sum of the individual contributions, but there were diminishing returns for increased total contribution. In other words the common benefit function $B(x' + x)$ satisfied $B'' < 0$. Under these assumptions, there is a convergence-stable level of contribution to the common good, and this level of contribution is also an ESS. If instead the common benefit is an accelerating function of the total contribution ($B'' > 0$) over part of the range of benefits, the level of contribution can converge to a fitness minimum under adaptive dynamics. Evolutionary branching will then occur, resulting in a polymorphic population in which some individuals make a high contribution and others a low contribution (Doebeli et al., 2004).

Morozov and Best (2012) consider a model of the evolution of virulence in a pathogen when the infected species is also subject to mortality from predation. In this model the probability that a predator catches an infected prey increases with the level of virulence of the pathogen. Virulence thus affects predator success and hence predator density. This then feeds back to affect prey numbers and pathogen transmission rates. Morozov and Best show that this feedback can result in evolutionary branching in the level of pathogen virulence, with some pathogens having low and others high virulence.

For an asexual species, evolutionary branching produces two separate lines. In some sense this is speciation, although for clonal species all lines are separate and can all be regarded as species. For a sexually reproducing species to evolve into two distinct species there must be some form of reproductive isolation that prevents the interbreeding of the emerging species. If the population is well mixed this must be brought about by the co-evolution of sufficiently strong assortative mating. Various models dealing with sexually reproducing populations suggest that speciation by evolutionary branching and assortative pairing is a real possibility (e.g. Doebeli and Dieckmann, 2000; Doebeli et al., 2005; van Doorn et al., 2009), although details of models have been questioned by Gavrilets (2005).

4.4 Replicator Dynamics

It is common in game theory to use a replicator equation as a possible dynamical foundation for the analysis (e.g. Weibull, 1995; Hofbauer and Sigmund, 1998; Broom and Rychtàr, 2013). The equation deals with the dynamics of a fixed number of strategies, or genetic types, in a single large population. Let $n_i(t)$ be the number of individuals using strategy x_i, $i = 1, \ldots, k$, and $v_i(t) = n_i(t)/n(t)$ the relative frequency of x_i, where t is time measured in generations and $n(t)$ is the total population size. The equation follows from a definition of fitness $w_i(t)$ as the per capita rate of increase of the strategy:

$$w_i(t) = \frac{1}{n_i(t)} \frac{dn_i(t)}{dt}. \tag{4.15}$$

Note that in general this fitness depends on the composition of the population at time t. It is related to but not the same as invasion fitness described in Section 2.4, because it goes beyond the study of the invasion of a rare mutant into a resident population. From the definition we get

$$\frac{dv_i(t)}{dt} = v_i(t)\big(w_i(t) - \bar{w}(t)\big), \; i = 1, \ldots, k, \tag{4.16}$$

where $\bar{w} = \sum_i v_i w_i$ is the average fitness in the population. Equation (4.2) in Box 4.1 is the important special case of only two strategies, one of which is present at low frequency.

If the strategies x_i can be treated as real-valued traits, it follows that

$$\frac{d\bar{x}(t)}{dt} = \mathrm{Cov}\big(w.(t), x.(t)\big), \tag{4.17}$$

where $\bar{x} = \sum_i v_i x_i$ is the population mean trait and $\mathrm{Cov}(w., x.) = \sum_i v_i (w_i - \bar{w})(x_i - \bar{x})$ is the population covariance of the trait x_i with fitness w_i. This is a version of the celebrated Price equation (Frank, 1995). The derivation of the equations is left as Exercise 4.8.

The replicator and Price equations have the advantage of a certain generality, in that they follow from a definition of fitness of a strategy as the per-capita rate of increase. They are thus helpful in giving an understanding of how selection operates. However, they do not in themselves solve the difficulties of a population dynamical analysis of a polymorphic population and, in the simple versions given here, they do not deal with issues of population structure and multilocus genotype–phenotype maps. Although opinions vary, one can say that the development of adaptive dynamics has been more strongly oriented towards handling such difficulties.

For pairwise interactions in a large population, in the form of a symmetric matrix game with discrete actions, the replicator equation becomes

$$\frac{dv_i(t)}{dt} = v_i(t)\big(W(x_i(t), \bar{x}(t)) - W(\bar{x}(t), \bar{x}(t))\big), \tag{4.18}$$

where x_i is a mixed strategy of individuals of type i, $\bar{x}$ is the population mean strategy, and $W(x_i, x_j)$ is the expected payoff for strategy x_i against x_j, which is computed from the payoff matrix of the game. As an example, for the Hawk–Dove game we would have $x_i = (p_i, 1 - p_i)$, where p_i is the type's probability of using Hawk, and $W(x_i, x_j)$ is given by eqs (3.4). It is shown by Hofbauer and Sigmund (1998) that an ESS for the matrix game corresponds to a dynamically stable population mean strategy $\bar{x} = \bar{x}^*$. This is an interesting result, but one should keep in mind that the replicator dynamics is a special kind of dynamics that need not apply to real populations. In general it need not be the case that an ESS corresponds to a stable point for realistic kinds of evolutionary dynamics, for instance for the dynamics of continuous multidimensional traits (see Section 6.1).

The replicator equation represents a highly idealized view of evolutionary changes. For game theory in biology, we think it is preferable to stay somewhat closer to the population-genetic underpinnings, which can be done using adaptive dynamics. The overwhelmingly most productive application of evolutionary dynamics to game theory has been and still is the study of the invasion of a rare mutant into a resident population. A valuable aspect of that approach is that it limits attention to the simpler, but still challenging question of the invasion of a rare mutant, rather than attempting to describe a full evolutionary dynamics, which is better left to population genetics. Another valuable aspect is that one can readily include such things as diploid genetics, interactions between relatives, correlations between traits, and metapopulation dynamics in the description of the invasion process.

4.5 Games Between Relatives

The study of invasion of rare mutant strategies into a resident population is a leading theme in this chapter. Our approach to games between relatives also develops this idea. For these games, a rare mutant strategy can have an appreciable chance of interacting with other mutant strategies, and need not interact only or predominantly with resident strategies. Relatedness can thus lead to positive assortment of strategies (see below), but games between relatives also include other situations. One example we have encountered is dispersal to reduce kin competition (Section 3.10). Another example is the interaction between parents and offspring about parental investment, where the players of the game are from different generations and can, depending on the precise circumstances, have partly diverging evolutionary interests (see below). The evolutionary analysis of parent-offspring conflicts was initiated by Trivers (1974) and has subsequently been given much attention.

In the original formulation of kin selection by Hamilton (1964), the concept of inclusive fitness was used to study interactions between relatives. In principle the concept has wide applicability (Gardner et al., 2011), but care is sometimes needed for a correct interpretation (Fromhage and Jennions, 2019). However, we will not make use of it here. Let us just note that a main idea of the inclusive fitness approach is to assign fitness effects to an 'actor', corresponding to the reproductive consequences of the action for the actor and the actor's relatives. Alternatively, instead of such an actor-centred approach, one can sum up all reproductive effects for a focal 'recipient' individual, and this is referred to as the direct fitness approach (Taylor and Frank, 1996; Taylor et al., 2007). Furthermore, a very straightforward approach that is related to direct fitness is to simply compute invasion fitness of a rare mutant in a resident population. This can be thought of as a 'gene-centred' approach and we use it here.

We start by examining the evolution of helping, which is perhaps the most studied issue for interactions between relatives.

4.5.1 The Evolution of Helping

Helping others can be costly, so that helping behaviour is not expected to evolve unless the cost of providing help is more than compensated for by some benefit. This was illustrated by the Prisoner's Dilemma game (Section 3.2) where we saw that the strategy of cooperation would not evolve. When pairs are formed from population members at random, defect is the only ESS in that game. There can, however, be reasons to expect a positive assortment where helping individuals are more likely than non-helpers to find themselves interacting with others who help. For example, suppose that offspring have limited dispersal, so that they mature close to their natal environment (population viscosity), and that individuals interact with those around them. Then individuals will be more closely related to those individuals with whom they interact than to average population members. Thus if a mutation arises that provides more help to others than the resident population strategy, then mutants will receive more help from others than residents because some of a mutant individual's relatives will also have the mutation. Population viscosity is likely to be the most widespread reason for positive assortment in nature, which was originally analysed by Hamilton (1964).

Non-random pairing can thus lead to cooperators doing better than defectors. Our analysis here follows that in McElreath and Boyd (2007). Consider a large population in which each individual either plays cooperate (C) or defect (D) in pair-wise interactions. When pairs are formed there may be a tendency for cooperators to be paired together and for defectors to be paired together. Let p_C be the probability that the partner of a cooperative individual is also cooperative and let p_D be the probability that the partner of a defector is cooperative. Then under positive assortment we would have $p_C > p_D$. In fact the correlation ρ between the level of cooperativeness of pair members can be shown to be exactly $p_C - p_D$ (Exercise 4.10). Conditioning on whether the partner is cooperative or not, the payoff to a cooperator is

$$W(C) = p_C(b - c) + (1 - p_C)(-c).$$

Similarly, the payoff to a defector is

$$W(D) = p_D b + (1 - p_D)0.$$

Thus

$$W(C) > W(D) \Leftrightarrow (p_C - p_D)b > c. \qquad (4.19)$$

This shows that if the correlation $\rho = p_C - p_D$ is sufficiently high cooperators can do better than defectors. Box 7.4 illustrates similar effects of correlation in a model in which traits are continuous.

In Box 4.3 the inequality in eq (4.19) is applied to interactions between relatives. Note that the right-hand inequality in eq (4.20) in Box 4.3 holds for all values of the proportion α of the C allele in the population, which means that the condition tells us when C is an ESS and when D is an ESS. Thus, C is an ESS when the right-hand

> **Box 4.3 Kin interactions and positive assortment**
>
> We consider a population of haploid individuals, each with one of two alternative alleles, C and D, at a focal locus. C individuals always cooperate in the Prisoner's Dilemma, D individuals always defect. A proportion α of the population has the C allele. The two alleles in the two members of a pair are said to be identical by descent (IBD) if they are both copies of the same allele present in a recent ancestor. The relatedness between pair members is taken to be the probability that their alleles are IBD. More generally, including for diploid organisms, the coefficient of relatedness is often defined as the probability that a randomly chosen allele at a locus in one individual has an IBD allele at the locus in the other individual. For example, in diploid organisms relatedness is 1 for identical twins, 0.5 for sibs of unrelated parents, and 0.125 for first cousins.
>
> In our population, assume that the average relatedness between pair members is r. Conditioning on whether partner's allele is IBD we have
>
> $$p_C = P(\text{alleles IDB}|\text{individual is C}) \times 1 +$$
> $$P(\text{alleles not IBD}|\text{individual is C}) \times \alpha$$
> $$= r + (1-r)\alpha.$$
>
> Similarly
>
> $$p_D = P(\text{alleles IDB}|\text{individual is D}) \times 0 +$$
> $$P(\text{alleles not IBD}|\text{individual is C}) \times \alpha$$
> $$= (1-r)\alpha.$$
>
> Thus $p_C - p_D = r$, so that eq (4.19) becomes
>
> $$W(C) > W(D) \Leftrightarrow rb > c \qquad (4.20)$$
>
> This relationship is referred to as Hamilton's rule.

inequality in eq (4.20) holds, and D is an ESS when the reversed inequality holds. These results are thus in accordance with Hamilton's rule.

4.5.2 Hawk–Dove Game Between Relatives

Let us now analyse the Hawk–Dove game, making the same assumptions about relatedness as in Box 4.3. Grafen (1979) originally analysed this game, using the gene-centred approach. With the payoffs from Table 3.2, we first examine the invasion of a mixed strategy of playing Hawk with probability p' into a resident population using p. Instead of eq (3.5) we get

$$W(p',p) = W_0 + r\left[p'^2\frac{1}{2}(V-C) + p'(1-p')V + (1-p')^2\frac{1}{2}V\right] +$$
$$(1-r)\left[p'p\frac{1}{2}(V-C) + p'(1-p)V + (1-p')(1-p)\frac{1}{2}V\right] \quad (4.21)$$
$$= W_0 + \frac{1}{2}V\left[1 + (1-r)(p'-p-\frac{C}{V}p'p) - r\frac{C}{V}p'^2\right].$$

Following the same logic as in Box 4.3, a player using the strategy p' has a probability r of interacting with relatives who also use p', and a probability $1 - r$ of interacting with random members of the population who use p. The fitness gradient (eq (4.4)) is then

$$D(p) = \frac{1}{2}V\left[1 - r - (1+r)\frac{C}{V}p\right].\tag{4.22}$$

Solving the equation $D(p^*) = 0$, we get

$$p^* = \frac{(1-r)V}{(1+r)C},\tag{4.23}$$

if $V/C \leq (1+r)/(1-r)$. Compared with the game for unrelated individuals, this probability of playing Hawk is smaller than V/C, which can be seen as a tendency towards cooperation. For $V/C > (1+r)/(1-r)$, the equilibrium is $p^* = 1$. As illustrated in panel (a) of Fig. 4.9, for $r > 0$, the equilibrium is a strict maximum of the payoff $W(p,p^*)$, thus satisfying the condition (ES1) for evolutionary stability. Note that this differs from the case $r = 0$, for which all strategies are best responses to the ESS.

Even though relatedness has the effect of decreasing the evolutionarily stable intensity of fighting, very dangerous fighting between closely related individuals still occurs in nature. As illustrated in panel (b) of Fig. 4.9, if two newly emerged honey

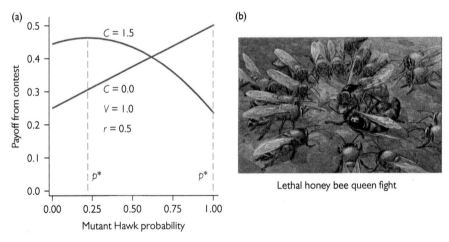

Lethal honey bee queen fight

Fig. 4.9 (a) Illustration of the payoff to a mutant strategy p for two ESSs for the Hawk–Dove game between relatives. In both cases $r = 0.5$ and $V = 1.0$. With a more valuable future ($C = W_0 = 1.5$) beyond the contested resource (e.g. the inheritance of a colony), the ESS is $p^* = 0.222$, as given in eq (4.23). If the future beyond the contest has no value ($C = W_0 = 0.0$), a fight to the death is the ESS. (b) Illustration of a fight to the death between two newly emerged honey bee queens. The worker bees look on and will accept the winner as the new queen of the colony. Painting by Hashime Murayama. Reused with permission from the National Geographic Image Collection.

bee queens meet, they will fight to the death. The explanation lies in the colony life cycle of these insects. When a colony has grown to a certain size, the previous queen leaves the colony with part of the workers, in order to found a new colony (swarming). At this point in time, the colony has reared a number of new queens, but only one of them can inherit the colony (cf. lethal fighting in Section 3.5). The first new queen to emerge will try to locate the other queen-rearing cells and kill their inhabitants, but if two or more new queens emerge and meet, they fight to the death (Gilley, 2001). Honey bee queens are typically highly related, with $r = 0.75$ if they are full sisters (same father), and $r = 0.25$ if they are half sisters. The example illustrates that even simple game theory models can deliver surprising and valid predictions.

4.5.3 Parent–Offspring Conflict

Interactions between parents and offspring involve such issues as how much parents should invest in a given clutch of offspring compared with future clutches, and how the investment in a clutch should be distributed among the individual offspring. Offspring begging for food is a kind of negotiation between offspring and parents, and between the offspring. Here we study a simpler situation, where parents provide a fixed amount of resources to a clutch, and the question is how this amount is distributed among the offspring. We will compare the case where one chick (the oldest) is dominant and can decide on the proportion p of resources to itself with the case where a parent decides the proportions. In many species of birds, the chicks in a clutch compete intensely over food and the oldest chick often dominates its younger siblings (Mock et al., 1990).

Inspired by Parker et al. (1989), we examine the simplest possible situation in which a clutch consists of two siblings, with the older chick being dominant. We assume that the survival to adulthood of a chick that gets a proportion p of the parental resources is

$$s(p) = 1 - \exp(-a(p - p_{\min})) \tag{4.24}$$

for $p \geq p_{\min}$ and zero otherwise. The parameter a is related to the total amount of resources invested and $p_{\min}$ is the smallest proportion required for a chick to have a chance to survive. If the dominant chick gets the proportion p with survival $s(p)$, the other chick will get $1 - p$ with survival $s(1 - p)$. In panel (a) of Fig. 4.10 we illustrate this chick survival function, together with the evolutionarily stable allocations if they are determined by the dominant chick (dashed lines) or by a parent (dotted line).

Figure 4.10b shows a great egret parent with chicks. Sibling competition is fierce in these birds (Mock et al., 1990), but their breeding characteristics do not agree in detail with our model here. In great egrets the observation is that the parents do not interfere in sibling fights.

To derive the allocations illustrated in Fig. 4.10, we assume that individuals are diploid. For each of the two cases of dominant chick control and parental control, we examine the payoff (proportional to invasion fitness) of a rare mutant p in a resident population using q. When the mutant is rare, it occurs overwhelmingly as an allele in a heterozygote genotype. Furthermore, at most one of the mother and the father of a clutch is a mutant heterozygote. As a simplification, we assume that generations

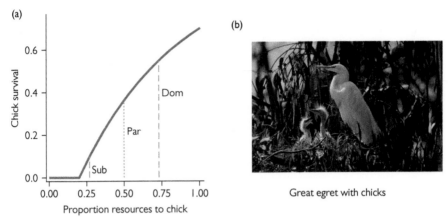

Fig. 4.10 (a) Chick survival as a function of the proportion of parental resources allocated to the chick. The curve is given by eq (4.24) with $p_{min} = 0.2$ and $a = 1.5$. The dotted line shows equal investment in each of two chicks and is evolutionarily stable for parental control of the allocation. The dashed lines show the evolutionarily stable allocation to the dominant and subordinate chicks if the dominant chick is in control. The relatedness between chicks is $r = 0.5$. (b) A great egret parent with chicks. Original photo by 'Mike' Michael L. Baird. Published under the Creative Commons Attribution 2.0 Generic license (CC BY 2.0).

are non-overlapping. For each of the two cases, we look for the change from one generation to the next of the frequency of the mutant (gene-centred approach). We can then use as payoff the expected number of adult mutant offspring per breeding attempt (clutch) where one of the parents is a mutant. If the dominant chick is in control of the allocation, the payoff is

$$W_{dom}(p,q) = \frac{1}{4}\left[\left(s(p) + s(1-p)\right) + s(p) + s(1-q) + 0\right] \quad (4.25)$$
$$= \frac{1}{2}\left[s(p) + \frac{1}{2}s(1-p) + \frac{1}{2}s(1-q)\right].$$

We arrive at this expression by averaging over the four equally likely possibilities for offspring genotypes when one of the parents (mother or father) is a mutant heterozygote. The first possibility is that both dominant and subordinate chicks are mutants, the second is that only the dominant chick is mutant, the third is that only the subordinate chick is mutant, and the fourth, giving a payoff of 0, is that neither chick is mutant. The fitness derivative (eq (4.4)) for this case is

$$D_{dom}(q) = \frac{1}{2}\left[s'(q) - \frac{1}{2}s'(1-q)\right], \quad (4.26)$$

so that a condition for evolutionary equilibrium is $s'(q) = \frac{1}{2}s'(1-q)$. The condition entails that the marginal change in survival for the dominant chick is twice that of the subordinate. Using eq (4.24), the solution is

$$q = p^* = \frac{1}{2} + \frac{\log(2)}{2a},$$

which is illustrated by the rightmost dashed line in Fig. 4.10a. Similarly, if one parent, for instance the mother, is in control of the allocation, the payoff is

$$W_{par}(p,q) = \frac{1}{4}[s(p) + s(1-p)] + \frac{1}{4}[s(q) + s(1-q)]. \qquad (4.27)$$

We arrive at this expression by noting that with probability $\frac{1}{2}$ the mother is a mutant, in which case the allocation is p, with each chick having the probability $\frac{1}{2}$ of being mutant, and with probability $\frac{1}{2}$ the father is a mutant, in which case the allocation is q, again with each chick having the probability $\frac{1}{2}$ of being mutant. The fitness derivative for this case is

$$D_{par}(q) = \frac{1}{4}[s'(q) - s'(1-q)]. \qquad (4.28)$$

A condition for evolutionary equilibrium is $s'(q) = s'(1-q)$, with $p^* = 0.5$ as a solution, and this is illustrated by the dotted line in Fig. 4.10a.

One thing to note about this example is that the mutants for the two cases should be thought of as mutants at different loci. Genes influencing the behaviour of a dominant chick are likely to be different from genes influencing parental feeding behaviour. For parent–offspring conflicts, this will be the general situation, and it provides a recipe for how to study games between relatives. For the analysis, one should thus examine the invasion fitness of mutants of different genes, for instance genes for offspring behaviour vs. genes for parental behaviour, taking into account the population structure produced by relatedness.

4.6 Exercises

Ex. 4.1. Consider the coordination game with payoffs given by Table 3.3. Let $W(p',p)$ be the expected payoff to a mutant that drives on the right with probability p' when the resident strategy is to drive on the right with probability p. Show that $W(p',p) = (1-2p')(1-2p)$. Verify that $p^* = 0.5$ is a Nash equilibrium. For any mutant strategy $p' \neq p^*$ verify that conditions (ES1) and (ES2)(i) hold, but that condition (ES2)(ii) does not hold, so that p^* is not an ESS.

Ex. 4.2. Consider the analysis in Box 4.1, except now assume that there are discrete non-overlapping generations, so that $n(t)$ and $n^*(t)$ are the numbers of mutants and residents, respectively, in generation t. Assume that numbers now change according to $n(t+1) = K[w_0 + w(\epsilon(t))]n(t)$ and $n^*(t+1) = K[w_0 + w^*(\epsilon(t))]n^*(t)$, where $\epsilon(t) = n(t)/(n(t) + n^*(t))$ and w_0 and K are constants. Shown that

$$\epsilon(t+1) - \epsilon(t) = \frac{\epsilon(t)(1 - \epsilon(t))}{\bar{W}}\Delta,$$

where $\bar{W} = (1 - \epsilon(t))(w_0 + w^*(\epsilon(t))) + \epsilon(t)(w_0 + w(\epsilon(t)))$.

Ex. 4.3. Consider a two-player game with two actions, labelled u_1 and u_2, in which a strategy specifies the probability of taking action u_2. Let a_{ij} denote the payoff to an individual that chooses action u_i when partner chooses action u_j. Define $\alpha = (a_{11} + a_{22}) - (a_{12} + a_{21})$ and assume that $\alpha \neq 0$. Suppose the payoffs are such that there is a Nash equilibrium p^* lying strictly between 0 and 1. Show that we can write the payoff function as $W(p',p) = K(p) + \alpha(p - p^*)p'$, where $K(p)$ is a function of p. Show that $\alpha < 0$ if and only if each of the following three statements hold: (i) the unique best response to each pure strategies is the other pure strategy, (ii) p^* is an ESS, (iii) p^* is convergence stable.

Ex. 4.4. Show that the conditions given in Section 4.1 for a strategy to be an ESS for a playing-the-field game are equivalent to the conditions for a two-player game to be an ESS, when the game is actually a two-player game. That is, assume that conditions (ES2)(i) holds, and show that this means that conditions (ES2')(ii) and (ES2)(ii) are equivalent when the game is a two-player game.

Ex. 4.5. Consider the simple sex-allocation game of Section 3.8 with no mortality during maturation. Show that the Nash equilibrium strategy, $p^* = 0.5$, satisfies condition (ES2') and is hence an ESS.

Ex. 4.6. Consider Exercise 3.10 on temperature-dependent sex determination in which the trait under selection is the threshold temperature. Take the payoff $W(x',x)$ to be the expected number of matings obtained by the offspring of a female with threshold x' when the resident threshold is x. Derive an expression for $W(x',x)$. Hence show that the trait $x^* = 20 + 10/\sqrt{3}$ is the unique best response to itself and is hence an ESS. Show that the strength of selection $D(x) = \frac{\partial W}{\partial x'}(x,x)$ is given by $D(x) = \frac{N}{20\beta(x)}(1 - 3\beta^2(x))$ where $\beta(x) = (x - 20)/10$. Verify that x^* is convergence stable and hence a CSS.

Ex. 4.7. In a two-player game, payoffs are given by $W(x',x) = x'x - (x' + \frac{4}{75}x'^3)$ where strategies lie in the range $x \geq 0$. Find the best response function $\hat{b}(x)$ and the three ESSs for this game. Write down the equation of adaptive dynamics and hence show that the smallest and largest ESSs are convergence stable, while the middle ESS is not.

Ex. 4.8. Derive the continuous-time replicator dynamics in eq (4.16) from the definition of fitness in eq (4.15), by introducing that $n_i(t) = v_i(t)n(t)$. Verify that the average fitness $\bar{w}$ is zero if the total population size is stationary, i.e. $dn(t)/dt = 0$. Also, derive the Price equation (4.17) from the replicator equation.

Ex. 4.9. The Rock–Paper–Scissors game is a two-player game in which each contestant chooses one of three actions: rock (R), paper (P) or scissors (S). Payoffs for a generalised version of this game are given by Table 4.1. Here $a > 0$ and $b > 0$. The standard form of the game has $a = b$.

Table 4.1 Payoffs in a generalized Rock–Paper–Scissors game.

Payoff to focal		Action of partner		
		R	S	P
Action of focal	R	0	a	$-b$
	S	$-b$	0	a
	P	a	$-b$	0

The unique Nash equilibrium strategy for this game is to choose each action with probability $\frac{1}{3}$. Show that this Nash equilibrium is an ESS when $a > b$ and that there is no ESS when $a \le b$.

Consider the continuous replicator dynamics in eq (4.16) when the population consists of a mixture of the three pure strategists. Let $v_R(t)$ be the proportion of R strategists at time t, with $v_S(t)$ and $v_P(t)$ defined similarly. Show that $\bar{w} = (a - b)M$ where $M = v_R v_S + v_S v_P + v_P v_R$. Show that $\dot{v}_R(t) = v_R(t)[av_S(t) - bv_P(t) - (a - b)M(t)]$. Set $y(t) = v_R(t)v_S(t)v_P(t)$. Show that $\dot{y}(t) = (a - b)(1 - 3M(t))y(t)$. Let $a > b$. Show that in this case $v_i(t) \to \frac{1}{3}$ as $t \to \infty$ for $i = R, S, P$ provided that $y(0) > 0$. Investigate the limiting behaviour when $a = b$ and when $a < b$. (See e.g. Hofbauer and Sigmund (2003).)

Ex. 4.10. Consider the Prisoner's Dilemma with positive assortment (Box 4.3). Let α be the proportion of cooperators in the population. Choose a randomly selected population member and set $X_1 = 0$ if this individual is a defector and $X_1 = 1$ if the individual is a cooperator. Similarly, let X_2 be 0 or 1 depending on whether the partner of the individual is a defector or cooperator, respectively. Show that $\text{Var}(X_i) = \alpha(1 - \alpha)$ and that $\text{Cov}(X_1, X_2) = \alpha(p_C - \alpha)$. By conditioning on the X-value of partner, derive the identity $\alpha = (1 - \alpha)p_D + \alpha p_C$. Hence show that the correlation between X_1 and X_2 is $\rho(X_1, X_2) = p_C - p_D$.

5

Learning in Large Worlds

Many things that are important for individuals are variable. Among the examples are properties of an individual and its social partners, such as size, strength, and similar qualities, and characteristics of the environment, such as the availability of resources. Individuals might then need to learn about their particular circumstances, rather than relying on innate precepts. For game theory there are two major ways to conceptualize such learning. The first is to assume that individuals have innate representations of the probability distributions of variable features and use experiences to update the priors to adequate posteriors (cf. Section 3.12). We refer to this widely used Bayesian perspective as a 'small-worlds' approach. It presupposes that individuals have innate representations of the states of the world, over which prior distributions are defined, as well as representations of the state dynamics. For this to be reasonable the number of states should in some sense be small. The second way is to assume that individuals have much more limited innate precepts, for instance only about what should be sought and what should be avoided, perhaps guided by emotions like pleasure and pain. This can be referred to as a 'large-worlds' approach. It may give rise to seemingly adaptive behaviour also in situations where an individual does not have an innate model of the world, for instance because the world, including the decision-making machinery of social partners, is too complex. One can use the term model-free learning to refer to this approach.

An advantage of the large-worlds perspective is that it can be combined with investigation of the properties of model-free learning that are evolutionarily favoured in particular situations. We can for instance study the evolution of an individual's initial, unlearned tendencies to perform actions, how much the individual pays attention to different classes of stimuli, and the magnitude of the primary rewards that guide learning. As we illustrate in this and later chapters, the approach can be helpful as a method of analysis also for situations (games) that individuals encounter on a day-to-day basis, while living and interacting in social groups.

We present a sequence of examples illustrating both different biological applications and important elements of learning models. In Section 5.1 we introduce our basic learning approach, which is actor–critic learning (Box 5.1), using a stylized coordination game with a single state and two actions. We then use this approach to study learning in the Hawk–Dove game (Section 5.2), making the point that learning rates can evolve and that this can have important implications for the outcome of learning. How to model learning when actions vary along a continuum is dealt with in

Game Theory in Biology: Concepts and Frontiers. John M. McNamara and Olof Leimar,
Oxford University Press (2020). © John M. McNamara and Olof Leimar (2020).
DOI: 10.1093/oso/9780198815778.003.0005

Section 5.3, using contributions to a common good (Section 3.1) as an example. Social dominance and the formation of social hierarchies is an important and challenging problem for game theory. In Section 5.4 we introduce a learning model of this phenomenon, which is further extended in Section 8.6. The model includes the new element that individuals observe a continuously distributed variable, for instance representing relative size, before choosing an action. We end the chapter by giving an overview of learning approaches in game theory (Section 5.5), including discussing whether or not the learning dynamics will converge towards an endpoint for the game that is being learned how to play, such as a Nash equilibrium or an ESS.

In small worlds a natural question is whether learning rules are evolutionarily stable. For large-worlds models the corresponding question is whether different parameters or traits of learning mechanisms, for instance learning rates, are at an evolutionary equilibrium. We discuss this issue in greater detail in Section 8.8, after having presented a diverse set of examples.

5.1 Reinforcement Learning

For a large-worlds approach, one needs to assume some particular mechanism of model-free learning. Harley (1981) and Maynard Smith (1982) proposed learning dynamics for game theory in biology, which we describe in Section 5.5, together with other similar suggestions. Here we present our own favoured approach. Reinforcement learning, as described by Sutton and Barto (2018), was developed by computer scientists for machine learning and robotics. It was inspired by and is closely related to ideas in animal psychology and neuroscience. The approach now contributes to the rapid emergence of applications of artificial intelligence and provides an important basis for experimental investigations of learning in neuroscience. Reinforcement learning thus has major advantages: it is conceptually and computationally mature and it has the potential for biologically realistic descriptions of animal behaviour. The basic idea is that individuals explore through randomness in their actions and tend to increase the preference for actions that result in higher than so-far-estimated rewards (which applies to instrumental learning generally).

The overarching aim of reinforcement learning is to maximize perceived rewards over a suitable time scale. Learning is thus driven by rewards and is fundamentally different from evolutionary change, which is driven by success in survival and reproduction, which need not involve psychological processes such as perception of reward. For learning to be adaptive, it follows that individuals must possess innate, evolved mechanisms for detecting rewards, and these are referred to as primary rewards. From these, individuals can learn that other stimuli, or states, serve as predictors of primary rewards, and they are then called secondary rewards. The study of primary and secondary rewards is one of the main preoccupations of animal learning psychology, often discussed in terms of unconditioned and conditioned stimuli. Reinforcement learning also incorporates prediction of rewards by associating states with values that correspond to estimated future rewards. In this chapter we focus on the simplest case where the primary rewards are the payoffs from repeated plays of a one-shot game in a large population.

There are a number of different but related modelling approaches in reinforcement learning, many of which are described by Sutton and Barto (2018). Here we focus on one of them, actor–critic learning (see p. 94), but in the next chapter we also illustrate Sarsa learning (Box 6.5), which is another common approach.

For biological relevance, it is useful to specify what individuals might learn about in a given situation. For instance, in Section 3.11 we examined the Hawk–Dove game for individuals that vary in fighting ability and know their own ability. In practice, individuals might not know their own fighting ability in comparison with other individuals, but need to learn about it through interactions, and we study this kind of process in Sections 5.2 and 5.4. First, however, as a simpler beginning, we use the stylized example of the coordination game from Section 3.7. The analysis illustrates how individuals in a population can learn to coordinate their driving, on the left or on the right side of the road. An interpretation would be that individuals learn and adjust to how the majority of the population drive (Fig. 5.1).

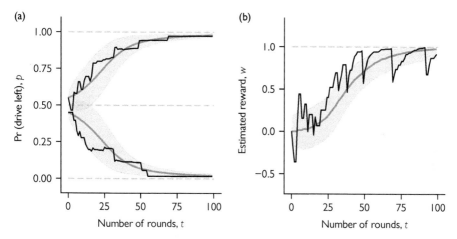

Fig. 5.1 Learning in a coordination game. In a large population ($N = 8000$) random pairs of individuals are repeatedly chosen to play a game with payoffs in Table 3.3. Individuals use actor–critic reinforcement learning as described in Box 5.1, with learning rates $\alpha_w = 0.2$ and $\alpha_\theta = 0.4$. The result is that they coordinate on either driving left or driving right. The outcome depends on the tendency to drive left or right in the starting population. The x-axes indicate the expected number of rounds of play per individual. (a) The population average (dark grey lines) $\pm$ one standard deviation (light grey shading) for two cases, with starting probabilities of driving left of $p = 0.45$ and $p = 0.55$. For each case, the learning trajectory is shown for one individual as a thin black line. (b) The median (dark grey) and the range from first to third quartiles (light grey) for the corresponding estimated reward from the game, for the case of starting at $p = 0.55$ (the other case is very similar).

5.1.1 The Actor–Critic Approach

A mechanism for learning in game theory needs both to build up estimates of the value for an individual of being in or reaching a certain state and to form preferences for the actions to perform in the state. Actor–critic learning implements these as distinct but linked processes. The learning dynamics are described in Box 5.1 for two-player games with two actions and in a large population where players are randomly paired. The value of being in the single state of the game, which in this case is the estimated reward from playing the game, is updated using the temporal difference (TD) error, as specified by eq (5.3) in Box 5.1 and illustrated in Fig. 5.1b. The TD error is the difference between actual and estimated rewards. The updating is similar to the much studied Rescorla–Wagner model of classical conditioning (Staddon, 2016), which describes the updating of an associative value. The interpretation of the Rescorla–Wagner model is that learning is driven by 'surprise', in the form of a discrepancy between perceived and estimated values. In reinforcement learning, the surprise is given by the TD error in eq (5.2). Based on findings in neuroscience the TD error is interpreted as a reinforcement signal that guides learning (Sutton and Barto, 2018), and it is a reason for the name 'reinforcement learning'. The TD error and the updating in eq (5.3) make up the critic component of actor–critic learning.

The actor component of actor–critic learning is an updating of action preferences, given by eq (5.5) and illustrated in Fig. 5.1a. The action probabilities are logistic functions of preferences, as in eq (5.1). The updating involves the product of the TD error from the critic component and the so-called eligibility of the action, defined in eq (5.4). The eligibility measures the relative change of the action probability with a change in preference. As seen from eq (5.4), if the current preference is such that the action performed has close to maximal probability, the eligibility will be small and the preference is changed very little, but if the action performed has smaller probability, the eligibility can be larger and the change in preference is then more sensitive to the TD error. In the general case of several actions, the corresponding quantity to eq (5.4) is referred to as an eligibility vector (Sutton and Barto, 2018). The actor–critic updating of the action preferences is a kind of policy gradient method, which means that the method is likely to perform well in terms of learning leading to higher rewards (Sutton and Barto, 2018). From the perspective of animal psychology, the actor component implements operant, or instrumental, conditioning (Staddon, 2016).

It is suggested that the actor learning process may correspond to a form of neural plasticity, so-called spike-timing-dependent plasticity (Roelfsema and Holtmaat, 2018; Sutton and Barto, 2018). However, compared with real neural processes, the learning mechanism in Box 5.1 might be a simplification. Neurons cannot implement negative firing rates, and for this reason excitatory and inhibitory influences on the preference for an action need to have separate mechanisms, as put forward by Collins and Frank (2014).

5.1.2 Average Actor–Critic Dynamics in a Population

We can use the individual learning dynamics from Box 5.1 to derive equations for the change of the population averages of the learning parameters w_{it} and θ_{it} (Fig. 5.1

Box 5.1 Actor–critic learning with a single state and two actions

We examine a simple case where individuals in a large population meet a sequence of randomly selected partners to play a game with two actions. The payoffs from the game are interpreted as rewards and individuals learn which actions to prefer in order to achieve high rewards. Because there are no payoff-relevant differences between rounds of the game (e.g. the identity and thus characteristics of the other player), there is only one state. In round t individual i chooses its first action with probability p_{it}, which is expressed in terms of a preference θ_{it} for the action, as follows:

$$p_{it} = \frac{1}{1 + \exp(-\theta_{it})}. \qquad (5.1)$$

The probability is a logistic function of the preference. Learning means that the preference is updated based on rewards from using the action. Individual i has an estimate w_{it} of the reward at the start of round t. At the start of the next round, the individual has observed the actual reward R_{it} from the recent play of the game, which depends on the actions used by both players, and can form the difference

$$\delta_{it} = R_{it} - w_{it} \qquad (5.2)$$

between estimated and actual rewards. This is referred to as the temporal-difference (TD) error. The TD error is used to update the estimated reward for the next round:

$$w_{i,t+1} = w_{it} + \alpha_w \delta_{it}, \qquad (5.3)$$

where α_w is a rate of learning. This represents the 'critic' of actor–critic learning. The TD error is also used to update the preference, in such a way that successful actions become more likely and less successful actions less likely. For this purpose, a derivative of the logarithm of the probability of performing the action, first or second, is used

$$\zeta_{it} = \frac{\partial \log \Pr(\text{action}|\theta_{it})}{\partial \theta_{it}} = \begin{cases} 1 - p_{it} & \text{if first action} \\ -p_{it} & \text{if second action} \end{cases} \qquad (5.4)$$

which follows from eq (5.1). The quantity ζ_{it} is referred to as the eligibility of the action used. The update of the preference ('actor') is then

$$\theta_{i,t+1} = \theta_{it} + \alpha_\theta \delta_{it} \zeta_{it}, \qquad (5.5)$$

where α_θ is a rate of learning, which need not be the same as α_w.

---

illustrates this kind of dynamics). For the average estimated reward, it follows from eq (5.3) that

$$\frac{d\bar{w}_t}{dt} = \alpha_w \bar{\delta}_t. \qquad (5.6)$$

For a learning steady state, we should then have $\bar{\delta}_t = 0$, or $\bar{R}_t = \bar{w}_t$. To derive a similar equation for the average preference $\bar{\theta}_t$, we first note from eq (5.4) that, given an individual's current action preference θ_{it}, the expected eligibility of the next action is equal to zero (e.g. from equation (5.4), $p_i(1 - p_i) - (1 - p_i)p_i = 0$). Such a relation follows generally by noting that the total probability, summed over actions, is equal to one. Using this and eq (5.5) we get that

$$\frac{d\bar{\theta}_t}{dt} = \alpha_\theta \operatorname{Cov}(\delta_{it}, \zeta_{it}),\tag{5.7}$$

where Cov denotes the population covariance between the TD error and the eligibility. For a learning steady state, we should then have $\operatorname{Cov}(\delta_{it}, \zeta_{it}) = 0$. We can compare this with the Price equation (4.17). Although these equations describe fundamentally different processes, natural selection vs. actor–critic learning, they are both helpful in providing intuitive understanding.

5.2 Learning and the Hawk–Dove Game

We now examine learning in the Hawk–Dove game (described in Section 3.5) from a few different perspectives. As a starting point we study an example where all individuals in the population have the same fighting ability, but they are unaware of the value V of winning and the cost C of losing a Hawk–Hawk interaction, and instead learn about these from the rewards they obtain (Fig. 5.2). The outcome of actor–critic learning is that the population average probability of using Hawk approaches the ESS

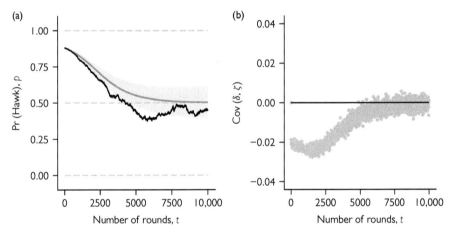

Fig. 5.2 Learning in a Hawk–Dove game. In a large population ($N = 8000$) random pairs of individuals are repeatedly chosen to play a Hawk–Dove game, as described in Section 3.5. If both choose Hawk, each has a 50% chance of winning, gaining $V = 0.5$, and the other then losing $C = 1.0$. Individuals use actor–critic reinforcement learning as described in Box 5.1, with learning rates $\alpha_w = 0.02$ and $\alpha_\theta = 0.02$. The x-axes indicate the expected number of rounds of play per individual. The result is that the population average probability of playing Hawk gradually approaches the ESS probability $p^* = V/C = 0.5$. (a) The population average (dark grey line) $\pm$ one standard deviation (light grey shading) probability of playing Hawk. The learning trajectory is shown for one individual as a black line. (b) The population covariance between the TD error and the eligibility, from eq (5.7), shown as one light grey point at each time step (the points overlap).

strategy of $p^* = V/C$, with considerable variation in individual strategies (Fig. 5.2a). For the rather low rates of learning in this example, the approach to a learning equilibrium is quite slow, taking several thousand rounds per individual, which is confirmed by the low covariance between the TD error and eligibility (Fig. 5.2b), from eq (5.7).

The situation in Fig. 5.2 is biologically somewhat artificial. While individuals might well learn about the value of winning and the cost of fighting, learning about the individual's own fighting ability in relation to others is likely to be a more prominent phenomenon. As a first illustration of this we examine a situation where individuals vary in fighting ability q, as described in Section 3.11 and Fig. 3.9, but they are unaware of their own q and can learn about it from wins and losses in Hawk–Hawk interactions against randomly selected opponents.

Two contrasting illustrations of the outcome of such learning appear in Fig. 5.3, with a large number of rounds per individual and rather low rates of learning in Fig. 5.3a, and fewer rounds per individual and higher, evolved rates of learning

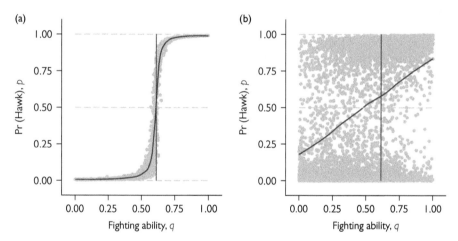

Fig. 5.3 Learning outcomes in Hawk–Dove games with variation in fighting ability q, with either fixed low or evolved rates of learning. At the start of a generation an individual's value of q is drawn from a uniform distribution on the interval $[0, 1]$. In a large population ($N = 8000$) random pairs of individuals are repeatedly chosen to play a Hawk–Dove game, as described in Section 3.11, with the probability of winning a Hawk–Hawk interaction given in eq (3.18). Individuals use actor–critic reinforcement learning as described in Box 5.1. The panels show the probability of using Hawk as a function of fighting ability, after a specified number of rounds of learning. The action probabilities are shown as light-grey points (overlapping), with non-linear local (LOESS) regression fits shown as dark grey lines, and the ESS threshold (from Fig. 3.9) as a vertical line. (a) There are many rounds of interaction (5×10^4) per individual, with rather low learning rates $\alpha_w = 0.02$ and $\alpha_\theta = 0.02$. (b) Each generation there are on average 100 rounds of interaction per individual and the learning rates are allowed to evolve over many generations, with payoffs equal to one plus the average reward from a game, until reaching an evolutionary equilibrium with $\bar{\alpha}_w = 0.111$ and $\bar{\alpha}_\theta = 2.164$.

in Fig. 5.3b. We see that with many rounds and slow learning, the ESS threshold strategy is a good approximation to the learning outcome (Fig. 5.3a). There are, however, good biological reasons to consider the evolution of learning rates, and also that learning should sometimes be selected to be fast. In animal psychology, learning rates are assumed to be influenced by the salience, in the sense of perceived intensity or noticeability, of a conditioned stimulus (Staddon, 2016). Furthermore, the reinforcing effects of rewards and penalties can vary in strength and influence learning (Berridge and Robinson, 1998; Wise, 2004), and it is reasonable that these effects can change in evolution. For our Hawk–Dove game, this would mean that individuals evolve to perceive wins and losses in a fight as highly striking or salient, and therefore make larger updates to their action preferences, as in eq (5.5) in Box 5.1. The reason this is adaptive is that fights are costly, making it important for individuals to relatively quickly move towards using mainly Hawk or mainly Dove, already after a limited amount of experience. As shown in Fig. 5.3b, the consequence is that a fair number of individuals with fighting ability q below the ESS threshold still tend to use Hawk, and vice versa for Dove. This can be seen as a trade-off between the cost of obtaining information and the benefit of using actions that are suited to the individual's characteristics.

We can also investigate the consequences of such higher rates of learning with a greater number of rounds per individual. We find that a threshold-like learning equilibrium has not quite been reached even after 1000 rounds per individual (Fig. 5.4a),

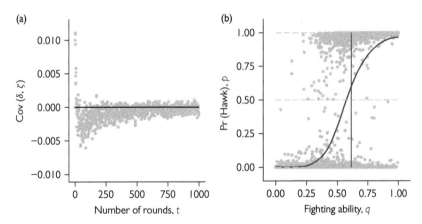

Fig. 5.4 Learning in a Hawk–Dove game with variation in fighting ability. There are many rounds per individual (1000) and the rates of learning are fixed at the average evolved values from Fig. 5.3b ($\alpha_w = 0.111$ and $\alpha_\theta = 2.164$). An individual's value of q is drawn from a uniform distribution on the interval $[0, 1]$. (a) The population covariance between the TD error and the eligibility, from eq (5.7), shown as one light grey point at each time step (the points overlap). (b) The learning outcome. The action probabilities are shown as light-grey points (overlapping), with a non-linear local (LOESS) regression fit shown as a dark grey line, and the ESS threshold (from Fig. 3.9) as a vertical line.

but that the individual actions have come somewhat closer to the ESS threshold strategy (Fig. 5.4b). We conclude that it need not be realistic to assume that individual learning reaches or even comes very close to an ESS of a one-shot game.

Note that for the actor–critic learning in the examples above, individuals do not directly learn about their own fighting ability q, in the sense of forming an actual estimate of this value, but instead simply learn which action to use based on the rewards they perceive. It is, however, possible to allow for more sophisticated learning by introducing several states, or stimuli, that individuals can take into account (see Sections 5.4 and 8.6).

We can compare the results in Fig. 5.3b with a small-worlds model analysed by Fawcett and Johnstone (2010). These authors studied the Hawk–Dove game in a large population with somewhat different assumptions than here: there were only two levels of fighting ability, either high or low, and instead of interactions over a period of some given length they used a continual turnover of individuals. In any case, the ESSs they computed shows qualitative similarity to our results in that they found rather rapid learning.

5.3 Learning in a Game of Joint Benefit of Investment

Actions that vary along a continuum are often needed for a good description of animal behaviour. The action might be the amount of effort invested into an activity, and individuals can modify their investments through reinforcement learning. We use the example of contributing to a common benefit at a cost (Section 3.1) to illustrate actor–critic learning with continuous actions. For simplicity we examine a case where random pairs of individuals play investment games similar to that in Fig. 3.1, but we assume that individuals vary in quality q, which influences their cost of investment. As illustrated in Fig. 5.5a, the common benefit (positive reward) from actions u_i and u_j is

$$B(u_{sum}) = b_0 + b_1 u_{sum} - \tfrac{1}{2} b_2 u_{sum}^2, \tag{5.8}$$

where $u_{sum} = u_i + u_j$, and the cost to individual i (negative reward) is

$$C(u_i, q_i) = c_1 u_i (1 - q_i) + \tfrac{1}{2} c_2 u_i^2. \tag{5.9}$$

The individual quality q_i influences how steeply the cost increases with u_i.

Actor–critic learning with continuous actions is described in Box 5.2. If individuals do not know their quality, they can learn about q from the rewards they obtain. As seen in Fig. 5.5b, the resulting individual learning trajectories are rather noisy. Part of the reason is that an individual interacts with a sequence of random partners, who vary in quality and also in their particular learning experiences, which causes the net reward from a game to vary considerably.

We can examine how the learning rates α_w, α_θ and the parameter σ evolve in such a situation, by assuming that they are genetically determined traits of an individual. Such an assumption could be biologically realistic, because observations

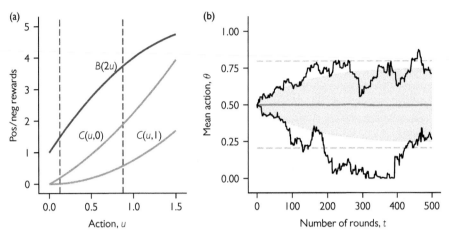

Fig. 5.5 Learning in an investment game. In a large population ($N = 8000$) random pairs of individuals play investment games with continuous actions. If i with quality q_i uses action u_i and meets j using u_j, the reward for i is $B(u_i + u_j) - C(u_i, q_i)$, with benefit and cost from eqs (5.8, 5.9), illustrated in (a). The dashed vertical lines show the ESS investments for $q = 0$ and $q = 1$, if an individual knows its q_i and meets another with q_j distributed uniformly on $[0, 1]$. (b) Individuals use actor–critic reinforcement learning as described in Box 5.2, with learning rates $\alpha_w = 0.04$, $\alpha_\theta = 0.004$ and exploration parameter $\sigma = 0.1$. The x-axis gives the expected number of rounds of play per individual and the population average (dark grey line) $\pm$ one standard deviation (light grey shading) of the action preference θ is shown. The learning trajectories are also shown for two individuals as thin black lines, the lower with $q \approx 0.1$ and the upper with $q \approx 0.9$. The dashed horizontal lines show the ESS investments for these values of q. Payoff parameters: $b_0 = 1$, $b_1 = 2$, $b_2 = 0.5$, $c_1 = c_2 = 1.5$.

Box 5.2 Actor–critic learning with continuous actions

For the simplest case with continuous actions, an individual's action preference θ is the mean of a normal distribution from which actions u are drawn. The standard deviation σ of the distribution measures the individual's tendency to explore around the mean. Thus, in round t individual i chooses its action from the distribution

$$P(u|\theta_{it}) = \frac{1}{\sqrt{2\pi\sigma^2}} \exp\left(-\frac{(u - \theta_{it})^2}{2\sigma^2}\right). \tag{5.10}$$

Just as for the case of two actions (Box 5.1), the estimated reward w_{it} is updated using the TD error $\delta_{it} = R_{it} - w_{it}$:

$$w_{i,t+1} = w_{it} + \alpha_w \delta_{it}. \tag{5.11}$$

The TD error is also used to update the preference, in such a way that successful actions become more likely and less successful actions less likely. A derivative of the log of the probability density of the action is used for this,

Box 5.2 *Continued*

$$\zeta_{it} = \frac{\partial \log P(u|\theta_{it})}{\partial \theta_{it}} = \frac{u - \theta_{it}}{\sigma^2}, \tag{5.12}$$

which follows from eq (5.10) and is referred to as the eligibility of the action. The update of the action preference is then

$$\theta_{i,t+1} = \theta_{it} + \alpha_\theta \delta_{it} \zeta_{it}. \tag{5.13}$$

Learning involves changes in both the estimated reward w_i and the action mean value θ_i and both are driven by the TD error δ_i. The learning rates α_w, α_θ and the magnitude of exploration σ should satisfy an approximate relation for learning to be reasonably efficient. Thus, from eqs (5.11, 5.13), and noting that $u - \theta_i$ in eq (5.12) has a magnitude of about σ, we ought to have

$$\alpha_w \Delta \theta \sim \frac{\alpha_\theta}{\sigma} \Delta w \tag{5.14}$$

for learning to cause the w_i and θ_i to move over approximate ranges Δw and $\Delta \theta$ during learning.

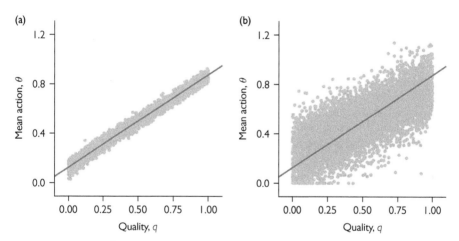

Fig. 5.6 Learning outcomes in the investment game with variation in individual quality q, with either fixed low or evolved rates of learning with an evolved exploration parameter. At the start of a generation an individual's q is drawn from a uniform distribution on the interval $[0, 1]$. In a large population ($N = 8000$) random pairs of individuals are repeatedly chosen to interact, with payoff parameters as in Fig. 5.5. The panels show the action preference θ as a function of quality, as light-grey points (overlapping) after a specified number of rounds of learning. The dark-grey lines show the ESS investment in a one-shot game where an individual knows its q, with qualities distributed uniformly on $[0, 1]$. (a) There are many rounds of interaction (10^4) per individual, with fixed low learning rates $\alpha_w = 0.002$, $\alpha_\theta = 0.0002$, and $\sigma = 0.10$. (b) Each generation there are on average 100 rounds of interaction per individual and the learning rates and exploration parameter are allowed to evolve over many generations, until reaching an evolutionary equilibrium with $\bar{\alpha}_w = 0.204$, $\bar{\alpha}_\theta = 0.007$, and $\bar{\sigma} = 0.098$.

of behavioural variation find that individuals vary in how sensitive they are to stimuli and how much they tend to explore (this represents so-called animal personality variation). We can implement this as three unlinked diploid loci, one for each of α_w, α_θ, and σ, and with individual reproductive success proportional to the mean reward during a generation.

As in the Hawk–Dove game analysed above, individuals face a trade-off between the advantages of slower and more careful learning (Fig. 5.6a) and the need to learn from a limited amount of experience (Fig. 5.6b). If we compare how well the learning outcome is approximated by an ESS strategy for a situation where an individual knows its q and plays another with q distributed uniformly on $[0, 1]$, we see that slow learning can, after very many rounds of interaction, come quite close to the ESS, whereas adaptive learning rates produce greater deviations from this ESS. This shows that an ESS of a one-shot game with known q can be a reasonable guide to learned behaviour, but there are sometimes quite noticeable deviations. As we have seen, these deviations can be favoured by natural selection. In any case, it is of interest to compare a learning outcome with a one-shot ESS prediction.

5.4 A Dominance Game

Dominance hierarchies are found in many group-living animals. It is an important and challenging problem for game theory to model social dominance, where individuals use aggressive displays and fights to establish who is dominant and who is subordinate in pairwise relationships. A useful perspective is that individuals learn from aggressive interactions to behave as dominant or subordinate towards another group member. Although the Hawk–Dove game could be used to model social dominance, we will use the assessment game from Section 3.12 because it contains elements that are present in real interactions. In this section we use the game to study aggressive behaviour in large groups where there is no individual recognition. In Section 8.6 we study the important problem of how individuals learn about the characteristics of other members in smaller groups.

Individuals vary in some quality q, which we assume is normally distributed in the population with mean μ_q and standard deviation σ_q. We might, for instance, have the logarithm of body size as q, or some other aspect of fighting ability. A real dominance interaction can consist of a sequence of several behaviours, including threat display and fighting. We simplify this to two stages. In the first stage, individuals observe some measure ξ of relative quality. For the interaction between individuals i and j in round t, we write the observation by i as

$$\xi_{ij}(t) = b_1 q_i - b_2 q_j + \epsilon_i(t), \tag{5.15}$$

where b_1 and b_2 determine the influence of the individual's own and the opponent's qualities on the observation and $\epsilon_i(t)$ is an error of observation, assumed to be normal with standard deviation σ. In the second stage, i and j perform actions. As a simplification, we assume there are only two actions, A and S, where A is an aggressive display or actual fighting behaviour and S is subordinate or submissive display.

The ultimate function of dominance behaviour is the division of resources, such as food or mates, but the behaviour often occurs in nature when there is no immediate conflict over a resource. In a sense, the conflict is about the dominance position. We implement this by assuming that immediately following the interaction in a round there is a probability that a resource becomes available, and we use V to denote the resulting expected fitness value. As in Section 3.12, if both i and j use action S, each has a 50% chance of obtaining an available resource (or, alternatively, they share it). If one uses A and the other S, the aggressive individual gets the resource, and this is the advantage of being dominant. If both use A, we assume that they are preoccupied with aggressive behaviour, and neither gets the resource (e.g. some other, unspecified agent gets the resource, or, if it is a potential mate, that the resource moves on). This makes it advantageous to establish dominance before a major part of the resources appear. There is a fitness cost of AA interactions, from exhaustion or injuries, given by

$$C\exp(-q_i + q_j). \tag{5.16}$$

The observation in eq (5.15) is statistically related to this cost.

5.4.1 Case of No Observations and No Learning

Before going into learning, is is instructive to perform a fitness-based analysis of a special case of the game, where there are no observations of relative quality and the game is played once between random individuals in a large population. If individuals are unaware of their own quality one can show (Exercise 5.4) that the evolutionarily stable probability of using A is given by

$$p_A^* = \frac{V}{V + 2C\exp(\sigma_q^2)}. \tag{5.17}$$

Alternatively, provided that an individual knows its q, we can consider threshold strategies $\hat{q}$ where the action A is used if $q > \hat{q}$. This is similar to the threshold strategies for the Hawk–Dove game in Section 3.11, where x is used to denote the threshold instead of $\hat{q}$, as illustrated in Fig. 3.9. The derivation of an equation for the evolutionarily stable threshold is dealt with in Exercise 5.5. Just as for the Hawk–Dove game one can show that the proportion of the population using the aggressive action A is smaller for the ES threshold strategy than for the ESS where individuals do not know their quality.

5.4.2 Learning and Primary Rewards

Reinforcement learning is driven by rewards, which can be either positive or negative. One way of applying learning to the dominance game is to assume that individuals have a primary motivation to perform the aggressive behaviour A. We can implement this by assuming that an individual perceives a reward v from performing A. This primary reward is genetically determined and can evolve, for instance to correspond to the resources obtained. In addition to this positive reward, there is also a perceived cost of an AA interaction. For simplicity, and to give a scale for the rewards, we assume the perceived cost is given by eq (5.16). Learning then proceeds by random pairs of individuals being drawn from a large group, making observations as in eq (5.15), and updating their actor–critic learning parameters as described in Box 5.3.

Box 5.3 Actor–critic learning with continuous observations/states

A randomly selected pair of individuals in a group play a game with two actions, which is repeated several times. Before choosing actions the individuals, i and j, make observations, ξ_{ij} for i and ξ_{ji} for j, which may be correlated with the expected rewards. We let the observations act as states. Individual i chooses its first action with probability $p_{it}(\xi_{ij})$. We modify the expression in eq (5.1) in Box 5.1 to

$$p_{it}(\xi_{ij}) = \frac{1}{1 + \exp\left(-(\theta_{it} + \gamma_{it}\xi_{ij})\right)}, \tag{5.18}$$

making the probability a logistic function of a preference that depends linearly on ξ_{ij} with slope γ_{it}. Individual i also has an estimated value,

$$\hat{w}_{it} = w_{it} + g_{it}\xi_{ij}, \tag{5.19}$$

at the start of the interaction. At the start of the next time step, the individual has observed the actual reward R_{it} from the recent play, which depends on the actions used by both players, and can form the TD error

$$\delta_{it} = R_{it} - \hat{w}_{it}. \tag{5.20}$$

This is used to update the learning parameters w_{it} and g_{it}:

$$w_{i,t+1} = w_{it} + \alpha_w \delta_{it} \frac{\partial \hat{w}_{it}}{\partial w_{it}} = w_{it} + \alpha_w \delta_{it} \tag{5.21}$$

$$g_{i,t+1} = g_{it} + \alpha_w \delta_{it} \frac{\partial \hat{w}_{it}}{\partial g_{it}} = g_{it} + \alpha_w \delta_{it}\xi_{ij},$$

where α_w is a rate of learning. For the actions we have the eligibilities

$$\zeta_{it} = \frac{\partial \log \Pr(\text{action})}{\partial \theta_{it}} = \begin{cases} 1 - p_{it} & \text{if first action} \\ -p_{it} & \text{if second action} \end{cases} \tag{5.22}$$

$$\eta_{it} = \frac{\partial \log \Pr(\text{action})}{\partial \gamma_{it}} = \begin{cases} (1 - p_{it})\xi_{ij} & \text{if first action} \\ -p_{it}\xi_{ij} & \text{if second action} \end{cases}$$

which are used to update the preference parameters. The updates of θ_{it} and γ_{it} are then

$$\theta_{i,t+1} = \theta_{it} + \alpha_\theta \delta_{it} \zeta_{it} \tag{5.23}$$

$$\gamma_{i,t+1} = \gamma_{it} + \alpha_\theta \delta_{it} \eta_{it},$$

where α_θ is a rate of learning.

5.4.3 Slow Learning Without Observations

As we have seen, for low learning rates the outcome of learning sometimes approximates an ESS of a one-shot game with the rewards as payoff. This possible correspondence is one way of using game theory to throw light on learning outcomes. As a starting point, we use this approach for the dominance game without the observations in eq (5.15). The corresponding one-shot threshold ESS, where A is played if $q_i > \hat{q}^*$, is worked out in Exercise 5.6. As illustrated in Fig. 5.7a, for fixed v and fixed low learning rates there is good agreement between the one-shot ESS and the learning outcome. As Fig. 5.7b shows, the learning is quite slow, in particular for individuals with values of q_i that are close to the threshold $\hat{q}^*$. Taking evolution into account, with slow learning we might expect the perceived reward v of performing A to evolve towards the ESS threshold value of a fitness-based analysis from Exercise 5.5. Exercise 5.7 deals with this correspondence.

However, as we noted for the Hawk–Dove game in Fig. 5.3, in reality we might expect learning rates to evolve to higher values, for which individuals more quickly settle on one action or the other. In such a case it is not so easy to work out analytically an evolutionary equilibrium value for v, but it can be found using individual-based evolutionary simulation.

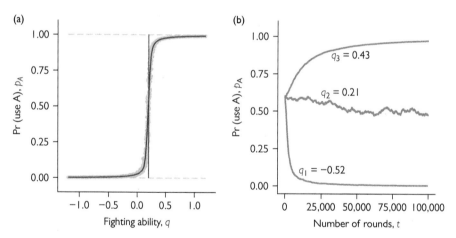

Fig. 5.7 Learning in the dominance game without observations. In a large population ($N = 8000$) random pairs of individuals are repeatedly chosen to play the dominance game with actions A and S. Learning follows Box 5.3 for the special case of $\xi_{ij} = 0$ and with small learning rates ($\alpha_w = \alpha_\theta = 0.02$). (a) After very many rounds of learning (10^5 per individual) the population has come close to a threshold strategy where A is played if $q > \hat{q} = 0.20$. This threshold corresponds to the ESS for the one-shot game analysed in Exercise 5.6. (b) Illustration of learning trajectories for three specific individuals, labelled with their qualities q_i. The x-axis gives the expected number of rounds of play per individual and the y-axis gives $p_A = 1/(1 + \exp(-\theta_{it}))$ from eq (5.18). The distribution of q in the population is normal with $\mu_q = 0$ and $\sigma_q = 0.5$ and the reward parameters are $v = 0.1$ and $C = 0.2$.

5.4.4 Evolution of Learning with Observations

If individuals make observations that are perfectly correlated with the cost of an AA interaction (i.e. if $b_1 = b_2$ and $\sigma = 0$ for eq (5.15)), the result of evolution can be that there is nothing to learn for an individual. This is because the observation ξ_{ij} provides perfect information about the rewards. The initial action preferences (at the start of a generation) can then evolve to be optimal for these rewards. Furthermore, when the observations ξ_{ij} and ξ_{ji} are perfectly correlated, we expect the endpoint of evolution to be a strategy where individuals play A if $\xi > 0$ and S if $\xi < 0$, and interactions are then either AS or SA.

We are, however, interested in the case where the correlation between an observation and the cost of an AA interaction is less than perfect. In such a situation evolution favours initial learning parameters (i.e. w_{i0}, g_{i0}, θ_{i0}, and γ_{i0}) such that all four kinds of combination of actions occur and individuals learn to adjust their action preferences, using the actor–critic approach from Box 5.3. An illustration of this appears in Fig. 5.8.

The learning works so that individuals with higher q_i tend to adjust θ_{it} and γ_{it} such that the value $\xi = \xi_0$ of the observation for which the preference is zero

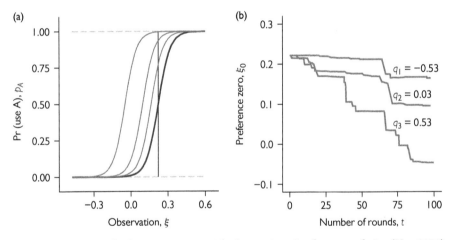

Fig. 5.8 Learning in the dominance game with observations. In a large population ($N = 8000$) random pairs of individuals are repeatedly chosen to play the dominance game. Learning follows Box 5.3. The distribution of q_i and the parameters b_1, b_2, and σ for ξ_{ij} are chosen such that, for random pairs, 50% of the variation in ξ_{ij} is due to variation in $q_i - q_j$. Learning rates and other genetically determined parameters, such as initial action preferences and the reward v from performing A, have values that are averages from evolutionary simulations over many generations. (a) Probability of performing A as a function of the observation. Initial logistic $p_A(\xi_{ij})$ (dark grey), and the final, learned logistic curves (light grey) for three individuals with different qualities q_i are shown. (b) Learning trajectories of the preference zero $\xi_{0t} = -\theta_{it}/\gamma_{it}$ for the three individuals (labelled with q_i). Parameters (see eqs (5.15, 5.16) and Box 5.3): distribution and observation of fighting ability, $\mu_q = 0$, $\sigma_q = 0.5$, $b_1 = b_2 = 0.707$, $\sigma = 0.5$; mean evolved values, $\alpha_w = 0.27$, $\alpha_\theta = 4.05$, $w_{i0} = 0.39$, $g_{i0} = 0.35$, $\theta_{i0} = -4.58$, $\gamma_{i0} = 20.66$, $v = 0.36$; payoff, $V = 0.31$, $C = 0.2$.

$(\xi_{0t} = -\theta_{it}/\gamma_{it})$ is shifted towards lower values. This value is referred to as the preference zero in Fig. 5.8b. The consequence of the shift is that individuals with higher q_i increase their tendency to perform the aggressive action A (in this example the correlation between q_i and the final learned ξ_0 is -0.51). Still, a major influence on the action preferences comes from evolution of the initial $p_A(\xi_{ij})$, shown in Fig. 5.8a, with learning producing adjustments to this logistic curve. The influence of learning of course depends on the strength of the correlation between ξ_{ij} and the cost in eq (5.16), being weaker for a higher correlation.

5.5 Approaches to Learning in Game Theory

The relative-payoff-sum (RPS) learning rule proposed by Harley (1981) and Maynard Smith (1982) appears to have been developed with the aim to make players approach an ESS. It makes use of discounted payoff sums for different actions u. For an individual who obtains a reward R_t in round t, we can write a total discounted payoff sum as $S_0 = r$ and

$$S_t = r + \sum_{\tau=0}^{t-1} \gamma^\tau R_{t-\tau},$$

for $t \geq 1$, where $\gamma \leq 1$ is a discount factor and $r > 0$ is referred to as a (total) residual payoff. With actions u_k, for instance u_1 and u_2, we can split the total S_t into components, $S_t = S_{1t} + S_{2t}$, where only those rounds where action u_k is used contribute to S_{kt}. In doing this, the residual should also be split in some way, as $r = r_1 + r_2$, so that

$$S_{kt} = r_k + \sum_{\substack{\tau=0 \\ u=u_k}}^{t-1} \gamma^\tau R_{t-\tau}.$$

The RPS learning rule is then to use action u_k in round t with probability $p_{kt} = S_{k,t-1}/S_{t-1}$. For this to always work one requires that the rewards and residuals are positive. A consequence of the rule is that actions that have yielded higher rewards, but also those that have been used more often, have a higher probability of being chosen by the learner.

   Although the RPS rule is not directly based on ideas about learning in animal psychology, it is broadly speaking a kind of reinforcement learning. In general, beyond the actor–critic approach described by Sutton and Barto (2018) and which we have used in this chapter, reinforcement learning can refer to any process where individuals learn from rewards that are somehow reinforcing, in a way that does not involve foresight or detailed understanding of the game situation. This broader interpretation has for instance been used by Roth and Erev (1995), and Erev and Roth (1998, 2014) in influential work on human behaviour in experimental games. Their basic reinforcement-learning model corresponds to the RPS learning rule.

A consequence of the RPS rule is that the ratio of the action probabilities is equal to a ratio of discounted payoff sums: $p_{kt}/p_{lt} = S_{k,t-1}/S_{l,t-1}$. For a discount parameter close to $\gamma = 1$, and after many rounds, the ratio of action probabilities is thus close to a ratio of summed rewards, which is sometimes referred to as a matching law (Herrnstein, 1970). This possible law has received much interest. If learning takes a population near a mixed ESS of a game, the average rewards of the actions will equalize, as in the Hawk–Dove game with $V < C$, and the matching law will then hold for population averages. This applies to any learning rule that takes a population near an ESS.

From a biological perspective, it is worth noting that the function of learning is not to reach a Nash equilibrium or an ESS of a one-shot game, but rather to be an adaptive tool for individuals to deal with variable circumstances. Originally, Harley (1981) and Maynard Smith (1982) argued that RPS is an evolutionarily stable learning rule, in the sense that it could not be invaded by another learning rule. It has since become clear that this claim is not correct (e.g. commentary by Krebs and Kacelnik in Maynard Smith, 1984), and it is limited to the long-term outcome of learning in a population. For instance, as we have seen in Sections 5.2 and 5.4, there are good reasons to expect learning rates to be tuned by natural selection, and the RPS rule says nothing about this.

Among the many variants of reinforcement learning, there are some that show qualitative similarity to the actor–critic approach. For instance, so-called Bush–Mosteller learning (Bush and Mosteller, 1955) assumes that individuals have an 'aspiration level' and a 'satisfaction', which can be compared with an estimated reward w_{it} and a TD error δ_{it}, and action probabilities are updated in a somewhat similar way to the action preferences in the actor–critic approach. In general, the various kinds of reinforcement learning are large-worlds approaches.

There are also learning rules that go beyond model-free reinforcement learning, by assuming that individuals understand that they are playing a particular game, although they might not know how other individuals decide on which actions to use. Fictitious play is a learning mechanism of this kind where individuals prefer actions that are the best responses to the actions they have observed others to use. It has been used as a dynamical foundation for game theory in economics (e.g. Fudenberg and Levine, 1998), and it has also been combined with aspects of reinforcement learning, for instance in the so-called experience-weighted attraction approach (Camerer and Ho, 1999). It is not known if animals actually use these more sophisticated learning rules. However, work in neuroscience has shown that model-free reinforcement learning oversimplifies the real situation (e.g. Langdon et al., 2018). Instead of a simple scalar TD error, as for instance used in actor–critic learning, animals encode more aspects of 'surprising' events than just a deviation from an estimated value.

5.5.1 Convergence towards an Endpoint of a Game

In the examples in this chapter on the Hawk–Dove, investment, and dominance games (Sections 5.2, 5.3, and 5.4), we found that low rates of learning produced learning outcomes near an ESS of a one-shot game, after many rounds of learning. The general question of whether learning will converge to a Nash equilibrium is much studied in economic game theory. The results are mixed, in the sense that there are special classes of games for which there is convergence, but also examples of non-convergence, such as cycling or otherwise fluctuating learning dynamics. Pangallo et al. (2019) randomly generated a large number of two-player games with two or more (discrete) actions per player, and examined the convergence properties for several classes of learning dynamics, including Bush–Mosteller reinforcement learning. They found that for competitive games, where gains by one player tend to come at a loss for the other, and with many actions, non-convergence was the typical outcome. In biology we are mainly interested in games that represent widespread and important interactions, and non-convergence of learning dynamics need not be common for these games. In any case, the relation between learning outcomes and game equilibria should always be examined, because game theoretic analyses add valuable understanding about evolved learning rules and learning outcomes.

5.5.2 Large and Small Worlds

The distinction between large and small worlds comes from general theories of rational human decision-making and learning under ignorance (Savage, 1972; Binmore, 2009; Huttegger, 2017). The large-worlds approach is a sort of questioning or criticism of the realism of the Bayesian small-worlds approach (Binmore, 2009). For game theory in biology, the decision-making processes of other individuals are among the most complex aspects of an individual's environment. So, for instance, in fictitious play individuals are assumed to understand the game they are playing, but they do not have an accurate representation of how other individuals make decisions. For this reason, fictitious play is a large-worlds approach, although it goes beyond basic reinforcement learning.

There are in fact rather few examples where game-theory models of social interactions have achieved a thoroughgoing small-worlds approach when there is uncertainty about the characteristics of other individuals. Some of these examples we present later in the book (Chapter 8). We argue that learning, including actor–critic reinforcement learning, will be especially helpful to study social interactions where individuals respond to each other's characteristics, as we develop in Section 8.6. It could well be that it is also a realistic description of how animals deal with social interactions. Further, for situations that animals encounter frequently and are important in their lives it could be realistic to assume, as we have illustrated in this chapter, that evolution will tune certain aspects of the learning process, including learning rates.

5.6 Exercises

Ex. 5.1. Derive the eligibility for actor–critic learning when there are discrete actions. Examine the case of more than two actions, numbered $k = 1, \ldots K$, with action probabilities for individual i at round t as

$$p_{ikt} = \frac{\exp(\theta_{ikt})}{\sum_l \exp(\theta_{ilt})}.$$

Derive updates to the θ_{ikt}, similar to eqs (5.4, 5.5). Show that with two actions and $\theta_{it} = \theta_{i1t} - \theta_{i2t}$, you get eq (5.5), except for a factor of 2 that might be absorbed into the learning rate.

Ex. 5.2. Derive the eligibility for actor–critic learning when there are continuous actions with a log-normal distribution (the analysis is similar to the one in Box 5.2).

Ex. 5.3. Work out the ESS result for a one-shot game shown in Fig. 5.6a, b.

Ex. 5.4. Investigate the one-shot assessment game where individuals are unaware of their own quality q and do not make observations of the kind given in eq (5.15). Verify that the evolutionarily stable probability of playing A is given by eq (5.17).

Ex. 5.5. Continuing from the previous exercise, now assume that individuals know their own quality q and use threshold strategies where action A is chosen if $q > \hat{q}$. Show that an evolutionarily stable $\hat{q}^*$ should satisfy the equation

$$V\Phi(\hat{q}/\sigma_q) = 2C\exp\left(-\hat{q} + \frac{1}{2}\sigma_q^2\right)\left[1 - \Phi(\hat{q}/\sigma_q - \sigma_q)\right],$$

where Φ is the standard normal cumulative distribution function.

Ex. 5.6. As in the previous exercise, assume that individuals know their own quality q and use threshold strategies where action A is chosen if $q > \hat{q}$ (and do not make observations ξ_{ij}). Investigate the one-shot game where the perceived reward v acts as payoff for performing the aggressive action A. Show that an evolutionarily stable $\hat{q}^*$ should satisfy the equation

$$v = C\exp\left(-\hat{q} + \frac{1}{2}\sigma_q^2\right)\left[1 - \Phi(\hat{q}/\sigma_q - \sigma_q)\right],$$

where Φ is the standard normal cumulative distribution function.

Ex. 5.7. Show that the evolutionarily stable thresholds $\hat{q}^*$ in Exercises 5.5 and 5.6 are equal when the perceived reward v of the action A and the fitness value of the resource satisfy

$$v = \frac{V}{2}(1 - p_A(\hat{q}^*)).$$

6

Co-evolution of Traits

The stability analyses in Chapter 4 deal with the evolution of a single trait, regarding other traits as fixed. That approach is perfectly reasonable when other traits do not interact with the focal trait, but is limiting and can miss important effects when there is a strong interaction. For example, in Section 6.6 we consider two traits: the degree of prosociality in a group and the dispersal rate between groups. It is perfectly possible to analyse the evolution of each of these traits in isolation, keeping the other trait fixed. However, the kin relatedness in groups depends on the dispersal rate and the degree of prosociality depends on the kin relatedness, so there is an interaction between the traits. As a consequence, when both are allowed to co-evolve there can be two distinct ESSs. At one ESS, behaviour within a group is cooperative and dispersal is low, and at the other behaviour is more selfish and dispersal is high. This result would not be so obvious if the traits were treated in isolation. In this chapter we consider how traits co-evolve when there is interaction between them.

Multidimensional stability is more complicated than stability in one dimension. Even verifying that a singular point is an ESS requires care since a singular point can be a maximum when approached from one direction but a minimum when approached from another. Both the relative speeds with which traits evolve and the genetic covariance between them can affect whether an ESS is convergence stable. We outline some of these issues in Section 6.1.

In two-player interactions the players can be in well-defined roles, with one player in role 1 and the other role 2 (Section 6.2). Furthermore, it may be reasonable to assume that each recognizes its own role. For example, when animals are territorial and an intruder challenges an established territory owner, it will often be reasonable to assume that the intruder is aware that the animal that it is challenging is the territory owner. When there are two clear-cut roles we are concerned with the co-evolution of two traits: behaviour in role 1 and behaviour in role 2. There is clearly an interaction between these traits: the best action as an intruder depends on the behaviour of the territory owner. When the traits are continuous the slope of the best response in one role to the trait value in the other role can be thought of as a measure of the strength of interaction of the two traits. These best response slopes are also important in determining the convergence stability of a trait combination. We end Section 6.2 by showing that the interaction based on roles can result in evolutionary predictions that are different from the analogous scenario without roles, using owner–intruder interactions as an example.

Game Theory in Biology: Concepts and Frontiers. John M. McNamara and Olof Leimar,
Oxford University Press (2020). © John M. McNamara and Olof Leimar (2020).
DOI: 10.1093/oso/9780198815778.003.0006

There can be two or more clear-cut roles also when individuals are playing the field rather than engaging in two-player interactions. The analysis of these cases is fairly similar to the one for two-player role asymmetry. As an example, in Section 6.3 we model the evolution of anisogamy when there are two distinct sexes. The relevant criterion for convergence stability (Box 6.2) can be expressed in a similar way to the corresponding criterion for a two-player role asymmetry.

Many traits outside of role asymmetries are naturally multidimensional, indicating that we ought to study the co-evolution between trait components. For example, Barta et al. (2014) emphasize that parental care involves several components such as feeding of offspring and nest defence, and thus cannot adequately be described by a one-dimensional trait. An individual's ability to perform a task, like care of offspring, and its tendency to engage in that task is another natural combination of traits. In Section 6.4 we study the co-evolution of such abilities and tendencies, including the ability to care and parental effort, with role specialization as a possible outcome. We then show that learning can promote individual specialization, in the form of correlations between abilities and behavioural tendencies, using scale-eating cichlids as an example (Section 6.5).

Even if different traits are not component parts of an overarching general activity, it can still be important to include them into a single evolutionary analysis. This will be the case when there is a notable fitness interaction between them. As we mentioned, in Section 6.6 we use the co-evolution of prosociality and dispersal to illustrate this possibility.

In evolutionary biology co-evolution often refers to reciprocal evolutionary changes in the traits of two or more species, occurring as a consequence of their interaction. Game theory has not contributed greatly to this field, but because of its general importance we devote Section 6.7 to it. We present an example of co-evolutionary convergence vs. advergence in Müllerian mimicry, using a different model of reinforcement learning and generalization (Box 6.5) compared with the models in the previous chapter. We also give a brief overview of evolutionary modelling of other species interactions, including the possibility of arms races.

6.1 Stability in More than One Dimension

When n traits co-evolve we can represent the strategy of an individual by the n-dimensional vector $\mathbf{x} = (x_1, x_2, \ldots, x_n)$, where x_i is the value of trait i. The payoff to a mutant with strategy $\mathbf{x}' = (x_1', x_2', \ldots, x_n')$ in a resident $\mathbf{x}$ population is denoted by $W(\mathbf{x}', \mathbf{x})$. The concept of a Nash equilibrium for this case is then a direct extension of that for one dimension given by eq (2.4). Specifically, the strategy $\mathbf{x}^* = (x_1^*, x_2^*, \ldots, x_n^*)$ is a Nash equilibrium if for all possible mutants $\mathbf{x}' = (x_1', x_2', \ldots, x_n')$ we have

$$W(\mathbf{x}', \mathbf{x}^*) \leq W(\mathbf{x}^*, \mathbf{x}^*). \tag{6.1}$$

Box 6.1 Multidimensional stability

Let $\lambda(\mathbf{x}',\mathbf{x})$ be the invasion fitness of a mutant strategy $\mathbf{x}' = (x_1', x_2', \ldots, x_n')$ in a resident population with strategy is $\mathbf{x} = (x_1, x_2, \ldots, x_n)$. The selection gradient $D_i(\mathbf{x}) = \frac{\partial \lambda}{\partial x_i'}(\mathbf{x}, \mathbf{x})$ indicates the strength and direction of selection at $\mathbf{x}$.

We examine the stability of a singular point $\mathbf{x}^*$, satisfying $D_i(\mathbf{x}^*) = 0$ for all i. A singular point could be a maximum, minimum, or a saddle (a maximum from one direction, but a minimum from another) as a function of the mutant strategy. The Hessian matrix $\mathbf{H}$ has elements $h_{ij} = \frac{\partial^2 \lambda}{\partial x_i' \partial x_j'}(\mathbf{x}, \mathbf{x})$. Taylor series expanding about the singular point gives

$$\lambda(\mathbf{x}', \mathbf{x}^*) \approx \lambda(\mathbf{x}^*, \mathbf{x}^*) + \sum_{i,j} h_{ij}(x_i' - x_i^*)(x_j' - x_j^*). \tag{6.2}$$

Thus if $\mathbf{H}$ is negative definite (i.e. $\sum_{i,j} h_{ij} z_i z_j < 0$ for all non-zero vectors $\mathbf{z}$), then $\mathbf{x}^*$ is a strict local maximum and is hence a local ESS.

Convergence stability. We use the multidimensional adaptive dynamics

$$\frac{dx_i}{dt}(\mathbf{x}) = m(\mathbf{x}) \sum_j c_{ij}(\mathbf{x}) D_j(\mathbf{x}), \tag{6.3}$$

where $\mathbf{C} = (c_{ij})$ is a genetic covariance matrix. Let $\mathbf{A}$ be the Jacobian matrix at $\mathbf{x}^*$, which has elements $a_{jk} = \frac{\partial D_j}{\partial x_k}(\mathbf{x}^*)$. Taylor expanding $D_j(\mathbf{x})$ to first order gives $D_j(\mathbf{x}) = \sum_k a_{jk}(x_k - x_k^*)$. Thus we can write the dynamics in the neighbourhood of $\mathbf{x}^*$ as $\frac{dx_i}{dt}(\mathbf{x}) = m(\mathbf{x}) \sum_k b_{ik}(\mathbf{x})(x_k - x_k^*)$, where $\mathbf{B}$ is the matrix product $\mathbf{B} = \mathbf{CA}$. Suppose that the eigenvalues of $\mathbf{B}$ have negative real parts. Then $\mathbf{x}$ will tend to $\mathbf{x}^*$ if these strategies are initially close, so that $\mathbf{x}^*$ is convergence stable. Conversely, if some eigenvalues of $\mathbf{B}$ have positive real parts, $\mathbf{x}$ will move away from $\mathbf{x}^*$ for certain initially close strategies.

Often the genetic covariance matrix $\mathbf{C}$ is not known. Leimar (2005, 2009) shows that if the matrix $\mathbf{A} + \mathbf{A}^T$ is negative definite then $\mathbf{x}^*$ is convergence stable for all covariance matrices. This property is referred to as strong convergence stability.

As before, this is a necessary but not sufficient condition for evolutionary stability. A sufficient condition for $\mathbf{x}^*$ to be a local ESS when traits are continuous is given in Box 6.1.

Convergence stability in higher dimensions is more subtle than in one dimension. One reason is that not only the direction of selection but also the rate of change of one trait relative to another can affect stability (Exercise 6.2). A second is that there may be a genetic correlation between traits. So for example, adult male size and adult female size may be influenced by the same genes, in which case selection to increase male size may also increase female size even if there is no direct selection on female size. Box 6.1 presents an analysis of convergence stability under adaptive dynamics in the multidimensional case.

6.2 Role Asymmetries

There are often well-defined roles in two-player interactions, with one player in each role and each player recognizing its own role. For example, in biparental care the roles are male and female. In a dispute over a territory one individual may be the current territory owner and the other an intruder. In such interactions the role of an individual may directly affect that individual's payoff. For example, possession of a territory may be more valuable to a territory owner than to an intruder if the owner has already invested time and effort in finding out where food sources in the territory are located. However, as the example of territory ownership presented below illustrates, even without payoff asymmetries, the mere existence of well-defined roles can alter evolutionary predictions.

Each individual might carry genes specifying the behaviour in each role, regardless of the current role of that individual. So, for example, females carry genes that specify what to do in the current situation and might also carry genes that specify what they would have done if male (e.g. if the genes are autosomal). Motivated by this, we define a strategy by a pair of traits (x_1, x_2) to be adopted in each role. Consider a rare mutant strategy (x'_1, x'_2) in a population with resident strategy (x_1, x_2). Since the mutant is rare, a mutant in role 1 is partnered by a resident in role 2. Let $W_1(x'_1, x_2)$ be the mutant's local payoff in this role. Similarly $W_2(x'_2, x_1)$ is the mutant's local payoff in role 2. At any time there are an equal number of individuals in each role, so in the simplest setting where roles are independent of the strategy, half of the mutants are in each role. If this is the case we can define a fitness proxy W as

$$W((x'_1, x'_2), (x_1, x_2)) = W_1(x'_1, x_2) + W_2(x'_2, x_1), \tag{6.4}$$

which we assume is a strong fitness proxy. Let us also assume that the genetic system is such that the evolution of one trait does not constrain the evolution of the other trait. Mutations that alter the value of one of the traits need then not alter the value of the other trait. For this case the best response to a resident strategy should simultaneously maximize the local payoffs W_1 and W_2. In particular, at a Nash equilibrium (x_1^*, x_2^*) we have

$$W_1(x'_1, x_2^*) \leq W_1(x_1^*, x_2^*) \quad \text{for all } x'_1, \tag{6.5}$$

and

$$W_2(x'_2, x_1^*) \leq W_2(x_2^*, x_1^*) \quad \text{for all } x'_2. \tag{6.6}$$

Suppose that for resident strategy (x_1, x_2) the local payoff $W_1(x'_1, x_2)$ has a maximum at $x'_1 = \hat{x}_1$. We refer to $\hat{x}_1$ as a role 1 local best response to the resident strategy, and similarly for a role 2 local best response $\hat{x}_2$. When local best responses are always unique we write them as $\hat{b}_1(x_2)$ and $\hat{b}_2(x_1)$. The Nash equilibrium condition can then be expressed as

$$x_1^* = \hat{b}_1(x_2^*) \text{ and } x_2^* = \hat{b}_2(x_1^*). \tag{6.7}$$

6.2.1 Convergence Stability with a Role Asymmetry

Suppose the trait in each role is continuous. Let the payoff to a rare mutant with strategy (x_1', x_2') in a population with resident strategy (x_1, x_2) be given by eq (6.4). Assume that for each x_2 the local payoff $W_1(x_1', x_2)$ has a strict maximum at $x_1' = \hat{b}_1(x_2)$ satisfying

$$\frac{\partial W_1}{\partial x_1'}(\hat{b}_1(x_2), x_2) = 0 \text{ and } \frac{\partial^2 W_1}{\partial x_1'^2}(\hat{b}_1(x_2), x_2) < 0. \qquad (6.8)$$

The local best response $\hat{b}_2(x_1)$ in role 2 has analogous properties. Let $x_1^* = \hat{b}_1(x_2^*)$ and $x_2^* = \hat{b}_2(x_1^*)$, so that (x_1^*, x_2^*) is an ESS. We analyse the convergence stability of this equilibrium in terms of the slopes of the local best response functions.

We first give an intuitive argument. Suppose that initially trait 1 has a value $x_1(0)$ that is close to, but not equal to x_1^*. Trait 2 then evolves to be the local best response to this trait 1 value, after which trait 1 evolves to be the best response to this trait 2 value, and so on. In this way we obtain a sequence $x_1(0), x_1(1), x_1(2), \dots$ of trait 1 values where $x_1(n) = \hat{b}_1(\hat{b}_2(x_1(n-1)))$. As Fig. 6.1 illustrates, the sequence of trait 1 values converges on x_1^* if the function B given by $B(x_1) = \hat{b}_1(\hat{b}_2(x_1)))$ has a derivative that satisfies $B'(x_1^*) < 1$. Conversely, if $B'(x_1^*) > 1$ the sequence moves away from x_1^*. Since $B'(x_1^*) = \hat{b}_1'(x_2^*)\hat{b}_2'(x_1^*)$ this suggests that

$$\hat{b}_1'(x_2^*)\hat{b}_2'(x_1^*) < 1 \implies (x_1^*, x_2^*) \text{ is convergence stable} \qquad (6.9)$$

$$\hat{b}_1'(x_2^*)\hat{b}_2'(x_1^*) > 1 \implies (x_1^*, x_2^*) \text{ is not convergence stable.} \qquad (6.10)$$

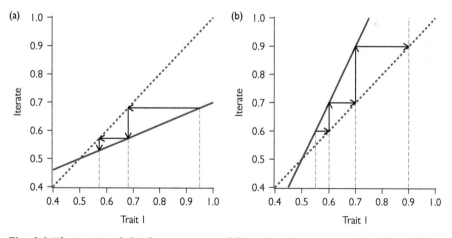

Fig. 6.1 The intuition behind convergence stability when there are two roles. The solid line gives a hypothetical function $B(x_1) = \hat{b}_1(\hat{b}_2(x_1))$. The dotted line is the 45° line. These lines intersect at $x_1^* = 0.5$. In (a) successive iterates approach x_1^*. In (b) successive iterates move away from x_1^*.

Of course evolution does not work by a series of steps in this way. However, the intuition is good because it can be shown that conditions (6.9) and (6.10) hold provided that there is no genetic covariance between the two traits (Exercise 6.5). In the special case in which the role asymmetry is purely a label, without payoff differences between roles, at a symmetric Nash equilibrium we have $x_1^* = x_2^* \equiv x^*$ and $\hat{b}_1'(x^*) = \hat{b}_2'(x^*) \equiv \hat{b}'(x^*)$. Thus (x^*, x^*) is convergence stable if $-1 < \hat{b}'(x^*) < 1$. Conversely, if either $\hat{b}'(x^*) < -1$ or $\hat{b}'(x^*) > 1$ then (x^*, x^*) is not convergence stable. Box 6.2 generalizes these results to playing-the-field situations.

6.2.2 Example: Territory Owner Versus Intruder

Suppose that a territory owner and an intruder contest the owner's territory. Each must choose either to play Hawk or Dove. The outcomes and payoffs to each are exactly as in the standard Hawk–Dove game (Section 3.5). In particular possession of this territory is worth V to both contestants and the cost of losing a fight is C. We assume that $V < C$, so that in the game without role asymmetries $p^* = V/C$ is the unique Nash equilibrium and is also an ESS (Section 4.1).

For this game there are two traits. Trait 1 is the probability, p_1, of playing Hawk when in the role of territory owner and trait 2 is the probability, p_2, of playing Hawk as an intruder. Suppose that the resident strategy is (p_1, p_2). Then the best trait 1 response (i.e. the best response as an owner) only depends on the resident trait 2 value (behaviour of intruders). Following the analysis of the standard Hawk–Dove game we see that it is given by $\hat{p}_1 = 1$ if $p_2 < V/C$, $\hat{p}_1 = 0$ if $p_2 > V/C$, and is any probability of playing Hawk if $p_2 = V/C$. The best trait 2 response is similarly $\hat{p}_2 = 1$ if $p_1 < V/C$, $\hat{p}_2 = 0$ if $p_1 > V/C$, and is any probability of playing Hawk if $p_1 = V/C$. At a Nash equilibrium p_1^* must be a trait 1 best response to p_2^* and vice versa. It can be seen that there are three Nash equilibria, $(1,0)$, $(0,1)$, and $(V/C, V/C)$. At the first Nash equilibrium owners always play Hawk and intruders always play Dove. Since the action in each role is the unique best response to the resident strategy, this Nash equilibrium is also an ESS. Similarly, the strategy $(0,1)$ that specifies play Dove as owner and Hawk as intruder is an ESS. At the third Nash equilibrium both owners and intruders play Hawk with probability V/C. In Section 4.1 we saw that the strategy $p^* = V/C$ is an ESS for the standard Hawk–Dove game. This was because, although mutants do equally well as residents against residents, once a mutation becomes common mutants play against other mutants and do less well than residents do against mutants. In the current game consider a mutation that, say, changes the aggressiveness when in the role of territory owner; i.e. a mutant's strategy is $(p_1', V/C)$ where $p_1' \neq V/C$. As before, mutants do equally well as residents against residents. Now, however, when two mutants play against one another, one is in the role of the intruder and plays Hawk with probability V/C just as residents do. Thus mutants do as well as residents even when they become common. Mutants can therefore invade by random drift and the Nash equilibrium is not an ESS.

In the above example the labelling of individuals as owner and intruder is the only difference from the standard Hawk–Dove game. This labelling dramatically changed the evolutionary prediction: the original unique ESS is no longer stable and instead there are two other ESSs. At each, behaviour is determined by a convention (such as play Hawk if owner), and once this convention is followed it does not pay for any population members to deviate.

Our reasoning above for why the mixed-strategy Nash equilibrium $(V/C, V/C)$ is not an ESS is an instance of a general argument put forward by Selten (1980). Selten shows that in any two-player game with role asymmetries there can be no mixed-strategy ESS; i.e. at an ESS no action can be chosen with a probability that lies strictly between 0 and 1. Despite the elegance of Selten's analysis and conclusion we believe the result is of limited applicability in biology. One reason is that in real games strategies often lie on a continuum, as for the threshold strategies in Section 3.11. It is then usually the case that all Nash equilibria are pure strategies so there certainly cannot be a mixed-strategy ESS. Another reason is that Selten's analysis assumes that game payoffs are fixed and do not depend on the strategy employed. To illustrate why this might not be the case, consider the territory owner versus intruder game. In that game V could be higher when ownership is respected as individuals tend to keep territories for longer. Also, C is the loss in expected future reproductive success as a result of losing a fight, and this expected future reproductive success depends on how easy it is to obtain a territory in the future. Thus both V and C depend on the resident strategy and both emerge from the analysis once the game is put into an ecological context, rather than being specified in advance. As a consequence, it turns out that mixed-strategy ESSs are possible (Section 9.5). Similarly, in Section 9.3 we consider the situation in which each of two parents have to decide whether to care for their common young or to desert. Again, putting the game into an ecological context means that payoffs emerge and a mixed-strategy ESS is possible.

6.2.3 *Real Owner–Intruder Contests*

Maynard Smith (1974) discovered that conventional settlement could be an ESS for owner–intruder contests and subsequent game-theory modelling confirmed this possibility (e.g. Maynard Smith and Parker, 1976; Hammerstein, 1981). Nevertheless, there is much field observation and experiments on owner–intruder interactions and these do not give much support for conventional settlement. For instance, based on observations of contests between spiders, Austad (1983) argued that cases of seeming territory owner's advantage most likely were consequences of asymmetries in resource value. As we have discussed (Section 3.5), the Hawk–Dove game is often an oversimplified representation of real contests, where individuals gain information about each other's size, strength, and motivation. Game theory analyses that take such factors into account show that there can be ESSs where owners and intruders use different local strategies even if there are no payoff asymmetries, such that one role is more persistent in fighting than the other (Leimar and Enquist, 1984; Enquist and Leimar, 1987). These ESSs occur when the resource value is low relative to the fitness costs of fighting. For higher resource value in relation to costs there is only one ESS, which is role-symmetric. This agrees qualitatively with the Hawk–Dove game. If we change our assumption in the analysis above to $V > C$, we find that the only ESS is to always play Hawk, regardless of the role. Because real owner–intruder conflicts are embedded into population processes and individual life-histories it is, however, not clear if the theoretical possibility of role-dependent local strategies without payoff differences between roles is important in nature. In spite of this, role asymmetries like owner–intruder are important because the population distributions of fighting abilities or resource values can differ between roles. For instance, stronger contestants can accumulate as owners (Leimar and Enquist, 1984). As a consequence, owners and intruders might have non-symmetric prior distributions of their relative fighting ability, with owners on average having higher fighting ability, and this can influence how persistent they are in fighting.

6.3 The Evolution of Anisogamy

Commonly, sexually reproducing species have two mating types, with sexual reproduction involving the formation of a zygote by the fusion of a gamete from one type with a gamete from the other type. Such mating systems can be classified as either isogamous or anisogamous (Fig. 6.2). Isogamy is common in extant unicellular eukaryote organisms and is liable to be ancestral in sexually reproducing species (Maynard Smith, 1982; Lessells et al., 2009). Anisogamy is the predominant system in multicellular organisms and appears to have evolved independently several times. Pathways by which this transition could occur and factors that select for small and large gametes are summarized by Lessells et al. (2009). One explanation for the evolution of anisogamy is as an adaptation to prevent competition between cytoplasmic symbionts (e.g. Hurst and Hamilton, 1992). A second, which we explore here, involves a trade-off. Due to resource limitation small gamete size results in more gametes and hence more matings, but the resulting zygote is smaller and hence less

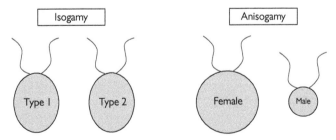

Fig. 6.2 Sexual reproduction in species with two mating types. Reproduction is isogamous if the gametes from the two types are of equal size. In anisogamy the gamete of one type is larger than the other, in which case this type is defined as female and the other type as male. Mating involves the fusing of gametes and only occurs between different types.

fit. There have been many evolutionary models of this trade-off starting with that of Parker et al. (1972). Here we largely follow the analysis of Matsuda and Abrams (1999), incorporating elements from the model of Lehtonen and Kokko (2011) (see also Bulmer and Parker, 2002).

6.3.1 Gamete Size Versus Number

We model a large population of a hypothetical marine organism that has two mating types and an annual cycle of reproduction. At reproduction there is one adult of each type per unit volume of water. Each type releases gametes into the water, which form a well-mixed mating pool. When two gametes of different types meet they fuse to form a zygote. If there are n_1 type 1 gametes and n_2 type 2 gametes per unit volume then a type i gamete fuses with probability $M_i(n_1, n_2)$. This probability does not depend on the size of the gamete. For each mating type, each adult has a total resource R to expend on gamete formation. If n gametes are released then each has size $x = R/n$. The size of the zygote is the sum $x_1 + x_2$ of the sizes of the two fusing gametes. The probability of survival to adulthood of the zygote is an increasing function $f(x_1 + x_2)$ of its size.

In this model the mating rates M_1 and M_2 must be related since every zygote is formed by the union of gametes of the different types. Thus the total number of type 1 gametes that fuse must equal the total number of type 2 gametes that fuse, so that $n_1 M_1(n_1, n_2) = n_2 M_2(n_1, n_2)$, where n_i is the number of type i gametes per unit volume. Houston and McNamara (2006) refer to the consistency condition that every offspring has one type 1 parent and one type 2 parent as the Fisher condition. The condition is important in models of mating systems and parental care (Queller, 1997; Houston and McNamara, 2006; Houston et al., 2005; Kokko and Jennions, 2008; Jennions and Fromhage, 2017) and its consequences for the evolution of anisogamy have been highlighted by Lehtonen and Kokko (2011). For computations we have used the mating probabilities $M_1(n_1, n_2) = n_2/(100 + n_1 + n_2)$ and $M_2(n_1, n_2) = n_1/(100 + n_1 + n_2)$. For these functions the probability that a gamete fuses with a gamete of the opposite type is never 1 but tends to 1 as the number of gametes of the opposite type increases. There is thus sperm limitation (Lessells et al., 2009). There is also sperm competition (Lessells et al., 2009) since the probability of fusion decreases with the number of gametes of the same type.

Let the resident strategy be (x_1, x_2); i.e. residents produce gametes of size x_1 if they are type 1 and gametes of size x_2 if they are type 2. Let a rare mutation in this population have strategy (x_1', x_2'). Then the expected total reproductive success of a mutant that is type 1 is

$$W_1(x_1'; x_1, x_2) = M_1(R/x_1, R/x_2)\left(R/x_1'\right) f(x_1' + x_2), \qquad (6.11)$$

Similarly, the expected total reproductive success of a mutant of type 2 is

$$W_2(x_2'; x_1, x_2) = M_2(R/x_1, R/x_2)\left(R/x_2'\right) f(x_1 + x_2'). \qquad (6.12)$$

We take invasion fitness to be

$$\lambda((x_1',x_2'),(x_1,x_2)) = W_1(x_1';x_1,x_2) + W_2(x_2';x_1,x_2) \qquad (6.13)$$

The selection pressure on trait i is $D_i(x_1,x_2) = \frac{\partial W_i}{\partial x_i'}$, where this derivative is evaluated at $x_i' = x_i$ (cf. eq (4.4)). If isogamy is a local ESS there is a gamete size x^* such that the reproductive success of each trait is (locally) maximized at x^*. This first requires $D_1(x^*,x^*) = 0$ and $D_2(x^*,x^*) = 0$, which are both equivalent to

$$x^* f'(2x^*) = f(2x^*) \qquad (6.14)$$

(Exercise 6.3). We also require these turning points to be maxima, which holds if the second derivative $\frac{\partial^2 W_i}{\partial x_i'^2}$ evaluated at $x_i' = x_i$ is negative for each i. These latter conditions are satisfied provided that $f''(2x^*) < 0$. Not every zygote reproductive success function f gives rise to an isogamous ESS, as Exercise 6.3 illustrates.

In order to investigate convergence stability of an isogamous ESS we first note that eq (6.12) can be used to analyse how the direction of selection on gamete size 2 depends on the resident gamete size 1. This direction of selection and the resulting 'best response' curve are shown in Fig. 6.3b. In this anisogamy example the local payoff for a mutant in one role depends on the resident behaviour in that role as well as the resident behaviour in the other role, whereas in the analysis of Section 6.2.1 there is no dependence on resident behaviour in the same role. This means that the derivation of the condition for convergence stability in terms of the product of the slopes of the best response functions no longer applies. Nevertheless, the condition can be extended (Box 6.2) and shows that the isogamous ESS is not convergence stable since the 'best response' curve has slope $\hat{b}'(0.5) = -2.9924$ at the isogamous ESS.

To analyse the rate of change of trait values under adaptive dynamics we assume that the genetic correlation matrix between trait 1 and trait 2 is proportional to the identity matrix (Box 6.1) so that we have $\frac{dx_i}{dt} = KD_i$, where K is a positive function of the resident strategy. As Exercise 6.4 shows, under these dynamics an isogamous ESS cannot be convergence stable. Fig. 6.3c illustrates typical dynamics when there is an isogamous ESS and initial gamete sizes are small, with one type of gamete (type 1) slightly larger than the other. Both gamete sizes first increase and approach the isogamous ESS, although the type that is initially larger remains larger. Evolutionary change is slow in the neighbourhood of this ESS, but eventually the two gamete sizes co-evolve with the initially larger type evolving to be bigger still and the initially smaller type evolving to be very small again.

Parental care is widespread, although not necessarily common, throughout the animal kingdom. When there is care in fish it is usually the male that does the caring, but overall female care predominates. Attempts have been made to relate the predominance of female care to the fact that the female (by definition) invests more in each gamete (e.g. Trivers, 1972). This debate continues with formal evolutionary models used to shed light on the issue (e.g. Fromhage and Jennions, 2016), although conclusions seem to depend on the modelling assumptions made.

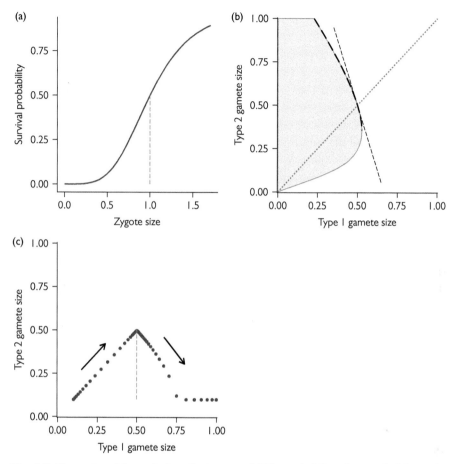

Fig. 6.3 Illustration of the evolution of anisogamy. (a) The probability of survival of the zygote as a function of zygote size. For this function ($f(s) = s^4/(1 + s^4)$) an isogamous ESS exists at which each gamete has size 0.5 units (so that the zygote is unit size). (b) Region (shaded grey) in which the selection pressure is to increase type 2 gamete size for given resident type 1 gamete size. The upper boundary to this region (dashed) is a 'best response' curve in that it is the size to which the type 2 gamete would evolve for given fixed resident type 1 gamete size, assuming type 2 gamete size started equal to that of type 1. The 45° line (dots) and tangent to the 'best response' curve (dashed straight line) are also shown. (c) Adaptive dynamics when initial gamete sizes are 0.1001 and 0.1 for types 1 and 2, respectively. The timescale of evolutionary change is determined by setting $K = 10^{-3}$. Points then correspond to combinations of gamete sizes at times $t = 0, 1, 2, \ldots$, with the arrows indicating the direction of evolution. The size of gametes is constrained to be at least 0.1 units. $R = 10^2$.

Box 6.2 Convergence stability when roles play the field

Let invasion fitness be given by eq (6.13). Then

$$D_1(x_1,x_2) \equiv \frac{\partial \lambda}{\partial x_1'}((x_1,x_2),(x_1,x_2)) = \frac{\partial W_1}{\partial x_1'}(x_1;x_1,x_2).$$

Assume there exists a curve $\tilde{b}_1(x_2)$ such that for each x_2 we have $D_1(x_1,x_2) > 0$ for $x_1 < \tilde{b}_1(x_2)$ and $D_1(x_1,x_2) < 0$ for $x_1 > \tilde{b}_1(x_2)$. Thus if x_2 were held fixed, trait x_1 would evolve to $\tilde{b}_1(x_2)$. We assume that $\tilde{b}_2(x_1)$ is similarly defined with analogous properties. Let (x_1^*,x_2^*) satisfy $x_1^* = \tilde{b}_1(x_2^*)$ and $x_2^* = \tilde{b}_2(x_1^*)$. We investigate the convergence stability of (x_1^*,x_2^*).

We follow the terminology of Box 6.1. Since $D_1(x_1,x_2)$ is assumed to be decreasing at $x_1^* = \tilde{b}_1(x_2^*)$ we have $a_{11} = \frac{\partial D_1}{\partial x_1}(x_1^*,x_2^*) < 0$. Similarly, $a_{22} < 0$. Thus the trace of the matrix **A** is negative. We also have $D_1(\tilde{b}_1(x_2),x_2) = 0$. Implicitly differentiating this function with respect to x_2 and evaluating the expression at $x_2 = x_2^*$ gives

$$a_{11}\tilde{b}_1'(x_2^*) + a_{12} = 0. \tag{6.15}$$

Similarly

$$a_{22}\tilde{b}_2'(x_1^*) + a_{21} = 0. \tag{6.16}$$

Putting these equations together, the determinant of **A** is given by $\Delta = a_{11}a_{22}(1 - \tilde{b}_1'(x_2^*)\tilde{b}_2'(x_1^*))$. Thus if $\tilde{b}_1'(x_2^*)\tilde{b}_2'(x_1^*) < 1$ then $\Delta > 0$ and both eigenvalues of **A** are negative (Exercise 6.1). Furthermore, if the covariance matrix **C** is diagonal then **B** = **CA** has negative eigenvalues, so that (x_1^*,x_2^*) is convergence stable. Similarly if $\tilde{b}_1'(x_2^*)\tilde{b}_2'(x_1^*) > 1$ then (x_1^*,x_2^*) is not convergence stable.

See Abrams et al. (1993) for a similar analysis.

6.4 Evolution of Abilities and Role Specialization

Within a population, organisms usually vary in their ability to perform specific tasks. They can also vary in their behavioural tendencies to perform a task. A possible outcome is then that co-evolution gives rise to two or more distinct types, where individuals of one type combine high ability and tendency to perform one task, with individuals of another type combining high ability and tendency to perform a different task (e.g. Roff, 1996). Learning might also contribute to role specialization, which we study in Section 6.5. The combination of co-adapted traits characterizing a type is sometimes referred to as a syndrome. Among the examples are dispersal dimorphisms (Roff and Fairbairn, 2007), with dispersing and sedentary types that each are adapted to their tasks, and mating type polymorphisms (Gross, 1996; Engqvist and Taborsky, 2016), for instance fighter and sneaker males (Sections 1.2 and 3.11). An individual's type can be genetically determined, but is often strongly influenced by environmental cues during development. Disruptive selection entails frequency dependence and promotes role specialization, by favouring distinct types. The frequencies of the types are stabilized by minority types performing better. Nevertheless, while variation in abilities and behavioural tendencies are ubiquitous, in many cases there is only one type present in a population, with stabilizing selection acting on its traits. For instance, often all males of a species acquire matings in a similar manner. One should however keep in mind that within-population polymorphism through co-evolution of traits is a widespread phenomenon, as are mating types (Gross, 1996) and other kinds of specialization. We now analyse the important instance in which the task is parental care.

6.4.1 Co-evolution of Parental Effort and Ability to Care

In our previous analysis of biparental care we assumed that the benefit to the young is an increasing but decelerating function of the sum of the parental efforts and that each parent pays a cost that is an increasing and accelerating function of its own effort (Section 3.4). Under these assumptions, parents are predicted to expend equal effort when both have the same cost functions. Here we examine a situation where parental effort and the ability to care co-evolve (McNamara and Wolf, 2015), in which case the outcome can be that care is not equally divided between parents. If the female and male have abilities θ_f and θ_m and expend efforts x_f and x_m, then the female and male payoffs are

$$W_f(x_f, x_m; \theta_f) = B(x_f + x_m) - K(x_f, \theta_f) - C(\theta_f) \tag{6.17}$$

$$W_m(x_f, x_m; \theta_m) = B(x_f + x_m) - K(x_m, \theta_m) - C(\theta_m). \tag{6.18}$$

The common benefit function B is assumed to be an increasing and decelerating function of the sum of the parental efforts. The cost to a parent has two components. The cost $C(\theta)$ of having ability θ increases with this ability. The cost of effort $K(x, \theta)$ is an increasing and accelerating function of effort for each ability θ. However, the cost of a given level of effort decreases with ability. Exercise 6.6 provides specific examples of these functions.

We are concerned with the co-evolution of the four traits θ_f, θ_m, x_f, and x_m. In order to gain insight into this process we consider what would happen if evolution always resulted in the ability of each parent being the best given its level of effort. This assumption reduces the problem to the standard form of a parental effort game in which just x_f and x_m co-evolve, but with a new 'effective' cost function $\tilde{C}(x)$. Box 6.3 derives the relationship between the form of this effective cost function and the slope of the best response function. When ability does not evolve and the cost of care is an accelerating function of effort the slope of the best response function is greater than -1 (but less than 0). This means that the symmetric ESS may be the only ESS (Fig. 6.4a). Furthermore, by the results of Section 6.2 the ESS is convergence stable. In contrast when ability is allowed to co-evolve with effort, the effective cost $\tilde{C}(x)$ may be a decelerating function of effort even though $K(x, \theta)$ is accelerating for each fixed θ. When this is so, the best response curves must cross at least three times (Fig. 6.4b, cf. Houston and Davies, 1985), and the symmetric ESS is not convergence stable. As Exercise 6.6 illustrates, whether or not $\tilde{C}$ is decelerating depends on the strength of interaction of ability and effort on cost.

Box 6.3 Stability of parental effort and ability

Assume that the ability of each parent is the best given the effort expended, so that a parent that expends effort x has ability $\hat{\theta}(x)$ where

$$K(x, \hat{\theta}(x)) + C(\hat{\theta}(x)) = \min_{\theta}[K(x, \theta) + C(\theta)]. \tag{6.19}$$

The effective cost of effort x is then $\tilde{C}(x) = K(x, \hat{\theta}(x)) + C(\hat{\theta}(x))$. Equations (6.17) and (6.18) thus reduce to

$$\tilde{W}_f(x_f, x_m) = B(x_f + x_m) - \tilde{C}(x_f) \tag{6.20}$$

$$\tilde{W}_m(x_f, x_m) = B(x_f + x_m) - \tilde{C}(x_m). \tag{6.21}$$

Now suppose that there is an ESS with equal efforts $x_f^* = x_m^* \equiv x^*$ by both parents. To find this ESS we differentiate the payoff and set this derivative to zero to find the best response $\hat{b}(x)$ of one parent to effort x of the other parent:

$$B'(\hat{b}(x) + x) - \tilde{C}'(\hat{b}(x)) = 0. \tag{6.22}$$

The ESS is then given by $\hat{b}(x^*) = x^*$. We will assume that $\tilde{C}''(x^*) - B''(2x^*) > 0$ to ensure that payoffs have a strict maximum at x^*.

To investigate convergence stability we differentiate eq (6.22), rearrange and set $x = x^*$ to give

$$\hat{b}'(x^*) = \frac{B''(2x^*)}{\tilde{C}''(x^*) - B''(2x^*)}. \tag{6.23}$$

$B''(2x^*) < 0$ by assumption. We have also assumed $\tilde{C}''(x^*) - B''(2x^*) > 0$. It follows that

$$\tilde{C}''(x^*) > 0 \implies -1 < \hat{b}'(x^*) < 0 \tag{6.24}$$

$$\tilde{C}''(x^*) < 0 \implies \hat{b}'(x^*) < -1. \tag{6.25}$$

The above analysis implicitly makes the strong assumption that ability evolves on a much faster timescale than effort. Although this is not realistic, the approach gives insight, and the results are consistent with evolutionary simulations in which ability and effort are allowed to both evolve on similar timescales (McNamara and Wolf, 2015). Figure 6.5 illustrates a typical simulation. Note, however, that an exact comparison of the analysis based on $\tilde{C}$ and an evolutionary simulation is difficult. In the analytic approach costs functions are given. The evolutionary simulation is based on a model of the entire life of each individual. Such models only specify consequences, such as the dependence of the probability of death on effort. The cost of death is not specified in advance. Since this cost is the loss of future reproductive success, and future success depends on the strategy of the mutant and residents, costs only emerge in a consistent manner once the ESS has been found (cf. Chapter 9).

When ability and effort co-evolve and there is disruptive selection, one sex evolves to be good at care and does most of the care, while the other sex evolves to be poor at care and does little care, as illustrated in Fig. 6.5. Since costs and benefits are

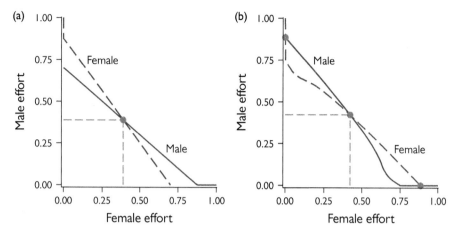

Fig. 6.4 The best parental effort of one sex as a function of the effort of the other sex, with and without the co-evolution of ability. (a) Ability constrained to be $\theta = 0.5$ for all population members. In this case there is a unique convergence stable ESS at which the female and male expend equal efforts. (b) Ability is the best given the effort expended. In this case the equal effort ESS is not convergence stable. There are two convergence-stable ESSs; at one the male expends zero effort, at the other the female expends zero effort. $B(z) = 2z - z^2$ for $0 \leq z < 1$ and $B(z) = 1$ for $z \geq 1$. $K(x, \theta) = 0.5v + v^2$ where $v = x/(1 + 2\theta)$. $C(\theta) = \theta^2/(1 + 10\theta)$.

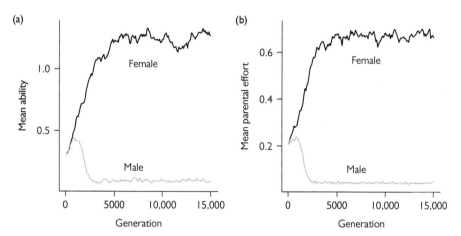

Fig. 6.5 The co-evolution of parenting ability and parental effort. (a) Mean ability. (b) Mean parental effort. Initially the sexes have equal ability and effort. $B, K,$ and C as for Fig. 6.4.

the same for both sexes, then whether it is females or males that do most of the care depends on initial conditions. However, once there are outside factors such as uncertainty of paternity, this has a strong influence on which sex does the majority of care (McNamara and Wolf, 2015).

6.5 Learning and Individual Specialization

Learning to specialize in certain behaviours is one possible reason for animal personality variation (Wolf and Weissing, 2010). A further phenomenon is that if individuals with certain abilities learn to make efficient use of the ability, this can give rise to disruptive selection on traits related to ability. As a consequence, the differences between individuals both in their ability and their behaviour can increase. We explore this using an example where there is random determination of a trait related to ability.

Scale-eating cichlids feed on scales of other fish in Lake Tanganyika. They attack victim fish from behind, either from the left or right, attempting to bite off scales. Their success in such an attack is influenced by the mouth asymmetry. For instance, an individual with a mouth skewed to the right is more efficient in attacking victims on the left flank. We follow Takeuchi and Oda (2017) in referring to such individuals as 'lefties', and individuals with mouths skewed to the left, making them more efficient at attacking the right flank, are 'righties'. The phenomenon was first studied by Hori (1993). Populations of scale-eating cichlids show a range of mouth asymmetries (Fig. 6.6), including individuals with forward-facing mouths. Although there are several studies on the issue, there is still some uncertainty whether the distribution of mouth asymmetry is bimodal or unimodal (Raffini and Meyer, 2019). How the mouth asymmetry of adult scale eaters is determined is also much discussed. The current view is that, although both genetic variation and plasticity (from experiences by juveniles of attacking victims) influence the asymmetry, the trait is mainly determined by random developmental variation (Raffini and Meyer, 2019).

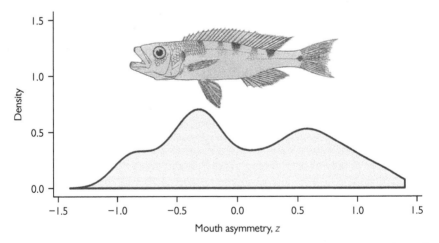

Fig. 6.6 Data on mouth asymmetry in scale-eating cichlid fish, *Perissodus microlepis*. The species is endemic to Lake Tanganyika. The figure shows a kernel-smoothed distribution of data on a mouth-asymmetry index from fig. 1 of Van Dooren et al. (2010). Illustration of *Perissodus microlepis* by George Albert Boulenger (in the public domain).

We can study the evolution of mouth asymmetry as the co-evolution of two traits, the extent of asymmetry and the direction of the skew. Thus, if z is mouth asymmetry, with $z < 0$ indicating lefties (with mouth skewed to the right) and $z > 0$ indicating righties, we can study random phenotype determination using two traits: the extent $|z|$ of asymmetry and the probability $q = q_1$ of being a lefty. Box 6.4 describes a model involving these traits.

Box 6.4 Mouth asymmetry in scale-eating cichlids

For a scale-eating individual with mouth asymmetry z, the basic success of an attack from left ($k = 1$) and right ($k = 2$) is

$$b_k(z) = \exp\left[-\frac{(z-\mu_k)^2}{2\sigma^2}\right], \tag{6.26}$$

with $\mu_2 = -\mu_1$. As a result of learning individual i has a probability p_{ik} to attack from direction k, with population average $\bar{p}_k$. The victims use the distribution of attacks on themselves to guard against attacks from a particular direction. Let $e(\bar{p}_k)$ measure the effect of the defence on the success of attacks from direction k. We assume

$$e(p) = v_0 - v_1\left(p - \tfrac{1}{2}\right) - \tfrac{1}{2}v_2\left(p - \tfrac{1}{2}\right)^2, \tag{6.27}$$

with $v_0 > 0$, $v_1 > 0$ and $0 < v_2 < 2v_1$, which is decreasing and satisfies $pe(p) + (1-p)$ $e(1-p) < e(\tfrac{1}{2})$ for $p \neq 0.5$. Thus, victims are better at defending if attacks are mostly from one direction (the general requirement is that $e(p)$ should be decreasing and concave). The success of an attack depends both on the mouth asymmetry of the attacker and on the defence:

$$s_k(z,\bar{p}_k) = b_k(z)e(\bar{p}_k). \tag{6.28}$$

The learned probability for an individual with asymmetry trait z of attacking from the right-hand direction is a sigmoid function of the difference

$$s_{21}(z,\bar{p}_2) = s_2(z,\bar{p}_2) - s_1(z,1-\bar{p}_2) \tag{6.29}$$

in success between right and left attacks:

$$p_2 = f(s_{21}(z,\bar{p}_2)) = \frac{1}{1 + \exp\left(-\kappa s_{21}(z,\bar{p}_2)\right)}. \tag{6.30}$$

Note that $p_1 = 1 - p_2 = f(-s_{21}(z,\bar{p}_2)) = f(s_{12}(z,\bar{p}_1))$, where we used $s_{12}(z,\bar{p}_1)$ defined in analogy with eq (6.29). Let the resident population of scale eaters contain a proportion q_1 with trait z_1 and a proportion $q_2 = 1 - q_1$ with trait z_2. There is a consistency requirement

$$\bar{p}_2 = q_1 f(s_{21}(z_1,\bar{p}_2)) + q_2 f(s_{21}(z_2,\bar{p}_2)), \tag{6.31}$$

with an equivalent formulation for $\bar{p}_1$. Using this equation, we can find a value for $\bar{p}_2 = p_2(z_1,z_2,q_2)$, given z_1, z_2, and q_2. From this we can find the payoff for a mutant individual with mouth asymmetry z':

$$W(z',z_1,z_2,q_2) = W_0 + \left(1 - f(s_{21}(z',\bar{p}_2))\right)s_1(z',1-\bar{p}_2) \tag{6.32}$$
$$+ f(s_{21}(z',\bar{p}_2))s_2(z',\bar{p}_2).$$

Furthermore, it is known that scale eaters do not have innate attack preferences but learn to prefer attacking victims from the left or from the right from their experience of which direction gives a higher success (Takeuchi and Oda, 2017). The speed of learning might then influence the evolution of mouth asymmetry. The victim fish also learn to defend against attacks from a particular direction.

In field experiments, Indermaur et al. (2018) showed that the defences influence the attack success of the scale eaters, with a higher success, and thus less efficient defence, with equal numbers of lefties and righties. This shows that it is more difficult for victims to defend against attacks when these are equally frequent from the left and the right. The model in Box 6.4 takes both learning by scale eaters and defence by their victims into account.

An example of an analysis of this model appears in Fig. 6.7. With the chosen parameters and with no learning, so that a scale eater continues to attack equally often from the left and the right, the evolutionary outcome is a population of symmetric individuals ($z = 0$), whereas with learning the result is a developmental

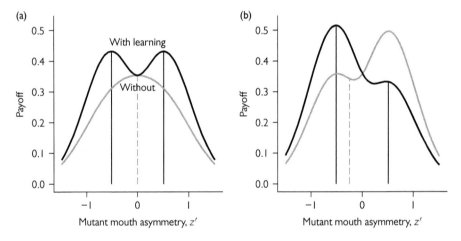

Fig. 6.7 Illustration of the effect of learning on the evolution of mouth asymmetry in scale-eating cichlids. (a) The light grey curve shows the payoff to mutant mouth asymmetry z' (from eq (6.32)) when there is no learning ($\kappa = 0$ in eq (6.30)). The resident population has forward-facing mouths (vertical light grey dashed line), and this mouth asymmetry has the highest payoff and is thus an ESS. With learning (dark grey curve, $\kappa = 4$) the evolutionary outcome is instead a mouth asymmetry of magnitude of $|z| = 0.51$ and equal chances of developing into skew to the right or left ($q_1 = q_2 = 0.5$), shown by the vertical dark grey lines. (b) Two cases of non-equilibrium situations with learning ($\kappa = 4$). The dark grey curve shows the payoff in a resident population with an equilibrium magnitude of asymmetry of $|z| = 0.51$ but a probability $q_2 = 0.9$ of being a righty. In this situation mutants with a lower probability of developing into a righty are favoured. The light grey curve shows mutant payoff in a monomorphic but skewed resident population ($|z| = 0.25$ and $q_1 = 1$). Parameters for the model in Box 6.4: $\mu_2 = -\mu_1 = 0.5$, $\sigma = 0.6$, $v_0 = v_1 = 0.5$, $v_2 = 0.75$, $\kappa = 0$ or $\kappa = 4$.

strategy that randomizes between skew to the right and the left (light vs. dark grey curves in Fig. 6.7a). Deviations from a 50–50 randomization are selected against, as illustrated in Fig. 6.7b. The dimorphic equilibrium is not reached by convergence of monomorphisms towards $z = 0$ followed by evolutionary branching, because $z = 0$ is not convergence stable. Instead, monomorphic evolution is directed away from this point, as illustrated by the light grey curve in Fig. 6.7b. A mutational jump between right- and left-skewed mouth is needed to reach a dimorphism, corresponding to a right–left developmental switch. Such switches could well occur in real organisms.

Only the outcome of learning is represented in Box 5.1, in eq (6.30). With individual-based simulation we can instead implement actor–critic learning by scale eaters, which is illustrated in Fig. 6.8. The starting point of the simulation is a monomorphic population with $z = 0$, but a dimorphic developmental strategy also evolves for other starting populations (e.g. monomorphic $z = -0.25$; not shown).

We looked for equilibria in a space of randomized strategies, but for this example we might instead have studied genetically polymorphic populations. In general, the circumstances favouring a randomized developmental strategy vs. a genetic poly-morphism differ (Leimar, 2005), but with frequency-dependent selection in a large random-mating population these can be alternative outcomes, as we pointed out for

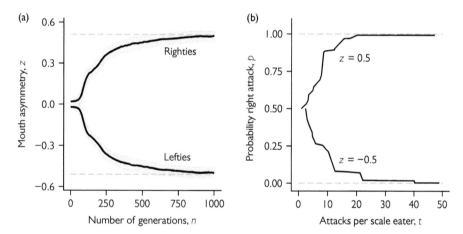

Fig. 6.8 Individual-based simulation of the evolution of mouth asymmetry in scale-eating cichlid fish. Attack success and defence follow the model in Box 6.4, but individuals use actor–critic learning (as in Box 5.1) when choosing whether to attack from the left or the right. The rewards are given by eq (6.28) and there are 100 attacks per scale eater per generation. There are two traits determining mouth asymmetry, the extent $|z|$ of asymmetry, and the probability q of developing a left skew (suitable for a righty), each coded by a diploid unlinked locus. The simulated population consists of 8000 scale eaters and 8000 victims, and each victim uses its current distribution of attacks to adjust its defence according to eq (6.27). (a) The evolution of mouth asymmetry over 1000 generations, starting from a population with forward-facing mouths. (b) Illustration of learning trajectories for a lefty and a righty in a dimorphic population. Learning rates are at an approximate equilibrium: $\alpha_w = 0.02$ and $\alpha_\theta = 4.8$.

the Hawk–Dove game (Section 3.5). For a finite population one can show that, because of genetic drift, the randomized outcome is favoured by selection (Maynard Smith, 1988), although the effect is weak in large populations.

Similar specialization as in the scale-eating cichlids can occur in other situations. In many social foraging groups some individuals are producers and others are scroungers (Fig. 1.1). Producers independently search for food sources which scroungers then attempt to exploit. The payoffs in each role are thought to be frequency dependent: the advantage of scrounging increases with the proportion of producers in the group. This producer–scrounger game has been extensively analysed (Barnard and Sibly, 1981; Giraldeau and Caraco, 2000; Giraldeau and Dubois, 2008). There is empirical evidence that individuals show some consistency over time in whether they are producers or scroungers (Fig. 1.1), although there is also flexibility (Giraldeau and Dubois, 2008; Reader, 2015; Aplin and Morand-Ferron, 2017b). In this respect, previous experience may be important: individuals may become better at a role as a result of learning, so that once a role is adopted an individual tends to keep the role. If there is variation between individuals in ability for the roles, we might also expect disruptive selection and the evolution of specialists for each role; some individuals would be better at producing and tend to produce, while others are better at scrounging and tend to scrounge. More generally, the co-existence of different personalities occupying different roles in groups may be maintained by frequency-dependent effects (e.g. Bergmuller and Taborsky, 2010). Groups would then tend to be composed of individuals that are good at the roles they perform.

6.6 Co-evolution of Prosociality and Dispersal

Dispersal, in the form of movement from the natal site to another breeding site, is an important life history trait. There is considerable evidence that within a group the propensity to disperse is correlated with other traits. In particular, high dispersal is often linked with a reduction in helping and other prosocial behaviours (Hochberg et al., 2008; Cote et al., 2010; Wey et al., 2015). One reason for this link could be that individuals that are not prosocial are more likely to be punished by other group members and disperse to avoid this punishment (Hochberg et al., 2008). We might also expect a positive association between prosocial behaviour and low dispersal probability that is mediated through the effect of relatedness. Individuals with a low dispersal probability are liable to be more closely related to other group members as an adult and during breeding, and this selects for prosocial behaviour. Conversely high dispersal results in low relatedness and selects for more selfish behaviour. This interdependence of dispersal and kin structure is modelled by Mullon et al. (2018). In their model groups occupy patches that are linked by dispersal. Individuals within a group interact pairwise in a game that yields resources that are used to enhance reproduction. In this interaction individuals can either behave prosocially and con-tribute to their partner or act selfishly. They show that there are regions of parameter space for which disruptive selection occurs. Evolution then results in a polymorphic

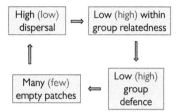

Fig. 6.9 Co-evolutionary feedback loop acting in the model of group defence and dispersal.

population in which there are two morphs: one morph has high prosocial behaviour and low dispersal while the other has more selfish behaviour and high dispersal.

In the model of Mullon et al. (2018) the two morphs co-exist. Co-existence is possible because of frequency-dependent competition between morphs: as the proportion of the low-dispersal morph increases, the competition for sites on a patch increases and the higher-dispersal morph has a relative advantage. In the model we develop below, the prosocial trait is the effort expended in defence of the whole group against an outside threat. The feedback that stabilizes a morph acts through the environment (Fig. 6.9). Thus rather than two morphs co-existing, our model results in one of two alternative morphs occupying the whole environment.

6.6.1 Model: Group Defence and Dispersal to Empty Areas

We develop a dispersal model in which N areas, each of which may contain a group of individuals, are linked by dispersal. Each area has K food territories. Each individual must occupy its own territory in order to survive. Thus the group occupying an area has maximum size K. At the beginning of each year each territory owner produces one offspring. It then dies with probability m, otherwise it retains its territory until next year. The offspring either remains in its natal area (with probability $1 - d$) or immediately disperses to a randomly selected area (with probability d). Those offspring that remain compete (scramble competition) for the territories that have become vacant due to the death of previous owners. Those individuals that disperse to a non-empty area die, since local group members oppose them. If a dispersing individual is lucky enough to disperse to an empty area it competes for a territory (again scramble competition) with others that also dispersed to this area. Those individuals that gain a territory are founder members of a new group. Individuals that do not gain territories immediately die without leaving offspring. Each year a threat to the whole group in an area occurs with probability β. If a threat appears, each group member decides how much effort to expend in group defence (its prosocial trait). Let u_i be the defence effort of the individual on territory i (set to zero if there is no individual on the territory). The probability the whole group is wiped out is a function $H(u_1, u_2, \ldots, u_K)$ that decreases with the defence efforts of the individuals. If the group is successful in defending against the threat, individual i dies with probability $h(u_i)$ that increases with its group defence effort.

In this model there is a trade-off—to defend the group or save oneself. We might expect that when the dispersal probability is high, members of a group are not highly related and expend little effort on group defence. Consequently, there will be many areas where the group has been wiped out, and hence many opportunities for dispersing offspring to find empty areas, so that dispersal remains high. Conversely, when the dispersal probability is low, individuals will be related and much more prosocial, putting more effort into group defence. This results in few empty patches and selects for low dispersal (Fig. 6.9). Figure 6.10 shows a case in which high initial

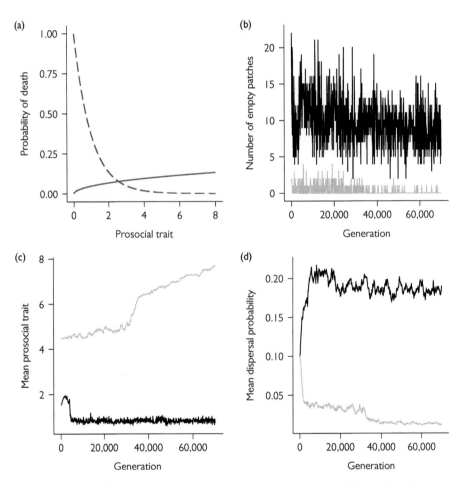

Fig. 6.10 The co-evolution of dispersal rate and prosociality. (a) Probability H that the group is wiped out when threatened as a function of the root mean square value of the prosocial trait values of group members (dashed line), and the probability h of individual death following a successful defence (solid line). (b), (c), and (d) show results of two evolutionary simulations based on these mortality functions. The darker lines correspond to starting traits $u = 1.5$ and $d = 0.1$. Lighter lines to starting values $u = 4.5$ and $d = 0.1$. [$H(u) = e^{-T}$, where T is the root mean square value of $u_1, u_2, \ldots, u_K$. $h(u) = 1 - e^{-0.05u^{0.5}}$. $m = 0.2$, $\beta = 0.4$.]

levels of group defence lead to even higher levels evolving, accompanied by rare vacant areas and low dispersal rates. In contrast, low initial levels of group defence lead to lower levels evolving accompanied by more empty areas and higher dispersal. Overall this suggests that when dispersal and defence effort are allowed to co-evolve there may be two different stable endpoints. Whether this is so depends very much on the functions H and h; often there is just one endpoint.

6.7 Co-evolution of Species

As we mentioned, in evolutionary biology co-evolution usually refers to reciprocal evolutionary changes in the traits of two or more species, as a consequence of their interactions. This is a large and complex area of investigation, in which game theory has only played a minor role. The species interactions can involve harm, as for parasites and hosts or predators and prey, or they can be mutualistic such that they benefit each of the species involved.

Here we illustrate the latter possibility using a model of the evolution of Müllerian mimicry. In this mimicry two or more species that are aposematic (i.e. unprofitable as prey, with striking appearances that signal their unprofitability) evolve to become similar in their appearance (Fig. 6.11), which benefits the prey because of reduced costs of predator education. The species interact through predators, who can learn more quickly to avoid them as prey by generalizing between their appearances. The first hypothesis about this form of mimicry evolution was put forward by Müller (1878) and it played a role in early discussions about evolution by natural selection. Müller suggested that learning by predators to avoid unpalatable prey worked such

Variable burnet moth   Nine–spotted moth

Fig. 6.11 Example of Müllerian mimicry in two insect species, the variable burnet moth *Zygaena ephialtes* (left) and the nine-spotted moth *Syntomis phegea* (right). In the Northern European part of its distributional range, *Z. ephialtes* displays red markings on a black background but in some regions of Southern Europe the markings are instead white, with some yellow (left). It is thought that this change in appearance by *Z. ephialtes* is the result of evolution to mimic the more numerous insect *S. phegea*. The species are unpalatable to their predators, including to insectivorous birds, who learn to avoid attacking such prey. Original photos by Hectonichus, Genova, Italy. Published under the Creative Commons Attribution – Share Alike license (CC BY-SA 3.0 and CC BY-SA 4.0).

that a predator would attack a given number of prey after which it would stop attacking. It then follows that by having the same appearance, two species can divide the cost of this predator education between them, and that the gain in per capita prey survival would be greater for a species with lower population density.

In Box 6.5 we present a reinforcement learning model that shows some similarity to Müller's original assumptions about learning. The learning mechanism works by updating an estimate of the value of performing a certain action (like attack) in a given situation (like a particular prey appearance). The preference for attack is assumed to be proportional to the estimated value (eq (6.33)), which is simpler but possibly less biologically realistic than actor–critic learning (Box 5.1). The learning model in Box 6.5 also includes generalization between different prey appearances, using a Gaussian generalization function (eq (6.35)). Let us now consider the evolution of prey appearances, while the predator learning rule is unchanging. Generalization

---

Box 6.5 Sarsa learning with generalization

The Sarsa model of learning (Sutton and Barto, 2018) is widely used in animal psychology. It works by updating an estimated value Q of choosing an action in a given state. Here we examine a case with continuous state ξ, where a predator discovers a potential prey with appearance ξ and must decide whether or not to attack. As a simplification we assume that the no-attack choice always results in zero reward, which is already estimated by the predator, so we only need to model the learning of the value of attacking. Further, we assume that the predator has much experience of attacking normal, palatable prey, having already estimated a value q_0 of such an attack, so we can limit ourselves to learning from attacks on unpalatable prey. Before any such attacks the estimated value is thus $Q_0(\xi) = q_0$. Let ξ_t, $t = 1, 2, \ldots$ be the appearances of successively attacked prey, and let $t = 0$ denote a naive predator. Following t, the predator has estimates $Q_t(\xi)$ and the probability of attacking a prey with appearance ξ is

$$p_t(\xi) = \frac{1}{1 + \exp(-\beta Q_t(\xi))}. \tag{6.33}$$

The parameter β sets how sharply $p_t(\xi)$ changes from close to 1 to close to 0 as $Q_t(\xi)$ goes from positive to negative. The values are updated using the difference between actual and estimated rewards from attacking ξ_t:

$$\delta_t = R_t - Q_t(\xi_t). \tag{6.34}$$

In the updating of values the predator generalizes from the appearance ξ_t of the prey in the recent attack to a different appearance ξ using a Gaussian generalization function

$$g(\xi, \xi_t) = \exp\left(-\frac{(\xi - \xi_t)^2}{2\sigma^2}\right). \tag{6.35}$$

The update is then given by

$$Q_{t+1}(\xi) = Q_t(\xi) + \alpha \delta_t g(\xi, \xi_t), \tag{6.36}$$

where α is a rate of learning.

influences mimicry evolution by causing predators to treat prey with similar appearances in similar ways. This is illustrated in Fig. 6.12a, showing the probability of survival for mutants with different appearances ξ, which could be a measure of prey colouration, when there are two resident populations with average appearance $\xi = 2$ and $\xi = 8$, respectively.

From the perspective of game theory, the best response in this situation is a mutant with appearance near the average of the population with highest population density, which is $\xi = 8$ in this example. The light grey arrows in the figure illustrate evolutionary changes for mutants from the less numerous population that could invade. This might suggest that mimicry can only evolve through major mutational changes. There is, however, an alternative evolutionary scenario. Note that the survival peaks around the population means in Fig. 6.12a are slightly asymmetric, with a skew in the direction of the other population. The skew is a consequence of generalization by predators. It will only be noticeable if the width of the generalization function (σ in eq (6.35)) is not too small compared with the separation of the population appearances. Such gradual evolution of mimicry is illustrated in Fig. 6.12b, using individual-based simulation.

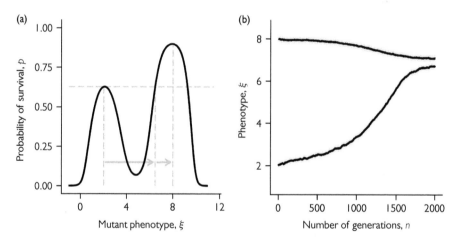

Fig. 6.12 Illustration of Müllerian mimicry evolution when predators learn and generalize according to the model in Box 6.5. There are $N_p = 100$ predators and two prey populations, one of size $N_1 = 1000$ and initial average appearance $\xi = 2.0$, and a larger one of size $N_2 = 5000$ and initial average appearance $\xi = 8.0$. The rate of detection of prey is 0.02 per predator per prey, over a season of unit duration. Reproduction is sexual with a simplified quantitative genetics inheritance, where the offspring phenotype is equal to the mid-parent plus a random normal deviate with standard deviation of 0.1. (a) The probability of survival of different mutant appearances for the initial situation. The light grey arrows indicate mutational changes of individuals in the first prey population that would be selected for. (b) Individual-based simulation of mimicry evolution for the two populations. The dark grey lines show the mean appearances and the grey shading indicates ± 1 SD. Learning parameters in Box 6.5: $\alpha = 0.6$, $\beta = 6.0$, $\sigma = 2.0$, $q_0 = 1.0$, and $R = -1.0$.

Fisher (1927, 1930) was the first to suggest such a possibility, initiating a debate about saltational vs. gradual changes in evolution (see Balogh and Leimar, 2005; Ruxton et al., 2008; Balogh et al., 2010; Leimar et al., 2012, for recent work on this matter). A related question is whether Müllerian mimicry evolution makes some species similar to the original appearance of other species, which is referred to as advergence, or if there is instead gradual co-evolutionary convergence to an intermediate appearance, somewhere in between the original appearances.

As is seen in Fig. 6.12a, saltational evolution will result in something close to advergence. However, gradual evolution can also approximate advergence (Fig. 6.12b), if there are pronounced differences in population density between prey species. From what is known about mimicry evolution in nature, it seems that advergence is the typical outcome (Mallet, 1999). The example in Fig. 6.11 illustrates advergence, such that *Z. ephialtes* has become similar in appearance to *S. phegea* without any noticeable change in the appearance of the latter (see Balogh et al., 2010; Leimar et al., 2012, for more background about this example).

Generalization over continuous stimuli, as represented by eq (6.35), has the potential to influence the evolution of signalling provided there is learning of the consequences of the signal by receivers. The phenomenon of stimulus generalization is much studied in animal psychology, with Gaussian generalization functions, often referred to as generalization gradients, giving a reasonable approximation to a wide range of experimental findings (Ghirlanda and Enquist, 2003). In machine learning, the term kernel-based methods is often used instead (Bishop, 2006; Sutton and Barto, 2018), where generalization functions appearing as in eq (6.36) are referred to as kernels. The model in Box 6.5 combines traditional ideas in animal psychology with those in machine learning. The approach can readily be adapted also to actor–critic learning, for instance for modelling of the kind described in Box 5.3.

6.7.1 Parasite–Host and Predator–Prey Interactions

The term co-evolution was introduced by Ehrlich and Raven (1964), although as we have seen the basic idea has a longer history. Ehrlich and Raven (1964) studied host–parasite interactions, specifically herbivorous insects (insects that feed on plants), arguing that adaptations of parasites to exploit their hosts and of hosts to defend against parasites have emerged in a stepwise pattern of many reciprocal evolutionary changes. A great volume of research on the topic has revealed that the evolutionary processes of hosts and parasites are complex, with important influences of phenomena like parasites switching between hosts over evolutionary time and hosts adapting to broad ranges of parasitic species. The concept of evolutionary stability has not proven very powerful in clarifying host–parasite relations. Instead it appears that outcomes similar to winners and losers in an arms race can occur, where on the one hand certain parasites successfully overcome host defences, while on the other hand the host defences fully succeed against other potential parasites.

Predator–prey interactions is another area where co-evolution has been much discussed, including using the perspective of evolutionary stability (Roughgarden,

1978; Takada and Kigami, 1991; Abrams et al., 1993; Abrams, 2000). The hunting and capture traits of a predator species can co-evolve with the anti-predator traits of its prey. For instance, suppose that each of the two species has a single trait, where the selection on trait values in one species only depends on the mean trait value in the other species. Then the mean trait value in one species will have a (delayed) response to the mean trait value in the other species. Since the fitness of a mutant of one species depends on the mean trait value of the other species, it depends indirectly on the mean trait value of its own species. As a result of feedback cycles of this type, co-evolution between a group of species can result in a subset of species having trait values minimizing fitness (Roughgarden, 1978; Takada and Kigami, 1991; Abrams et al., 1993; Abrams, 2000). Also, predator–prey interactions can have complex evolutionary outcomes with cycles both in the population dynamics and in the evolutionary dynamics (Abrams, 2000). Overall, the study of the evolution of these interactions is better characterized as evolutionary population biology rather than as an application of game theory in biology.

6.8 Concluding Comments

We have argued that in many systems it can be very limiting to consider traits in isolation. Mating systems provide many further examples. For instance, we might expect parental effort and divorce strategies to affect one another. If low effort by one parent leads to divorce by the partner, the evolution of parental effort will depend on divorce strategies in the population. Conversely, whether it is worth divorcing a partner depends on the likely parental efforts of alternative mates, so that the evolution of divorce strategies depends on parental effort strategies in the population. We analyse the co-evolution of the effort expended into a common good and the strategy for partner retention in Section 7.10, showing that the interaction of these traits can drive the evolution of high levels of cooperative behaviour.

Mutual mate choice provides another mating-system example. If a female is attempting to choose one male as her partner, then her choosiness can be expected to increase with her own attractiveness; i.e. increase with the number of males that will choose her. Thus the optimal mate choice strategy for females depends on that of males, and vice versa. This co-evolutionary situation is analysed by McNamara and Collins (1990). In the idealized case considered, they show that a class structure evolves for both sexes; pairs are only formed by males and females of the same class.

The co-evolution of male type and female mating preference can account for the evolution of such features as the male peacock tail, and provides a further spectacular example of the power of co-evolution. We analyse this Fisher runaway process in Section 10.5.

Some traits that we model as one-dimensional may really have different components. For example, in the model of parental effort of Section 3.4 the measure of effort is one-dimensional. However, as Barta et al. (2014) emphasize, care may involve more than one component with each parent allocating efforts between these components.

In particular, a parent must allocate its effort between provisioning and guarding the young. When care has two such dimensions, Barta et al. (2014) find that role specialization often evolves, with one parent concentrating on one of the tasks and the other parent concentrating on the other task. They also found that this promotes cooperation between the parents. In their model Barta et al. (2014) did not allow ability to evolve. It seems likely that had they done so, then there might be further disruptive selection leading to each sex evolving to be good at their specialized task (cf. Section 6.4).

The amount of variation in a trait may strongly affect the evolutionary direction of the trait via the effect of the variation on another trait. For example, suppose that the trait of interest is some measure of the degree to which an individual is cooperative in its interaction with others. The greater the variation in the trait, the greater the advantage in being choosy about which individuals to interact with. High choosiness then puts uncooperative individuals at a disadvantage since they are ignored. Thus co-evolution of cooperativeness and choosiness can result in high levels of the cooperative trait (Section 7.10).

Gaining information on the level of cooperative behaviour of others in the population may be valuable in deciding who to interact with and how to behave towards them. However, information gain takes time and may hence be costly. It is only worth paying this cost if there is sufficient variation in the population so that there is something to learn from observing individuals. If social sensitivity is worthwhile and evolves, this then exerts a selection on others to change their reputation by changing their behaviour, which then changes the selection pressure on being socially sensitive. The co-evolution of behavioural traits and social sensitivity is explored in Section 7.7.

6.9 Exercises

Ex. 6.1. Consider convergence stability in two dimensions. Let $\mathbf{B} = (b_{ij})$ be the matrix in Box 6.1. Let $T = b_{11} + b_{22}$ and $\Delta = b_{11}b_{22} - b_{12}b_{21}$ be the trace and determinant, respectively, of $\mathbf{B}$. Show that the real parts of the two eigenvalues of $\mathbf{B}$ are negative if and only if $T < 0$ and $\Delta > 0$.

Ex. 6.2. Let $\lambda((x_1', x_2'), (x_1, x_2)) = x_1'(x_1 + 2x_2) + x_2'(-x_1 + 3x_2) - (x_1'^2 + x_2'^2)$ be invasion fitness for a two-dimensional trait. Consider the stability of the singular point $x^* = (0,0)$. In the notation of Box 6.1 find the matrix $\mathbf{A}$. Suppose that covariance matrix $\mathbf{C}$ has elements $c_{11} = r_1 > 0$ and $c_{22} = r_2 > 0$, with off-diagonal elements zero. Thus the traits are genetically uncorrelated and r_1 and r_2 give the relative speeds of evolution of the two traits. Show that the singular point $x^* = (0,0)$ is convergence stable if $r_2 < r_1$ and is not convergence stable if $r_2 > r_1$.

Ex. 6.3. Consider the gamete size model of Section 6.3. Show that $D_i(x_1, x_2) = \frac{RM_i(x_1, x_2)}{x_i^2}[x_i f'(x_1 + x_2) - f(x_1 + x_2)]$. Hence show that eq (6.14) holds at an isogamous ESS. Show that eq (6.14) has a solution in the range $x^* > 0$ when $f(z) = e^{-1/z^2}$. Verify that $f''(2x^*) < 0$ so that x^* is an ESS. Show that eq (6.14) has no solution when $f(z) = z/(1+z)$. For further insight into why these cases differ, see the graphic representation of condition (6.14) in Maynard Smith (1982).

Ex. 6.4. Consider the gamete size model of Section 6.3. Let x^* given by eq (6.14). Let the genetic covariance matrix be proportional to the identity matrix. By Taylor series expansion about (x^*, x^*) show that the matrix coefficients describing the dynamics of the two traits in the neighbourhood of (x^*, x^*) (Box 6.1) can be written as $b_{11} = Hf''(2x^*)/x^*$ and $b_{12} = H[f''(2x^*)/x^* - f'(2x^*)/x^{*2}]$ for suitable constant H, with analogous expressions holding for b_{21} and b_{22}. Hence show that the two conditions $T = b_{11} + b_{22} < 0$ and $\Delta = b_{11}b_{22} - b_{12}b_{21} > 0$ cannot both hold, so that (x^*, x^*) is not convergence stable.

Ex. 6.5. Suppose there is a role asymmetry and that the trait in each role is continuous. Assume that there are unique local best response functions, with $\hat{b}_1(.)$ given by eq (6.8) and $\hat{b}_2(.)$ satisfying the analogous equation. Let $x_1^* = \hat{b}_1(x_2^*)$ and $x_2^* = \hat{b}_2(x_1^*)$, so that (x_1^*, x_2^*) is an ESS. (i) By implicit differentiation show that $\hat{b}_1'(x_2^*)\frac{\partial^2 W_1}{\partial x_1^2}(x_1^*, x_2^*) + \frac{\partial^2 W_1}{\partial x_1 \partial x_2}(x_1^*, x_2^*) = 0$, with an analogous equation holding for $\hat{b}_2'$. (ii) Deduce that $\hat{b}_1'(x_2^*)\hat{b}_2'(x_1^*) < 1$ if and only if the determinant of the Jacobian matrix $\mathbf{A}$ at this equilibrium (Box 6.1) satisfies $\Delta > 0$. (iii) Assume that the genetic covariance matrix $\mathbf{C}$ is diagonal. Show that the matrix $\mathbf{B} = \mathbf{C}\mathbf{A}$ has negative trace and a determinant that has the same sign as Δ.

Ex. 6.6. Consider the model of the co-evolution of parental effort and ability of Section 6.4. Let $K(x,\theta) = \beta(1-\theta)x + \frac{1}{2}\alpha x^2$ and $C(\theta) = \frac{1}{2}\theta^2$, where $0 < \beta < 1$ and $\alpha > 0$. Find the effective cost function $\tilde{C}(x)$ and verify that this function is concave if $\beta^2 > \alpha$. Let $B(z) = 2z - z^2$ for $0 \le z < 1$ and $B(z) = 1$ for $z \ge 1$. Set $\beta = 0.8$ and $\alpha = 0.14$. Find the best response functions in this case (remember that efforts have to be non-negative). Hence find the three ESS pairs of efforts.

7

Variation, Consistency, and Reputation

7.1 Variation has Consequences

In natural populations there is almost always a considerable amount of between-individual variation in behaviour as well as in other traits. For instance, for quantitative traits one finds that the standard deviation is often around 10% or more of the mean value (Houle, 1992). Furthermore, many estimates of the so-called repeatability of behaviours such as foraging, aggressiveness, and parental behaviour have been collected. The repeatability is the proportion of the variance in behaviour that is due to differences between individuals, and it ranges from around 0.1 to 0.8, with an average around 0.4 (Bell et al., 2009). For game theory this means that, first, we should not assume that variation is small and, second, that it is realistic to examine questions of consistency in behaviour and individual reputation.

One can conceptualize variation in behaviour as resulting from differences in the underlying strategy, in how the strategy is implemented, or from state differences. Strategy variation arises from underlying genetic variation and the implementation of a strategy is affected by developmental plasticity and noise. Concerning states, even if two individuals follow the same strategy, differences in previous experience give rise to state differences, so that the individuals may take different actions in the same current situation. Some phenotypic variation may thus be for adaptive reasons and some not, but regardless of its source, its existence has consequences for game theory.

Most models in the literature ignore that there may be large amounts of genetic variation at evolutionary stability. In considering the evolutionary stability of a Nash equilibrium the effect of variation due to mutation is certainly considered, but in contrast to what is the case in real populations, the amount of variation is assumed to be small; typically the fate of rare mutants is considered (Section 4.1).

In Section 3.11 we gave examples of state-dependent behaviour when there are state differences. In these examples the corresponding models with no state differences give similar predictions to the models with variation and a decision threshold when these latter models assume little variation in state, although there were some changes. For example, in the Hawk–Dove game variation in fighting ability reduces the frequency of fights and the payoffs of the alternative actions are no longer equalized at evolutionary stability: Hawks do better than Doves. As we will see, in other cases variation can produce even qualitative shifts in perspective.

Game Theory in Biology: Concepts and Frontiers. John M. McNamara and Olof Leimar,
Oxford University Press (2020). © John M. McNamara and Olof Leimar (2020).
DOI: 10.1093/oso/9780198815778.003.0007

All models are incomplete descriptions of the world that capture the essence of the biological issues of interest. They can be useful in helping to understand these issues, but they are always limited. It is important to understand what a given model can tell us, and what are its limitations. Models in which differences are ignored have provided important insights, but their limitations have not always been appreciated. Variation can completely change how we conceptualize games and change our predictions. Some of the major ways in which this occurs are illustrated in Fig. 7.1, and can be summarized as follows.

Explore all possibilities. A strategy is a rule that specifies the action taken for every possible state of the organism. However, if some states are never encountered it does not matter what an individual would have done in that state. The specification of what population members would have done had they entered the state can then drift, changing the selection pressure to enter the state. Variation can ensure that every state is reached with positive probability, removing this neutrality. This is a traditional idea in game theory about how variation can stabilize equilibria (Section 7.2).

Take a chance. Variation means that it can be optimal to take a chance and reject the current partner or option in the hope of obtaining a better one (Sections 7.3, 7.6, 11.2.2).

Signalling. When the best action of an individual depends on the attributes (e.g. ability, need, relatedness) of its partner then it will be advantageous to gain information on these attributes. But if the partner conveys this information via a signal, why should the signal be believed (Section 7.4)?

Reputation. Consistent differences lead to consistently different behaviour and hence to the possibility that reputations are formed and are informative. Once there are

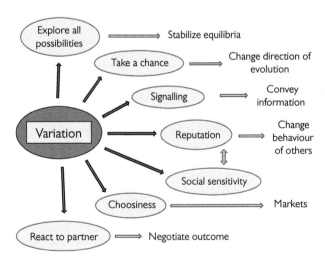

Fig. 7.1 Some of the major areas in which phenotypic variation is important, with an indication of the consequences of this variation.

reputation effects, individuals change their behaviour to change their reputation and hence change how others respond to them (Sections 7.5, 7.6).

Social sensitivity. When there are costs of observing the behaviour of other individuals, it will only be worth paying these cost if there is sufficient variation in behaviour so that there is something valuable to learn. Then the distribution of behaviours in the population determines the degree of sensitivity and hence the strength of the effect of reputation, which in turn feeds back to affect the distribution of behaviours. We examine this co-evolutionary feedback in Section 7.7.

Markets. Organisms often exchange commodities or services in a 'market'. The mean level of service is important in determining the behaviour of those receiving the service (Section 7.8). However, the variation in service can also be important. When there is no variation in a population there is no point in individuals being choosy about whom they interact with. In contrast, when prospective partners differ in their behaviour it often pays to be choosy about whom to partner with. Once choosiness evolves, there is then selection to change behaviour (and hence reputation) so as to increase the chances of being chosen. Variation and choosiness together lead to assortative pairing, which can select for population members to be more cooperative (Section 7.9). As an application of these ideas, in Section 7.10 we consider a model in which the commitment to a partner and choosiness co-evolve. As we demonstrate, the co-evolution of these traits can give rise to high commitments.

React to partner. When two individuals interact and there are individual differences, the behaviour of the partner will typically not be known in advance. It will then be worth changing one's own behaviour as the partner's behaviour is observed. Pair members can then be regarded as negotiating their behaviour with each other. We defer this topic until Chapter 8, where we show that the behaviour that is negotiated can be different from that when no negotiation is possible (Section 8.4).

In presenting the various examples of models in this chapter, we have two broad aims in mind. The first is a conceptual criticism of modelling practices in game theory that depend on an assumption that variation is small. The second is to illustrate that allowing for realistic amounts of variation and individual consistency makes it possible to model important biological phenomena that otherwise would be hard to understand (e.g. Section 7.10).

7.2 Variation and the Stability of Equilibria

Consider the trust game illustrated in Fig. 7.2. In this two-player game, one individual is assigned the role of Player 1 and the other Player 2. Player 1 must first decide whether to trust or reject Player 2. If Player 2 is rejected both players receive payoff $s > 0$ and the game ends. If Player 2 is trusted this player decides whether to cooperate with Player 1 or to defect. If cooperation is chosen each receives a payoff of r, where $0 < s < r < 1$. If Player 2 defects Player 1 gets a payoff of 0 and Player 2 a payoff of 1.

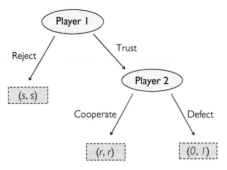

Fig. 7.2 Choices and payoffs in the trust game of Frank (1988). The pair of payoffs are for Player 1 and Player 2, respectively.

In this game it is clear that Player 2 should defect if trusted since $r < 1$. This means that Player 1 will get 0 if Player 2 is trusted and so should reject Player 2. Thus both players receive payoff s, whereas they could potentially have received the larger payoff of r. This game can be thus be seen as a version of the Prisoner's Dilemma game in which choices are made sequentially rather then simultaneously.

Suppose that the resident strategy is for Player 1 to reject Player 2 and for Player 2 to defect if trusted. Then the resident strategy is at a Nash equilibrium. It is not, however, an ESS. To see this consider the mutant strategy that specifies rejection when in the role of Player 1 but cooperation if trusted in the Player 2 role. This strategy has exactly the same payoff as the resident strategy even when common, and so will not be selected against. Formally, although condition (ES2)(i) holds, condition (ES2)(ii) does not. The problem here is that although the mutant and resident strategies differ in their choice of action in the role of Player 2, they never get to carry out this action as they are never trusted. In fact there is no ESS for this game.

Now suppose that in this resident population there is some mechanism that maintains variation in the action of Player 1 individuals, so that occasionally a Player 1 trusts its partner. This variation might be due to the occasional error, to mutation maintaining genetic variation, or any other source. When a Player 2 individual is trusted it gets to carry out its action. If the Player 2 individual is a mutant that cooperates it receives the payoff r, while a resident Player 2 defects and obtains the higher payoff of 1. Thus mutants of this kind are selected against. In other words, the occasional error by Player 1 stabilizes the Nash equilibrium against invasion by mutants.

A strategy is a rule that specifies the action taken for every possible state of the organism. In games in which there is a sequence of choices the state of an organism is usually defined by the sequence of actions that have occurred up until the present time. So, for example, in the trust game, being trusted is a state for Player 2. If in a population following a Nash equilibrium strategy some state is never reached, then there is no selection pressure on the action taken in that state. This means that mutations that change the action in this state alone are never selected against, so that the Nash equilibrium cannot be an ESS. If the Nash equilibrium is stabilized by occasional errors, as for the trust game, the resident strategy is referred to as a limit ESS. We return to this topic in Chapter 8 when we consider sequences of choices in more detail.

Variation can similarly remove the neutrality from signalling games (Section 7.4). Unless there is some possibility of receiving a particular signal there is no selection pressure on the response to that signal. Variation can ensure that all signals are possible.

7.3 Taking a Chance

Even if most outcomes of an action result in losses it may still be worth taking the action if some outcomes have a high gain. The following example illustrates this principle when there is variation in the level of cooperation of partners and it is worth taking a risk that the current partner is exceptionally cooperative. The example illustrates that when there is sufficient variation present in a population, evolved strategies may be far from that predicted by the standard theory based on resistance to invasion by rare mutants.

7.3.1 Taking Risks in a Finitely Repeated Prisoner's Dilemma Game

McNamara et al. (2004) consider a game in which two individuals play a sequence of rounds of the Prisoner's Dilemma game (Section 3.2) with one another. If in any round either player defects then the interaction ends and no more rounds are played. The idea is that individuals go off to seek more cooperative partners after a defection. This search phase is not explicitly incorporated into this model, although it is explicitly considered in a related model in Section 7.10. The interaction ends after the Nth round if it has not already done so. Here N is known to both players. A strategy for an individual specifies the number of rounds in which to cooperate before defecting. Each partner attempts to maximize its total payoff over the rounds that are played. For illustrative purposes we assume that each round has payoffs given by Table 7.1.

Note that if the last (i.e. Nth) round is played both players should defect. They will then each receive a payoff of 1 for this round. Now consider the $(N - 1)$th round. If

Table 7.1 Payoffs per round for the finitely repeated Prisoner's Dilemma model.

Payoff to focal		Action of partner	
		C	D
Action of focal	C	2	0
	D	5	1

a partner defects there will be no further rounds, so it is best to defect since $1 > 0$. If a partner cooperates then cooperation will give a reward of 2 from the current round and the individual will go on to get a reward of 1 from the Nth round. This is less than the payoff of 5 from defection. Thus it is best to defect whatever the action of the partner. A similar consideration applies to the partner, so both players should defect and the game will end with both players receiving 1. We can iterate backwards to earlier rounds in this way, deducing that both players should defect in the first round. This is the only Nash equilibrium for the game. Furthermore, since it is the unique best response to itself, it is an ESS.

Models in which two players play a number of rounds of the Prisoner's Dilemma against one another have become a testbed for the evolution of cooperation. These models assume that in any round there is always the possibility of at least another round, otherwise the type of backward induction argument presented above will rule out any cooperation. McNamara et al. (2004) incorporated a maximum on the number of round specifically because they wished to show that cooperation could still evolve when variation is maintained even though the standard arguments precluded it. Figure 7.3 illustrates the crucial role of variation. In each case illustrated the mean of the strategy values in the resident population is 5. The cases differ in the range of strategies that are present. When there is no variation, so that all residents have strategy 5, the best response of a mutant to the resident population is to cooperate for 4 rounds; i.e. defect before partner does. As the amount of variation about the mean of 5 increases it becomes best to cooperate for more rounds. To understand why this occurs suppose that the 9 strategies $1, 2, ..., 9$ are present in equal proportions. If 4 rounds of cooperation have passed, then partner's strategy is equally likely to be $4, 5, 6, 7, 8$, or 9. Thus the probability that partner will defect on the next round is only $\frac{1}{6}$. This makes it worthwhile to take a chance and cooperate for at least one

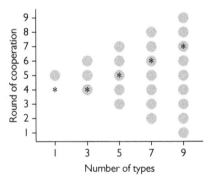

Fig. 7.3 Best responses in the repeated Prisoner's Dilemma in which there are a maximum of $N = 10$ rounds. Cases in which the resident population has 1,3,5,7, and 9 strategies are shown. In each case the strategies have a uniform distribution centred on the value 5. Best responses are indicated by a ∗. Payoffs on each round are given by Table 7.1.

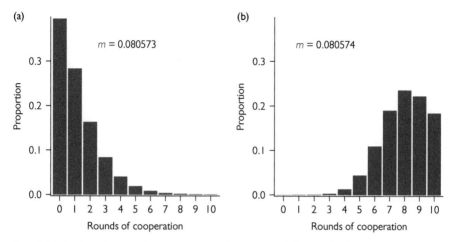

Fig. 7.4 The distribution of strategies at evolutionary stability in the finitely repeated Prisoner's Dilemma game with $N = 10$. Results for two mutation probabilities, m, are shown. Invasion fitness is proportional to 10 plus the total payoff in the game. Payoffs in each round are given by Table 7.1.

more round. This type of reasoning applies more generally. A formal analysis is set as Exercise 7.1.

In an evolutionary simulation, McNamara et al. (2004) considered a large asexual population with discrete non-overlapping generations in which the genetically determined trait is the number of rounds to cooperate before defecting. In each generation population members pair up to play the finitely repeated Prisoner's Dilemma game against each other. The number of offspring produced is proportional to a constant plus the payoff in this game. Offspring have the same trait value as their parent with probability $1 - 2m$, and have trait values $\max\{0, n - 1\}$ and $\min\{N, n + 1\}$ each with probability m. Figure 7.4 illustrates the effect of increasing m, which increases the amount of variation in the population. For low m evolution leads to a largely uncooperative population with some cooperation maintained by mutation. At a critical value of m there is a switch in the direction of selection as it becomes better to be more cooperative than the population mean. This results in the evolved population having a high level of cooperation.

In this example, sufficient variation results in a radical change in what evolves. In particular, at evolutionary stability the strategies present are not in Nash equilibrium, so the standard theory based on rare mutants is misleading. In the example, genetic variation is maintained through mutation, but could have instead been maintained by immigration. Phenotypic variation produced by developmental noise or errors in decision making produce similar effects, provided the total phenotypic variation contains an additive genetic component.

7.4 Signalling and the Handicap Principle

In this section we analyse the simplest form of a signalling game: the action–response game. In this game a signaller sends a signal to a receiver who then responds to the signal. We seek a signalling equilibrium. This is a Nash equilibrium at which the strategy of signallers is the best given the response strategy of receivers and the response strategy of receivers is the best given the strategy of signallers. To be concrete we focus on the following example.

7.4.1 Males Signal Parental Ability, Females Respond with Clutch Size

Suppose that a female and male bird have just paired. The female has to decide whether to lay one egg or two. Her best decision depends on the ability of the male to provision the nest. This ability, q, is either low (L) or high (H). All eggs survive unless the female lays two eggs when $q = L$. In this case b (≤ 2) eggs survive, and furthermore the female has to work harder to help the male, reducing her future reproductive success by K. The payoffs to the female for each of the four combinations of clutch size and male ability are given in Table 7.2. We assume that $b - K < 1$ so that the best action of the female is to lay one egg when $q = L$ and to lay two eggs when $q = H$.

It would be advantageous for the female to have information on the ability of the male. If ability is positively correlated with size then she can gain information from the male's size. Assuming that she can observe size, this cue is not something the male can control by his behaviour. A cue that is not under the control of the signaller (the male in this case) is referred to as an index. Female frogs prefer large mates. Since larger males can produce lower-frequency croaks, the deepness of the croak of a male is a cue of his size that cannot be faked and is an index (Gerhardt, 1994).

From now on we assume that there are no indices; the female bird has no cue as to the ability of her partner. Instead, the male transmits one of two signal s_1 or s_2 to the female. For example, he might bring her a small amount of food (signal s_1) or a large amount of food (signal s_2). We assume that producing signal s_1 is cost free, whereas the signal s_2 costs the male c_L if he has low ability and c_H if he has high ability. The payoff to a male is the number of surviving offspring minus any signalling cost. We examine circumstances in which there is a signalling equilibrium at which the male's signal is an honest indicator of his ability. That is, we seek a Nash equilibrium at which

Table 7.2 Payoff to the female in the male signalling game.

Payoff to female		Ability of male	
		L	H
Eggs laid	1	1	1
	2	$b - K$	2

males signal s_1 when having low ability and s_2 when having high ability, and females lay one egg when the signal is s_1 and lay two eggs when the signal is s_2. We describe two circumstances in which this is a Nash equilibrium.

7.4.1.1 The Handicap Principle

The idea that differential costs of signalling can maintain honesty is known as the handicap principle. This idea was first put forward in economics by Spence (1973), who applied it to the job market. He argued that an able person paid less of a cost to obtain a higher educational qualification than a less able person, so that the qualification acted as an informative signal of ability to employers. Independently, Zahavi used the idea of differential costs to explain the preference of females for mates that have exaggerated traits such as the large antlers of male deer and the elaborate tails of male peacocks (Zahavi, 1975, 1977). Zahavi argued that such traits lowered the survival of the male that possessed them. The trait thus acted as a test of male quality: an individual with a well-developed sexually selected character is an individual who has survived a test. By choosing a male with the most developed trait the female ensures that she obtains a mate of high quality. Zahavi's arguments for male characters as a handicap were purely verbal and many were not convinced by them until Grafen (1990) used a formal model to confirm that handicaps could work.

To illustrate the handicap principle in the clutch size model we set $b = 2$. Thus all eggs survive and all males prefer their mate to lay two eggs rather than one egg. Suppose that the resident female strategy is to lay one egg when the signal is s_1 and lay two eggs when the signal is s_2. Then a low-ability male would obtain payoff 1 if he signalled s_1 and payoff $2 - c_L$ if he signalled s_2. Thus if $c_L > 1$ his best signal is s_1. Similarly, if $c_H < 1$ a high-ability male does best to signal s_2. Thus if $c_H < 1 < c_L$ we have a signalling equilibrium at which males of each type are signalling optimally given the response strategy of females and females are responding optimally given the signalling strategy of males. This is also a separating equilibrium in that a female knows the ability of the male from his signal; in that sense the signal is honest. The honesty of signalling is maintained since high-ability males pay less of a cost to signal s_2 than low-ability males; i.e. this is an example of the handicap principle.

Since the work of Grafen (1990) a variety of models have investigated the effects of differential signalling costs and shown that these can result in a signalling equilibrium in which signals are informative even though they do not always provide perfect information (e.g. Johnstone and Grafen, 1993; Számadó, 1999; Bergstrom et al., 2002).

7.4.1.2 Costs of Deviation

In our example a low-ability male is less good at provisioning young. When two eggs are laid by the female this might result in complete nest failure, even when the female helps the male to care. Alternatively the female may just desert if after having laid two eggs she discovers that the male is of low ability. We may capture these scenarios by setting $b < 1$, so that if the female lays two eggs and the male is low ability then less than one offspring survives. We may also set $K = 0$ for illustrative purposes. We then seek a signalling equilibrium when there are no costs to male signalling; i.e. $c_L = c_H = 0$.

Assume that the female lays two eggs if and only if she receives the signal s_2. Then the best signal for a low-ability male is s_1 since $b < 1$ and the best signal for a high-ability male is s_2 since $1 < 2$. Given this signalling strategy females are following the best response strategy. So again we have a separating signalling equilibrium.

It was originally claimed that signals had to be costly to be honest (Zahavi, 1975; Grafen, 1990). However, as Enquist (1985) and Hurd (1995) point out, and our simple example illustrates, this need not be the case. In our example, if at the signalling equilibrium a low-ability male signals s_2 then the male pays the cost after the female lays two eggs. This is therefore an example of what Kirmani and Rao (2000) refer to as a default-contingent signal, meaning that a loss only occurs if the signaller defaults from his honest signal. For example, in an aggressive encounter between two conspecifics, a weak animal may have a cost-free choice between signalling submission or signalling that it will fight. The latter option may be advantageous if the opponent is weak and submits as a result of the signal. This advantage may, however, be outweighed if it turns out that the opponent is strong and the signal results in a fight (cf. Enquist, 1985).

We have considered a simple example of an action–response game, but other more complex signalling games are possible. For examples of these and a general overview of signalling issues, see Silk et al. (2000), Hurd and Enquist (2005), and Számadó (2011).

7.5 Reputation

Many individual differences persist over a lifetime or a significant part of it. Some are due to genetic variation, some a result of experience during development. In mammals and birds some state differences such as size are relatively fixed at adulthood. Others such as dominance status can persist as a result of behaviour of the individual and others. When animals can recognize other individuals and remember their past behaviour persistent differences lead to the feedback loop illustrated in Fig. 7.5. The logic of this feedback is as follows:

1. Individual differences lead to different behaviours.
2. Because the behaviour of each individual tends to persist, its past behaviour will be correlated with its current behaviour, so that the reputation of the individual, i.e. some measure of its past behaviour, is predictive of its current behaviour.
3. It will then be advantageous for population members to base their own current behaviour on the reputation of those they are interacting with.
4. This will select for individuals to change how others interact with them by changing behaviour so as to modify their own reputation.

We illustrate this feedback loop using the model in Box 7.1 that was developed by McNamara and Doodson (2015). In this model individual differences are due to differences in quality that might, for example, correspond to foraging ability. Each

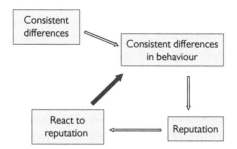

Fig. 7.5 Individual differences and individual recognition lead to the establishment of reputation, and the response of others to this leads to selection to change reputation by changing behaviour.

population member meets a series of different partners and plays a single round of a game with that partner. In a round each decides how much to contribute to the common good of the pair (Section 3.1), with higher quality individuals paying less of a cost for a given contribution. This leads to higher quality individuals tending to contribute more to the common good. Since quality differences persist, the reputation of an individual (a measure of their average contribution in the past) is positively correlated with their current contribution. In this model it is valuable to be able to predict this current contribution since the contribution of an individual that maximizes its payoff from the current round depends on its partner's contribution.

McNamara and Doodson (2015) contrast the case when there is no reputation with the case where individuals can form and respond to reputation. In the no-reputation case an individual's contribution to the common good of a pair is fixed, depending only on their quality. When reputation is possible, population members recognize others and observe the contributions of each population member in interactions with previous partners. Their own contribution with their current partner can then depend on their own quality and their partner's reputation (Box 7.1).

In this interaction, the degree to which an individual adjusts its own contribution as a function of the reputation of a partner (l, see Box 7.1) is a key trait. Results depend on whether the best response function for a single round of the game is increasing or decreasing, which depends on how the common benefit depends on the contributions of both partners. When the best response function is increasing, the evolved mean contribution to the common benefit is higher when there is reputation than when there is no reputation. The mean payoff is also higher. When the best response function is decreasing, the mean contribution and mean payoff are both lower when there is reputation than when there is no reputation. These results are illustrated in Fig. 7.6. To understand the results we first note that the predicted contribution of the current partner increases with their reputation. Thus in the case where the best response function is increasing, individuals evolve to increase their own effort

---

Box 7.1 Reputation based on quality variation

The quality of each population member is set during development and persists throughout life. Each individual knows their own quality. Individuals meet a series of different partners chosen at random. With each partner they play a single round of a game in which each contributes to the common good of the pair. The single-round payoff to an individual with quality q that contributes u when partner contributes v is $W_q(u, v) = B(u, v) - C_q(u)$. The common benefit $B(u, v)$ increases with both u and v. The cost $C_q(u)$ increases with the contribution u, and for given u decreases with quality q. A strategy specifies how much to contribute on each round as a function of own quality and the information available. The invasion fitness of a strategy is proportional to the average payoff obtained in following the strategy over many rounds.

When there is no reputation, and hence no information about each new partner, at evolutionary stability the contribution of an individual in each round is just a function of its own quality (Exercise 7.3).

When reputations are formed the previous contributions of an individual are summarized by a reputation, R, that is updated after each interaction according to $R' = \beta R + (1 - \beta)Z$, where Z is the individual's contribution in that interaction. Here β is a parameter satisfying $0 \leq \beta < 1$ that acts as a length of memory: when $\beta = 0$ the reputation is just the last contribution, whereas R approximates the average of all previous contributions when β is close to 1. It is assumed that an individual with quality q contributes

$$u = m + k(q - \mu) + l(r - m) \tag{7.1}$$

to the common good when the partner has reputation r. Here m is the individual's baseline effort, k specifies the responsiveness of the individual to deviations of its quality from the population mean quality μ, and l specifies its responsiveness to the deviation of the partner's reputation from the focal individual's own baseline effort. A strategy is thus specified by the vector $\mathbf{x} = (m, k, l)$. For simple quadratic benefit and cost functions the evolutionarily stable values of these parameters can be found analytically (McNamara and Doodson, 2015). For more complex functions they must be computed numerically.

---

with the reputation of the partner; i.e. l evolves to be positive. This then means that the greater the reputation of an individual the more a partner will contribute to the common good of the pair. As a result there is selection to increase reputation by increasing effort (m increases). The case of decreasing best response is analogous; l evolves to be negative and this then selects for individuals to lower their reputation by reducing m.

When the reputation loop illustrated in Fig. 7.5 operates an action has two effects: (i) it affects the current payoff, and (ii) affects future payoffs by altering reputation and hence the behaviour of future partners. In both of the cases illustrated in Fig. 7.6 it is worth reducing the current payoff in order to gain more future rewards. In the increasing best response case individuals put in a higher contribution, and hence pay a greater immediate cost than is best for their quality, in order to raise reputation. In the decreasing best response case individuals put in a lower contribution, and hence reap a smaller benefit than is best for their quality, in order to reduce their reputation.

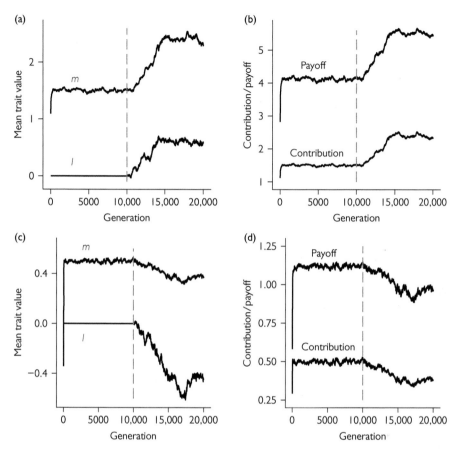

Fig. 7.6 Evolution without and with reputation in the model of Box 7.1. For the first 10,000 generations the parameter l in eq (7.1) is held fixed at $l = 0$, so that there is no response to reputation. After generation 10,000 l is free to evolve. (a) and (c) mean evolved values of the parameters m and l. (b) and (d) mean population contribution to the common good and mean population payoff per round. Benefit function $B(u, v) = 2(u + v) - \frac{1}{2}(u^2 + v^2) + Juv$. Cost function $C_q(u) = (1 - q)u + \frac{1}{2}u^2$. In (a) and (b) $J = 1$, so that there is a synergy between the contributions to the common good, resulting in an increasing best response function for a single round. In (c) and (d) $J = -1$, so that the common benefit depends on the sum of the contributions, resulting in a decreasing best response function. Quality is uniformly distributed between 0 and 1 and the memory parameter is $\beta = 0.8$.

7.6 Indirect Reciprocity

Alexander (1987) suggested that individuals will tend to help those who help others, a phenomenon he termed 'indirect reciprocity'. Alexander saw indirect reciprocity as important in human societies and possibly other highly social groups where everyone in a group is continually being assessed by others. He also argued that indirect reciprocity is fundamental to human moral norms. In this section we consider the logic behind indirect reciprocity, examining under what circumstances it can lead to high levels of helping in the population.

For definiteness, we assume that when two individuals interact each has the possibility of contributing a benefit, b, to the other at a cost, c, to themselves, so that they are playing a single round of the Prisoner's Dilemma with payoffs given by Table 3.1. Thus helping corresponds to the action of cooperation in this game. The simplest form of reputation is then based on the previous action of partner; we can define an individual's reputation as Good if it cooperated and Bad if it defected in its previous interaction. This form of reputation is an example of what Nowak and Sigmund (1998) call an image score. As we will see other forms of reputation formation are also possible.

The effects of reputation illustrated in Fig. 7.6 rely on the fact that the best action of a focal individual depends on its partner's current action. When this is so a partner's reputation partially predicts the partner's action and it is advantageous to respond to this reputation in order to improve the payoff from the current round. This then selects for population members to alter their behaviour, so altering their reputation, and hence the behaviour of future partners. This logic no longer works if the two-player game is the Prisoner's Dilemma. Of course this game has only two actions rather than a continuum of actions, but that is not the main conceptual difference. As we noted in Section 3.2, the Prisoner's Dilemma is an extreme game in that the best response to any action of a partner is always to defect with probability 1. This means that there is no immediate benefit from adjusting one's own behaviour to the reputation of a partner, so the feedback loop illustrated in Fig. 7.5 does not apply. Thus if an individual responds to a partner by cooperating this cannot be to improve current gains but can only be to improve future gains through changing the individual's reputation and hence the behaviour of future partners.

Nowak and Sigmund (1998) analyse a model in which individuals base their decision of whether to cooperate with an individual on the individual's image score. Three strategies are considered. ALLD is the strategy of always defecting, ALLC is the strategy of always cooperating, and the strategy DISC discriminates, specifying cooperation with a Good partner and defection with a Bad partner. Nowak and Sigmund (1998) show that high levels of cooperation can be maintained in a population comprising these three strategies. However, once errors are introduced into this model, the population rapidly evolves so that ALLD dominates (Panchanathan and Boyd, 2003). (See however, Berger and Grüne, (2016) who show that cooperation under image scoring can evolve when there are errors, provided the image score is not just a binary Good and Bad.) A major problem with the DISC strategy is that individuals gain a Bad reputation by defecting with a Bad partner. Consequently, in a population that comprises a mixture of ALLD and DISC, all DISC individuals rapidly acquire Bad reputations and cooperation ceases (Exercise 7.4). The weakness of using image scores is that this form of reputation takes no account of the reputation of the partner; the reputation becomes Bad after defecting with a Bad partner.

In image scoring and in the model of Section 7.5 the reputation of an individual only depends on the contribution made to a common good, it does not depend on the reputation of the partner who was helped by this contribution. Sugden (1986) introduced a form of reputation called 'standing' that applies to the Prisoner's Dilemma game and does have a dependence on the reputation of a partner. In his formulation the reputation of an individual is either Good or Bad. If an individual's reputation is Bad it becomes Good on cooperating. If an individual is Good it remains Good if either it cooperates or if it defects with a partner that is Bad; it only becomes Bad if the individual defects with a Good partner. Thus this form of reputation recognizes intent in the sense that it recognizes that defection with a Bad partner is justified and will not result in the reputation becoming Bad, whereas defection with a Good partner is unjustified and results in a Bad reputation.

Suppose that reputation is based on standing. Following Panchanathan and Boyd (2003) we consider a population in which there are three possible strategies: ALLD, ALLC and RDISC. The strategy ALLD specifies defection with all partners, ALLC specifies cooperation with all partners, and the strategy RDISC specifies cooperation if and only if the partner is Good. In Box 7.2 we analyse the payoffs to each strategy when each population member plays a sequence of rounds of the Prisoner's Dilemma with others. As is shown ALLD is an ESS and RDISC is an ESS provided there are sufficiently many rounds. This remains true when there are errors in which individuals occasionally fail to cooperate when this action is specified by their strategy (Panchanathan and Boyd, 2003), but RDISC might no longer be an ESS when there are errors of perception (Leimar and Hammerstein, 2001).

Ohtsuki and Iwasa generalize these results, analysing all forms of reputation dynamics and behavioural strategies when reputation is either Good or Bad (Ohtsuki and Iwasa, 2004, 2006). They identify combinations of reputation dynamics and evolutionarily stable behavioural strategies for which the average payoff per interaction is high. Eight such pairs were found, which they referred to as the leading eight. They found that elements of reputation dynamics that are common to the leading eight are that if their partner is Good individuals must cooperate in order to achieve a good reputation, whereas if their partner is Bad they do not lose a good reputation by defecting: properties that hold for reputation formation based on standing. Common elements of the behavioural strategy are that individuals cooperate with Good partners and punish Bad ones. Ohtsuki and Iwasa (2006) interpret these feature in terms of maintenance of cooperation, identification of defectors, justified punishment of defectors, and apology followed by forgiveness.

Further work considers the evolution of mixtures of strategies under replicator dynamics (Ohtsuki and Iwasa, 2007). The social norm, 'stern judging' (Kandori, 1992) is highlighted as leading to high levels of cooperation in this analysis and by considerations of the complexity of reputation formation (Santos et al., 2018). Under this social norm the reputation of an individual only depends on their last action; it is Good if they cooperated with a Good partner or defected with a Bad partner, and is Bad if they cooperated with a Bad partner or defected with a Good partner.

Box 7.2 Reciprocity based on standing

We assume a setting in which there is a sequence of rounds. In each round population members pair up and members of each pair play the prisoner's dilemma game with one another, where payoffs are specified by Table 3.1. Reputation is formed based on standing. The standing of all population members is initially Good. Population members play one of the three strategies, ALLD, ALLC, and RDISC. We analyse evolutionary stability in this population.

We first note that the (mean) payoff to an ALLC individual in the first round equals that to an RDISC individual. In subsequent rounds, both individuals have standing Good and are helped equally by others, but ALLC individuals have a lower payoff because they cooperate with ALLD individuals who have standing Bad. We will thus assume that strategy ALLC has been eliminated by selection.

Suppose that a proportion $1 - \rho$ of population members follow strategy ALLD and a proportion ρ follow strategy RDISC. In the first round the payoffs to individuals following strategies ALLD and RDISC are

$$w_D(1) = b\rho \text{ and } w_R(1) = b\rho - c, \tag{7.2}$$

respectively. In the second and all subsequent rounds all ALLD individuals have reputation Bad and all RDISC individuals have reputation Good. Payoffs per round are thus

$$w_D(2) = 0 \text{ and } w_R(2) = (b-c)\rho. \tag{7.3}$$

The total payoff under strategy ALLD is thus $W_D = b\rho$ and that under strategy RDISC is $W_R = b\rho - c + N(b-c)\rho$, where N is the mean number of rounds played after the first. Thus

$$W_R > W_D \iff \rho > \frac{1}{N}\left(\frac{c}{b-c}\right). \tag{7.4}$$

There is thus positive frequency dependence. As a result ALLD is always an ESS and the RDISC is an ESS provided $c < \left(\frac{N}{N+1}\right)b$. It is often assumed that after each round of the Prisoner's Dilemma there is at least one further round with probability w. In this case $N = w/(1-w)$, so that RDISC is an ESS if $w > \frac{c}{b}$. The condition ensures that there are sufficiently many rounds so that RDISC individuals, who pay a cost to maintain a Good reputation, have time to reap the benefits of this reputation in the future.

A general conclusion of the modelling work is that indirect reciprocity can produce high levels of cooperation in the Prisoner's Dilemma game, but that in order to do so the reputation of an individual cannot simply be based on their previous actions. Instead, reputation has to be formed in such a way as to also take into account the reputation of the recipients of these actions.

Some form of indirect reciprocity has strong empirical support in humans (e.g. Engelmann and Fischbacher, 2009). However, there is much less cooperation with non-kin in most other species than in humans. In particular, although there can be direct reciprocity, in which individuals return favours to others, there is only weak evidence for indirect reciprocity (see e.g. Molesti and Majolo, 2017). Thus despite the enormous theoretical effort devoted to indirect reciprocity, it is not clear if it is generally important in biology.

7.7 Differences Select for Social Sensitivity

If there is no variation in a population there is no advantage in observing the past behaviour of others. Once there is variation it can be advantageous to be socially sensitive, and observe the past behaviour of others in order to predict their current behaviour. However, the value of social sensitivity depends on the usefulness of the information obtained, which often depends on the amount and type of variation in the population. Thus when there are costs of sensitivity the range of behaviours in a population affects the optimal degree of sensitivity. Conversely, increasing the degree of sensitivity increases the importance of reputation and hence affects the range of behaviours. This section concerns this two-way interaction in models of the co-evolution of behaviour and social sensitivity.

Johnstone (2001) considers a model in which members of a large population play a long series of competitive rounds. In each round each population member is paired up with a different randomly selected opponent and the two contest a resource of value V. In this contest each can either play Hawk or Dove. Payoffs are as in the standard Hawk–Dove game (Section 3.5). It is assumed that $V < C$. There are three possible strategies in the population: (i) always play Hawk, (ii) always play Dove, and (iii) Eavesdrop. Individuals following the latter strategy have observed the outcomes of contests in the last round. If their current opponent gained the resource in the last round (either by playing Hawk or Dove) they play Dove, and if the opponent failed to gain the resource they play Hawk.

Let the frequency of pure Hawks, pure Doves, and Eavesdroppers in the population be f_H, f_D, and f_E, respectively. Johnstone (2001) shows that the proportions of these three types that gained the resource in the last round, p_H, p_D, and p_E, tend to limiting values that depend on the frequencies of the three strategies. This then allows the mean payoffs under the three strategies to be evaluated. Johnstone (2001) analyses the evolutionary outcome using replicator dynamics (Section 4.4), showing there is a unique stable attractor of the dynamics. At this equilibrium all three strategies co-exist.

If an opponent has gained the resource in the last round, this opponent is more likely to be a Hawk than the population frequency of Hawks. Thus by playing Dove against such an opponent an Eavesdropper avoids a costly Hawk–Hawk fight. Similarly, an opponent that failed to gain the resource in the last round is more likely to be a Dove than the population frequency of Doves, so that an Eavesdropper gains by playing Hawk. Eavesdroppers are inconsistent in their actions. Thus Eavesdroppers do not spread to fixation because as their proportion increases reputation becomes less predictive, so that eavesdropping becomes less valuable.

At equilibrium the proportion of interactions that results in fights is greater than in the standard Hawk–Dove game. This is because there is an extra premium for winning since those that have the reputation for having gained the last item do well against Eavesdroppers.

Wolf et al. (2011) present a model that generalizes the results of Johnstone (2001) to any symmetric two-player game with two actions that has a mixed-strategy ESS when reputation effects are not allowed. However, in contrast to the assumptions of Johnstone (2001) socially responsive individuals (Eavesdroppers) know the previous action of their current partner rather than just the outcome of this action. As with the case of the Hawk–Dove game, when eavesdropping is allowed at evolutionary stability the population is polymorphic with both responsive and unresponsive individuals present. Wolf et al. (2011) highlight that the presence of responsive individuals selects for consistency in others. So for example in the Hawk–Dove game if an unresponsive individual has been previously observed to play Hawk (Dove) a responsive individual plays Dove (Hawk) with this individual, so that it is best for the unresponsive individual to again play Hawk (Dove).

As the previous examples illustrate, increasing the number of socially sensitive individuals in a population may change behaviour so that it is not worth others being socially sensitive. This is particularly relevant when there are costs of observing others, as the trust model analysed by McNamara et al. (2009) illustrates. They consider the co-evolution of trustworthiness and social sensitivity in a version of the trust game illustrated in Fig. 7.2. In their model each population member plays a series of trust games against different randomly selected opponents. In each game one pair member is assigned (at random) as Player 1, the other as Player 2. Here we describe a special case of their model in which individuals in the role of Player 1 can pay a cost c to observe whether their partner was trustworthy the last time this individual was in the role of Player 2 and was trusted. In role 1 there are three local strategies. An unconditional acceptor (UA) accepts all Player 2s. An unconditional rejector (UR) rejects all Player 2s. Finally, a sampler (S) pays the observation cost, accepting the partner if they were trustworthy previously and rejecting them if they were untrustworthy. In role 2 the local strategy specifies the probability, p, of being trustworthy if trusted. The payoff to a strategy is the average net gain per round under the strategy.

The distribution of p determines which of the local Player 1 strategies is best. UR has a higher payoff than UA if and only if the mean of p satisfies $\mu < \frac{s}{r}$ (Box 7.3). The performance of S depends on both the mean and variance of p (Box 7.3). For a given mean, performance increases with the variance since the value of gaining knowledge increases with variation.

Figure 7.7 illustrates evolutionarily stable outcomes in three cases. When the rate of mutation of p is low there is insufficient variation to make it worth sampling. In this case $\mu < \frac{s}{r}$ and URs have the highest payoff. UR thus dominates the population with low levels of UA and S maintained by mutation from UR. For intermediate levels of mutation sampling is worthwhile. This selects for Player 2s to increase their p value so as to be trusted by samplers (Exercise 7.5), resulting in $\mu = \frac{s}{r}$. The payoffs of all

Box 7.3 Payoffs to Player 1 in the trust game

Consider a population where the Player 2 (P2) trait p has mean $\mu = E[p]$ and variance $\sigma^2 = \text{Var}(p)$. In this population, the payoff to an unconditional acceptor (UA) is $W_{\text{UA}} = E[pr] = \mu r$ and the payoff to an unconditional rejector (UR) is $W_{\text{UR}} = s$. Thus

$$W_{\text{UA}} > W_{\text{UR}} \Longleftrightarrow \mu > \frac{s}{r}. \tag{7.5}$$

A sampler accepts a P2 if and only if it is observed to be trustworthy on the one occasion they are observed. Suppose that Player 2 has trait value p. Then a sampler rejects this P2 (receiving payoff s) with probability $1 - p$ and accepts the P2 (receiving expected payoff pr) with probability p. Thus the sampler has mean payoff $w(p) = (1 - p)s + p^2 r$ in its interaction with this particular P2. Averaging over possible p values we have

$$W_{\text{sampler}} = (1 - \mu)s + (\mu^2 + \sigma^2)r - c. \tag{7.6}$$

Here we have used the identity $E[p^2] = \mu^2 + \sigma^2$.

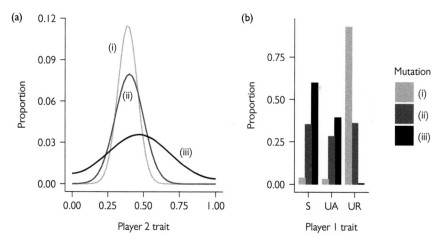

Fig. 7.7 Evolutionarily stable outcomes in the trust model. Three cases differing in the rate at which the probability Player 2 is trustworthy mutates are illustrated: (i) $\epsilon_2 = 0.001$, (ii) $\epsilon_2 = 0.02$, (iii) $\epsilon_2 = 0.6$. (a) The distribution of Player 2 trustworthiness. (b) The distribution of Player 1 traits. In the evolutionary simulation used to generate this figure the Player 2 trait (probability of trustworthiness, p) is assumed to take discrete values lying in the range $p = 0.0, 0.02, 0.04, \ldots, 0.98, 1.0$. Frequencies change according to discrete replicator dynamics with mutation. Player 1 traits mutate to another P1 trait with probability ϵ_1. p mutates to an adjacent value with probability ϵ_2. Parameters $s = 0.2, r = 0.5, c = 0.005, \epsilon_1 = 0.0002$.

three Player 1 strategies are equal and all co-exist. Finally, for high mutation samplers do best until μ evolves to be greater than $\frac{s}{r}$, and continue to increase until UA does as well. The population is then a mix of UA and S with low levels of UR maintained by mutation.

In this special case of the model of McNamara et al. (2009) mutation provides the variation that promotes sampling, and sampling drives up trustworthiness until the population is polymorphic for the local strategy in role 1. At this equilibrium the distribution of p is unimodal. In contrast, McNamara et al. (2009) consider cases in which Player 1s can observe the trustworthiness of a partner on more than one previous occasion. In some cases the distribution of p at evolutionary stability is bimodal. When this occurs, some Player 2s are very untrustworthy and rely on being accepted by UAs, whereas other Player 2s are much more trustworthy and gain by being accepted by samplers. In such cases the large variation in p maintains the occurrence of samplers and the occurrence of samplers with UAs maintains the variation in p.

7.8 Markets

Organisms often exchange commodities or services. Such biological markets have been highlighted by Ronald Noë and his co-workers (Noë and Hammerstein, 1994; Bshary and Noë, 2003; Hammerstein and Noë, 2016). A market often consists of two distinct trader classes, and this is the situation we consider in this section. For example, aphids and ants may form two classes in a market in which aphids provide ants with sugar-rich exudate and in exchange the ants provide the aphids with protection against predation. Cleaner fish and their clients trade with each other; the cleaner fish benefit by feeding on the ectoparasites of their clients and the client fish gain by having these parasites removed. Insects pollinate plants in exchange for nectar. Usually, the members of one class (or both classes) would benefit if they could get away with providing as poor service as possible. So for example, cleaner fish prefer to feed on the mucus of their clients rather than the ectoparasites, and do so if they can. A flowering plant would do better to produce less nectar if that did not affect the behaviour of pollinating insects. In such situations, whether it is possible to get away with poor service depends on the market setting. In this section and the following two sections we explore market forces and their consequences. As we will show the mean level of service given by one class is usually important. In many cases the variation in the quality of service is also important as the choosiness of recipients of service can be strongly affected by the range of options available.

Members of each trader class provide services to members of the other class. Figure 7.8 summarizes the relationship between the service given by members of one trader class and the response to this service of members of the other class. This dependence can be described as follows.

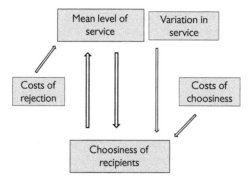

Fig. 7.8 The two-way interaction in a biological market. The service given by servicers selects for the choosiness behaviour of recipients. The choosiness of recipients selects for the service given.

Choosiness of recipients. Recipients of service can be choosy about the individuals that service them in a number of ways. (i) They might recognize those who would give poor service in advance and avoid these servicers. So for example, if a client fish has observed that a previous interaction between a cleaner and another client has ended in conflict, the cleaner is avoided (Bshary and Noë, 2003). (ii) Recipients may provide less in exchange or break off their interaction sooner if the service is poor. For example, a client is liable to terminate the interaction with a cleaner fish that feeds on mucus. An insect will tend to leave a flower earlier if there is little nectar. (iii) A recipient may avoid a servicer in the future if they have previously experienced poor service.

The degree to which recipients should be choosy depends on the costs and benefits of choosiness. Often recipients may be receiving positive benefits from a servicer, but these benefits are too few and the recipient should break off the interaction or avoid the servicer in the future in an attempt to find a better servicer. In the case of a foraging insect drinking the nectar from a flower, the rate at which it gains nectar will typically decrease as the amount of nectar remaining decreases. At what point should it leave the current flower and seek a new one? In the model we present below, this depends on the mean rate at which it can gain nectar from other flowers in the environment, which depends on the amounts of nectar in these flowers and the time taken to move between flowers. In the case of a client fish receiving bad service from a cleaner fish, should the client return to this cleaner the next time it requires cleaning, or should it seek a different cleaner? If all cleaners are the same there is clearly no point in being choosy. Thus choosiness is advantageous only if there is variation, so that there is a significant possibility of finding a better cleaner. The costs of rejecting a cleaner is also central. Finding a better cleaner may take time and involve energetic or mortality costs. Overall, the benefits of seeking a better partner must be weighed against the costs (e.g. Exercise 7.6).

Level of service. Recipient choosiness exerts a selection pressure to increase service. In particular, there is selection for servicers not to be 'outbid' by other servicers. For example in the nectar market model below, the critical rate of nectar intake at which a foraging insect should leave a flower increases with the nectar available at other flowers. Thus unless a plant keeps up with other plants in terms of nectar production, the flowers of this plant will receive short visits, resulting in less pollen transfer. The selection pressure to increase nectar production then depends on the loss due to reduced pollen transfer and the costs of producing more nectar. In the commitment model of Section 7.10 individuals have a choice between divorcing their partner or continuing to partner them in the following year. Divorce occurs if partners are not sufficiently committed to one another. If an individual is not committed it will tend to be divorced and enter a pool of single individuals. There is no immediate cost incurred as a result of the divorce. However, the individual will then pair up with a randomly selected member of the pool, and the commitment of pool members is below the population average since the pool contains all those individuals that have been divorced. Thus by being divorced an individual tends to re-pair with a poor partner.

We illustrate some of these ideas in a model of a market in which plants determine the nectar level in their flowers and insects exert choice through the time they spend at a flower. In this model, the main effects are driven by the mean level of service rather than the variation in service. Then in the following two sections we consider the situation in which the market involves a single species, with population members pairing and each pair member providing benefits to the other. In this situation, variation results in choosiness about partner, and this can result in a positive correlation between the benefits provided by pair members. As we show in Section 7.9 this assortment selects for cooperative behaviour in the population. Thus variation and choosiness can lead to cooperative behaviour, a phenomenon referred to as competitive altruism by Roberts (1998). In Section 7.10 we illustrate these ideas with a model in which each individual decides on its level of commitment to its partner. As we show, the commitment to partner (the service) and the strategy of partner divorce (the choosiness) co-evolve, resulting in high levels of commitment if individuals have the option to change partners many times during their lives.

7.8.1 A Model of the Nectar Market

Assume that a plant species relies on a particular insect for pollination. Visits by insects are rare and between visits a plant builds up the amount of nectar present in a flower until it is at some genetically determined level x. When an insect visits a flower it consumes nectar; its rate of consumption is $r(z)$ when the current amount of nectar in the flower is z. The rate of change in the amount of nectar during a visit is thus $\frac{dz}{dt} = -r(z)$. We assume that $r(z)$ is an increasing function of z, so that nectar is harder to extract as its level falls. The insect leaves when its instantaneous consumption rate falls to $r(z) = \gamma$, where γ can depend on the availability of nectar in the environment as a whole. The insect then moves on to another plant and again forages for nectar. The mean time taken to find another flower after leaving the previous flower is τ.

We assume that insects choose their leaving rate γ so as to maximize the average rate at which they obtain nectar. Plants gain a benefit $B(T)$ if the duration of an insect's visit is T. This is an increasing but decelerating function of the visit duration. They also pay a cost per visit equal to the amount of nectar consumed—this amount is equal to the amount of additional nectar that the plant must produce in order to replenish the amount present to its original level. The payoff to a plant equals the benefit minus cost.

To analyse this model assume that the resident strategy is for plants to provision flowers to level x. Let z_γ satisfy $r(z_\gamma) = \gamma$, so that z_γ is the amount of nectar left when an insects leaves a flower. Then the insect gains $x - z_\gamma$ nectar on each flower. Let $T(x, z)$ be the time taken to deplete a flower from its initial nectar level of x to level z. The mean rate at which the insect consumes nectar is $g(x, \gamma) = \frac{x - z_\gamma}{T(x, z_\gamma) + \tau}$. This mean rate is maximized by setting $\gamma = \gamma^*(x)$ where $\gamma^*(x)$ is the unique solution of the equation $g(x, \gamma^*(x)) = \gamma^*(x)$ (McNamara, 1985). Thus insects leave a flower when their instantaneous consumption rate falls to $\gamma^*(x)$, at which time the amount of nectar left in a flower is $z_{\gamma^*(x)}$.

Now consider a rare mutant plant that provisions to level x' when the resident provisioning level is x. An insect stays for time $T(x', z_{\gamma^*(x)})$ on a mutant flower. The payoff to the mutant is thus

$$W(x', x) = B(T(x', z_{\gamma^*(x)})) - (x' - z_{\gamma^*(x)}). \tag{7.7}$$

Figure 7.9 illustrates the effect of the resident strategy on the mean rate at which insects gain nectar and the best response of a rare mutant plant. As the resident provisioning level x increases an insect encounters more nectar on each flower and its instantaneous rate of consumption on arriving at a flower increases. As a consequence

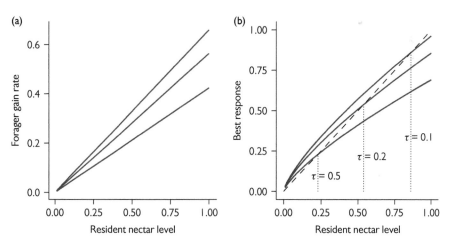

Fig. 7.9 Effect of the resident strategy (nectar provisioning level) in the model of a nectar market. (a) The mean consumption rate by an optimally foraging insect. (Lower curve $\tau = 0.5$, mid $\tau = 0.2$, top $\tau = 0.1$). (b) Best response to the resident strategy for each travel time. For each travel time, the unique CSS is the nectar level at which the best response curve crosses the (dashed) 45° line. $r(z) = z$ and $B(T) = 1 - e^{-3T}$—see Exercise 7.7.

its mean consumption rate $\gamma^*(x)$ under an optimal foraging strategy increases. Since this rate is also the instantaneous consumption rate at which it leaves each flower, the amount of nectar that remains on departure, $z_{\gamma^*(x)}$, also increases. This means that the duration of any visit of an insect to a rare x' mutant flower will decrease as x increases: in a market situation individuals do worse when their competitors offer more. For the function used to produce Fig. 7.9, these effects mean that the optimal nectar level of a rare mutant increases with the level of residents. As can be seen from the figure, for each value of the travel time τ there is a unique convergence-stable ESS (a CSS). This CSS increases as the travel time decreases. When τ decreases insects can afford to be more choosy, leaving when there is more nectar remaining. This intensifies the competition in the market, leading to higher 'bids' by competing plants.

7.9 Choosiness, Assortment, and Cooperation

We now consider markets in which there is just a single species, with population members acting both as service providers and recipients of service. Specifically, we assume that individuals pair up to perform a task that has potential benefits to the pair members, but where contributing to the common good has costs to the individual (cf. Section 3.1). Assume that there is variation in the level of contribution of pair members. We outline how the existence of variation in a market setting can lead to a positive correlation in the contribution of pair members. We then argue that the establishment of such a correlation selects for a greater contribution to the common good; i.e. greater cooperation between pair members.

In a market setting there are two main ways in which choosiness about a partner can lead to a positive correlation in the level of contribution of pair members. Both mechanisms rely on the fact that a pair forms or stays together if and only if both partners decide to do so.

1. Before a pair is formed individuals may have information about the likely contribution of prospective partners. This information might, for example, be as a result of reputation based on previous behaviour with others. Suppose also that in seeking a partner, individuals are able to reject one individual in favour of a more cooperative alternative. Then population members that are highly cooperative will tend to be rejected by few potential partners and so can afford to be choosy, whereas individuals that are uncooperative will be rejected by many and cannot afford to be choosy. As a consequence, highly cooperative individuals will tend to pair up with other highly cooperative individuals, leading to a positive correlation in the cooperativeness of members of a pair. McNamara and Collins (1990) illustrate this effect in the context of mutual mate choice.

2. Individuals may be able to break off their interaction with an uncooperative partner and seek another, more cooperative, partner. Then pairs will tend to stay together if both pair members are highly cooperative. In contrast, if a highly cooperative individual has an uncooperative partner it will 'divorce' the partner and seek a

better one. As a result, there is again a positive correlation in cooperativeness. McNamara et al. (1999a) illustrate this effect in the context of mating and divorce by both sexes.

Other processes have been suggested that could potentially lead to a positive assortment between non-relatives (e.g. Joshi et al., 2017).

The existence of a positive correlation exerts a selection pressure on the trait that is correlated. Box 7.4 presents an example in which, regardless of the source of the correlation, increased correlation leads to increased cooperation at evolutionary stability. Two cases are considered. In the first the variation in the contribution to the common good might be thought of as arising through developmental noise. In the second, there is no developmental noise but processes such as mutation maintain genetic variation. Both cases predict the same effect.

Box 7.4 A 'game' with assortative pairing

We consider a two-player game in which each of the two contestants contributes to the common good of the pair. The payoff to an individual that contributes u' when partner contributes u is

$$H(u',u) = u' + u + \frac{R}{2}(1 - u'^2),$$ (7.8)

where R is a positive parameter. This payoff is maximized at $u' = \frac{1}{R}$ regardless of the value of u.

Let U_1 and U_2 denote the contributions of members of a randomly selected pair. In both scenarios presented below we assume that the joint distribution of U_1 and U_2 is bivariate normal with correlation coefficient ρ.

Phenotypic variation. Assume that the actual contribution, u, of an individual with genetic trait x is drawn from a normal distribution with mean x and variance σ^2. Let a rare mutant have trait x'. The payoff to this mutant when the resident trait is x can be shown to be

$$W(x',x) = (1 + \rho)x' + (1 - \rho)x + \frac{R}{2}\left[1 - (\sigma^2 + x'^2)\right]$$ (7.9)

(Exercise 7.8). This is the payoff for an optimization problem; the optimal trait value is $x^* = \frac{(1+\rho)}{R}$.

Genetic variation. Now assume that the contribution of an individual with genetically determined trait x is just x. However, due to mutation or other sources of variation, the distribution of trait values in a population with mean trait value x is normally distributed with mean x and variance σ^2. A modification of the argument outlined in Exercise 7.8 then shows that the payoff to a mutant with trait value x' in a population with mean trait x is

$$W(x',x) = (1 + \rho)x' + (1 - \rho)x + \frac{R}{2}\left[1 - x'^2\right].$$ (7.10)

Thus, there is again selection to evolve to $x^* = \frac{(1+\rho)}{R}$, and if the genetic variation is generated by unbiased symmetric mutation the mean population trait will evolve to this value.

Although Box 7.4 provides a specific example, an increase in cooperativeness with the correlation between pair members is liable to be the norm. In Section 4.5 we noted that population viscosity could lead to a positive assortment between relatives, and that this could lead to the evolution of cooperative behaviour (Box 4.3). More generally, Fletcher and Doebeli (2009) emphasize that the evolution of altruism relies on a positive assortment between genotypes that provides help and phenotypic help. We now develop a model of commitment in which choosiness leads to a positive correlation in the commitment of pair members to one another, resulting in high levels of commitment.

7.10 Commitment

Around 90% of bird species are socially monogamous over breeding. By this we mean that a pair share parenting duties and remain together during the period of parental care. In many species this bond lasts for a single breeding season and the two individuals pair up with a different partner in the following year. In some long-lived birds the bond is more permanent and lasts for life. Pair-bond formation does occur in some mammal species, but is less common.

So what mental mechanisms are involved in pair-bond formation? Much of the work on this has been on the prairie vole, a mammal species found in the grasslands of North America. In prairie voles individuals have multiple litters with the same partner. Both partners are involved in care, sharing the nest and defending the territory. Pairs show what can be interpreted as affection to one another and appear committed to the partnership in that they sometimes do not take a new partner if their current partner dies. The empirical work on pair-bond formation in the prairie vole is summarized in McGraw and Young (2010). This work has established that receptors for oxytocin and vasopressin in the forebrain play a central role in pair-bond formation, acting through their effect on the dopamine-mediated reward pathways of the brain (Young, 2003; Lim et al., 2004). The hypothesis is that activation of these receptors in the reward centres results in the establishment of an association between the olfactory cues of the partner and the rewarding aspects of copulation. As a result the partner becomes highly rewarding. This form of conditioning is similar to that involved in drug addiction where a drug becomes highly rewarding.

Motivated by the above work we build a very basic model of the evolution of commitment in which an inflation factor can bias the reward system in favour of a partner. The model is asexual, and so is a long way from being a realistic model of pair-bond formation in breeding pairs. It may, however, capture a central evolutionary force that acts in a market setting to bring about commitment to a partner. The model is loosely based on that by McNamara et al. (2008).

7.10.1 A Model of the Co-evolution of Choosiness and Commitment

An asexual population has an annual cycle that has five phases: (i) Pairing, (ii) Resource gain, (iii) Reproduction, (iv) Possible divorce, and (v) Possible death. The details of each phase are as follows:

Pairing. Some individuals are already paired from the previous year, those that are not (i.e. they are single) pair up at random.

Resource gain. During resource gain, in each unit of time each member of a pair has the choice between being selfish by taking an outside option or cooperative by taking an option that helps both its partner and itself. The value of the outside option, r, varies with r drawn at random (uniformly) from the range $0 < r < R$. When making its choice an individual knows the current value of r. If the outside option is taken the individual gains the whole resource r: there is no contribution from this resource to the partner. The cooperative option yields a unit resource to each pair member. Individual are characterized by a threshold θ. An individual with this threshold takes the outside option if $r > \theta$, otherwise it helps the partner. Thus the proportion of time it helps the partner is $\rho = \frac{\theta}{R}$. The rates at which each pair member gains resources as a function of the ρ values of the pair is specified in Exercise 7.9.

Reproduction. Each population member reproduces asexually, where the number of offspring produced is large and is proportional to the mean rate at which resources were gained while paired. These offspring are subject to density-dependent competition before the start of the next pairing phase, with only sufficiently many surviving to compensate for the mortality of adults (see below). Surviving offspring are then adults and are single.

Divorce. Pair members decide whether to split up their pairing (divorce). Each individual has a divorce threshold d. Given that the partner commits a proportion of time ρ to helping the individual, then the partner is rejected if $\rho < d$. If either pair member rejects the other the pair divorce and each pair member becomes single, otherwise the partners attempt to stay together until the following year.

Death. The probability that each adult population member dies before the next round of the game is $1/n$, so that each population member plays an average of n rounds of the game over its lifetime. If an individual dies and its partner survives the partner is single at the beginning of the next pairing phase.

For this model, during the resource gain phase an individual maximizes its own rate of resource gain by taking an outside option if and only if the option yields a resource of value $r > 1$. This holds regardless of the behaviour of the partner. We allow for an individual to be biased in their decision to choose an option by assuming that each individual has an inflation factor θ. When choosing whether to help its partner or to take an outside option of resource value r, the outside option is chosen if and only if $r > \theta$. Thus an individual with factor θ behaves as if helping the partner has value θ to the individual, whereas its true value is 1. The resource gain rate is maximized when rewards are assigned their true value; i.e. $\theta = 1$. If an individual adopts a higher value of θ this results in a greater proportion of time ρ committed to its partner, which helps the partner and reduces the probability that the partner will divorce them, but at a resource cost to the individual.

In order to construct a model in which the propensity to help the partner and the divorce strategy co-evolve, we assume that an individual's inflation factor is given by $\theta = e^{\alpha}$ and their divorce threshold by $d = e^{\beta}/(1 + e^{\beta})$, where α and β are genetically determined parameters. Figure 7.10 illustrates evolved strategies.

When $n = 1$, so that individuals pair once and die after reproduction, the divorce strategy is irrelevant. In this case the mean value of the inflation parameter θ evolves to be close to 1, so that resources are assigned their true values.

When $n > 1$ individuals have the possibility to switch partners, and individuals that are not sufficiently committed are divorced by their partners. There are no explicit divorce costs, but population members that are single tend to be less committed to partners than those that have been retained by their partners, so that if an individual is rejected by its current parter the next partner is liable to be worse than average. There is thus selection to avoid being rejected, leading to the evolution of more committed behaviour. This then selects for individuals to be even more choosy, selecting for even higher values of commitment. This evolutionary process eventually settles down with θ significantly above the value 1 (Fig. 7.10a) and divorce thresholds being a little above the average level of commitment (Fig. 7.10b).

As n increases there are more opportunities to find and retain a good partner and the relationship with such a partner will tend to last longer. Thus increasing n increases the selection pressure to be choosy about the partner, so that evolved divorce thresholds increase (Fig. 7.10b). This results in an increase in the level of commitment, which is brought about by an increase in the inflation factor θ (Fig. 7.10a). Because pairs in which both pair members have high commitment tend to stay together, the correlation between the inflation factors of pair members is positive. As n increase, there is more time for assortative pairing to establish and this correlation increases

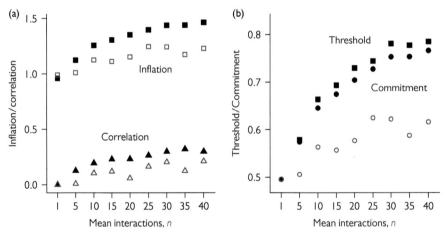

Fig. 7.10 Effects of the mean number of interactions (years) on evolved attributes in the model of the co-evolution of choosiness and commitment. (a) Mean inflation levels, θ (squares), and the correlation between the θ values of pair members (triangles). (b) Mean commitment level, ρ (circles), and mean divorce thresholds, d (squares). Filled symbols are at a high mutation rate on α, open symbols are at a low mutation rate: in (b) only the commitments levels are shown for the low mutation rate. Averages are over the last 10,000 years of an evolutionary simulation over many years. Population size 2^{14}. Maximum reward for an outside option, $R = 2$.

(Fig. 7.10b). In line with the predictions of Box 7.4, the mean rate of resource gain increases with increasing n (not shown).

In this model there would be no point in divorcing a partner if all other potential partners were the same. Furthermore, if we were to add divorce costs, it would only be advantageous to divorce the current partner if there is a reasonable probability of encountering a new partner who is significantly better. In Fig. 7.10 it is genetic variation in the parameter α, maintained by mutation, that gives rise to variation. The source of the variation is, however, largely irrelevant to the degree of choosiness that evolves. The amount of variation in the trait being chosen is crucial in most market models, with the degree of choosiness usually increasing with the amount of variation. This is illustrated in Fig. 7.10, where inflation levels, the correlation between pair members and the levels of commitment are all higher for the higher mutation case.

The market forces that lead to high commitment in the above model may well be important for many species, such as long-lived birds that retain their mate after breeding. For species that change their mate after breeding other forces must operate, although the enhancement or suppression of reward value might still be part of the biological mechanism.

In the above commitment model we incorporated a hypothetical psychological mechanism that affects the perception of rewards; the perceived value of earning rewards with the partner could be inflated (or suppressed). Inflation was determined by an inflation factor θ that we might think of as controlling the density of receptors in the brain of our hypothetical organism. Because this model is simple the exact details of the mechanism that brings about commitment may be unimportant in determining the level of commitment predicted. If we were to introduce complications, such as allowing the level of commitment to change in response to a change in the level of commitment of the partner, the details of the neural mechanisms that control behaviour might affect predictions. We believe it is important that future work in game theory does take into account known mechanisms (cf. McNamara and Houston, 2009). In particular, we need to understand the neural mechanisms that affect mental states such as commitment and trust, and how incorporating such mechanisms affects game-theoretical predictions. We further discuss the issue of incorporating mechanisms in Chapter 11.

7.11 Exercises

Ex. 7.1. Consider the finitely repeated Prisoner's Dilemma with a maximum of N rounds (Section 7.3). Let the payoff for a single round be given by Table 3.1. Let $F(n)$ be the probability that a randomly selected opponent cooperates on round n given that this opponent has cooperated on all previous rounds. Consider a mutant that maximizes its expected total payoff in this resident population. Let $V(n)$ be the expected total future payoff of the mutant from round n onwards, given that there has been cooperation on all previous rounds. Show that $V(n) = bF(n) + \max\{F(n) V(n+1) - c, 0\}$.

Suppose that in the resident population a proportion $\frac{1}{N+1}$ of individuals cooperates for n rounds before defecting (i.e. all strategies are equally likely) and that $c = 0.55b$. Evaluate $V(N)$, $V(N-1)$, and $V(N-2)$, and hence show that the best strategy for the mutant is to cooperate on the first $N-3$ rounds and then defect on round $N-2$.

Ex. 7.2. Let $b = 2$ and $c_H < 1 - 2\epsilon < c_L$ in the signalling model of Section 7.4. Suppose that there is signalling error, so that signal s_1 is received as s_2 with probability ϵ and s_2 is received as s_1 with probability ϵ. Let ρ be the proportion of high-ability males. Show that the male strategy of signalling s_2 if and only if high-ability and the female strategy of laying two eggs if and only if s_2 is received are a signalling equilibrium if $\frac{\rho\epsilon}{(1-\rho)(1-\epsilon)} < K - 1 < \frac{\rho(1-\epsilon)}{(1-\rho)\epsilon}$.

Ex. 7.3. In the model of McNamara and Doodson (2015) (Section 7.5) let $B(u, v) = 2(u+v) - (u+v)^2$ and $C_q(u) = (1-q)u + u^2$. Assume that quality (Q) is uniformly distributed between 0 and 1. Suppose that individuals do not know the reputation of opponents. Let $\hat{x}(q)$ be the contribution of an individual of quality q at a Nash equilibrium. Thus at this equilibrium the contribution of a randomly selected opponent is the random variable $Y = \hat{x}(Q)$. By noting that $E[B(x, Y)] - C_q(x)$ is maximized when $x = \hat{x}(q)$ show that $E[Y] = \frac{1}{4}$, and hence find the unique Nash equilibrium strategy.

Ex. 7.4. Consider a population in which each individuals play a sequence of rounds of the Prisoner's Dilemma game against different opponents (Section 7.6). Assume that reputation is an image score: Good if cooperated and Bad if defected with the previous partner. Let the proportions of ALLD, DISC, and ALLC individuals be ρ_D, ρ_{DISC}, and ρ_C, respectively (where $\rho_D + \rho_{DISC} + \rho_C = 1$). Initially ALLD individuals have reputation Bad and all other population members have reputation Good. Let $\alpha(n)$ be the proportion of population members that have a Good reputation after n rounds. Find an expression for $\alpha(n+1)$ in terms of $\alpha(n)$. Identify the limit of $\alpha(n)$ as n tends to infinity. Find the long-term average payoff per round for each strategy and identify the range of values of ρ_{DISC} for which the average payoff to ALLC exceeds that to DISC.

Ex. 7.5. (i) In the trust example, show that the local payoff to a P2 with trait value p is $V_{sampler}(p) = s + (1-s)p - (1-r)p^2$ when all P1s are samplers. Find the value, $\hat{p}$, of p maximizing the local P2 payoff when the population comprises a proportion β_{UA} of UAs and a proportion β_S of samplers.

(ii) Consider case (iii) in Fig. 7.7. In this case, at stability in the evolutionary simulation we have $E[p] = 0.46758$, $\text{Var}(p) = 0.21537^2$, $\beta_{UA} = 0.3947$, and $\beta_S = 0.5985$. Use the result of part (i) to show that $\hat{p}$ is close to $E[p]$. Use eq (7.6) of Box 7.3 to verify that UAs and samplers are doing approximately equally well at stability.

Ex. 7.6. In a biological market, an individual encounters a sequence of servicers and must choose just one. On encountering a server the value of service from this server is known and the individual must decide whether to choose this server or move on to the next. Once a server is rejected there is no possibility of later coming back to this server. The values of services are non-negative and are drawn independently from a distribution with probability density f. There is a search cost c of moving on to the next server. The individual behaves so as to maximize the expected service value minus the expected sum of the search costs. To do so a server is chosen if and only if their value is at least w. Justify the equation $w = -c + \int_0^\infty \max(w, v)f(v)dv$. Suppose that services are uniformly distributed between $1 - \delta$ and $1 + \delta$ where $0 \le \delta \le 1$. Also assume that $0 < c < \delta$. Show that $w = 1 + \delta - 2\sqrt{\delta c}$.

Ex. 7.7. Let $r(z) = z$ in the pollination model of Section 7.8. Let x be the resident strategy. (i) Show that an optimal forager leaves a flower when the amount of nectar left is $\gamma^*(x)$ where $\gamma^*(x) = \frac{x - \gamma^*(x)}{\log(x/\gamma^*(x)) + \tau}$. (ii) Let $B(T) = 1 - e^{-3T}$. Show that the best response of a mutant to resident strategy x is $\hat{b}(x) = 3^{\frac{1}{4}}(\gamma^*(x))^{\frac{3}{4}}$.

Ex. 7.8. In the model of Box 7.4 let x' be a rare mutant strategy in a resident x population. Note that if a mutant has actual contribution u' then the partners of this individual have the same distribution of contributions as that for a resident with actual contribution u'. Thus, from the properties of the bivariate normal distribution, $E(U|U' = u') = x + \rho(u' - x)$, where U is the contribution of the mutant's partner. Hence show that $W(x', x) = E(H(U', U))$ is given by eq (7.9).

Ex. 7.9. In the model of the co-evolution of choosiness and commitment, suppose that an individual commits a proportion ρ' of their time to the joint venture when partner commits ρ. Show that the rate at which resources are earned by the focal individual is $W(\rho', \rho) = \rho' + \rho + \frac{R}{2}(1 - \rho'^2)$.

8

Interaction, Negotiation, and Learning

8.1 Interaction over Time

Most of the two-player games in Chapter 3 have simultaneous moves, where both players choose their actions without knowing the action of the partner and do not change their choice once the action of the partner becomes known. These are referred to as one-shot, simultaneous, or sealed-bid games. We have already argued that real interactions are often not of this form (e.g. Section 3.5). Instead there is a sequence of decisions where each decision is contingent on the current information, and in particular on what has gone before. In this interaction over time there is often learning about the partner or the situation.

For example, consider the decision of parents whether to care for their young or to desert. McNamara et al. (2002) review evidence that the decisions of the two parents are often not independent. One of the cases they consider is that of the penduline tit. In the data reproduced in Fig. 8.1, out of $n = 130$ clutches the proportion in which the male cares is $p_m = (0 + 25)/130 = 0.19$. Similarly the proportion in which the female cares is $p_f = (0 + 67)/130 = 0.52$. If individuals are following a mixed-strategy Nash equilibrium with independent choices, then from these marginal proportions an estimate of the number of clutches in which there is biparental care is $n_{\text{bipar}} = np_m p_f = 130 \times (25/130) \times (67/130) = 12.9$. A comparison of observed values with those expected under independence provides strong evidence that decisions are not independent (the null hypothesis of independence can be rejected at $p < 0.0001$ in a chi-squared test). It is striking that there are no cases of biparental care. The obvious inference is that if one parent is caring the other notices this and leaves.

This chapter is concerned with the interaction process. By process we mean the sequence of actions and the information on which each action is based. As we show, one cannot predict the outcome of an interaction from the payoffs alone, but one must know the process. For example, just changing the order in which players choose their actions can completely change predictions (Section 8.2). Changing the order of choice has the effect of changing the information available to players. An individual can gain an advantage by giving reliable information to a partner. This might be achieved by individuals committing themselves to a course of action by deliberately handicapping themselves (Section 8.3), or through reputation effects. In an interaction individuals may differ in their abilities, energy reserves, or other aspects of state. Often the state of an individual is private information to that individual that

Game Theory in Biology: Concepts and Frontiers. John M. McNamara and Olof Leimar,
Oxford University Press (2020). © John M. McNamara and Olof Leimar (2020).
DOI: 10.1093/oso/9780198815778.003.0008

	Female care	Female desert	Male proportion
Male care	0 (12.9)	25 (12.1)	0.192
Male desert	67 (54.1)	48 (60.9)	0.808
Female proportion	0.515	0.485	

Fig. 8.1 Parental care behaviour in the penduline tit. The data from Persson and Ohrstrom (1989) show the number of clutches observed to have biparental care, male-only care, female-only care, and biparental desertion. The proportion of clutches with male care and female care are calculated from these data, together with the expected numbers (in brackets) under the four forms of care, given independent decisions by the parents. Drawing of penduline tits by Jos Zwarts. Published under the Creative Commons Attribution-Share Alike 4.0 International license (CC BY-SA 4.0).

is partly revealed by its actions. The interaction may then be a process in which the individuals gain partial information about one another, although it may be in the interests of each to hide information. So for example, it can be advantageous to a parent to appear less able to care than is actually the case, to force the mate to increase its care effort in order to compensate (Sections 8.4 and 8.5). It may even be the case that individuals do not know their own state and must learn about that, for instance when establishing dominance relations in a social group. This and other learning scenarios are considered in Sections 8.5, 8.6, and 8.7. Learning mechanisms can be especially helpful to study interactions over time when individuals respond to each other's actions that in turn are influenced by individual characteristics. Such a large-worlds perspective (Chapter 5) could also be realistic in terms of the traits and mechanisms assumed. We end the chapter (Section 8.8) with a general discussion of game theory for interaction over time, including the role of behavioural mechanisms and how one can think about evolutionary stability for large-worlds models.

8.2 Information and the Order of Choice

Consider two parents that have produced young together. Suppose that each can either care for the young (C) or desert (D). The advantage of caring is that the young do better, which benefits both parents. However, care is costly and reduces prospects for remating. For illustrative purposes, let the payoffs to the two parents be given by Table 8.1. This payoff structure might be appropriate if the male pays a greater cost for uniparental care than the female and also gains less from desertion, perhaps because the local population is male biased. We compare two processes by which the actions are chosen. Suppose first that each parent chooses its action without knowledge of the

Table 8.1 Payoffs to the two parents when each can either care (C) or desert (D).

Payoff to male		Female action			Payoff to female		Male action	
		C	D				C	D
Male action	C	6	3		Female action	C	6	4
	D	5	1			D	7	3

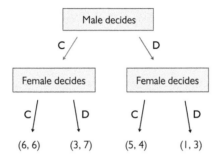

Fig. 8.2 The decision tree for the care versus desertion game when the male chooses first. Payoffs are ordered as (male, female).

decision of the other parent. Each then sticks to its decision once the decision of the other becomes known. In game theory this scenario is known as simultaneous moves. From Table 8.1 it can be seen that whatever the action of the female the male does best if he cares. Given that the male will care the female's best action is to desert. Thus the only Nash equilibrium is for the male to always care and the female to always desert.

Now suppose that the male chooses his action first. The female then chooses her action knowing the choice of the male. A game with this information structure, where one player makes its decision knowing that of the partner, is known as a Stackelberg game. We can represent this scenario by the game tree in Fig. 8.2. Note that the payoff structure is exactly as in Table 8.1. We can find a Nash equilibrium by working backwards. If the male cares the female will desert and the male will get payoff 3. If the male deserts the female will care and the male will get payoff 4. Thus the male should desert and the female will then care. The Nash equilibrium found by working backwards in this way is called a subgame perfect equilibrium (see Section 8.3).

If we compare the outcomes of the two processes we see they predict opposite forms of care. In the simultaneous choice case the male cares, in the sequential choice case he deserts, despite the fact that his best action is to care for either of the fixed actions of the female. This example shows that, by itself, the payoff structure is not enough to predict evolutionary outcomes; we must also specify the decision process. The evidence in Fig. 8.1 suggests that the penduline tits do not make simultaneous decisions. The process is not, however, liable to be as simple as one of the parents choosing first and the other reacting to this choice (Griggio et al., 2004; van Dijk et al., 2007). For example, during foraging trips each may assess the possibilities to remate if they desert, so that actions may also be contingent on this information. Furthermore, there is evidence that the female attempts to hide the fact she has already produced eggs from the male (Valera et al., 1997). This may allow her to gain an advantage by being the first to choose.

In comparing the simultaneous and sequential choice scenarios it is not the time ordering that is of paramount importance, rather it is the information structure. In the simultaneous case neither knows the decision of the other when their choices are made, in the sequential the female knows the male's choice. This is because it has already happened but, as we will see, credible threats (Section 8.3) can be another reliable source of information as to the (future rather than past) action of the partner. In our desertion game if the male chooses first the payoffs to the male and female at the subgame perfect equilibrium are 5 and 4, respectively. If the female chooses first they are 3 and 7. Thus each would prefer to choose first, so that each can gain an advantage by reliably informing their partner of their choice. This illustrates that in a game an individual can put a partner at a disadvantage by giving reliable information. Note that the partner should not ignore such information as this will typically be worse still. For example, a female that always deserts regardless of the male's decision would do worse than a female that deserts only if either the male has decided to care or if she is first to choose.

For some payoff structures, especially where it is best to anti-coordinate, it is better to choose second than choose first. In other words it can be better to receive information than to give it (McNamara et al., 2006b). The Rock–Paper–Scissors game (Exercise 4.9) is usually played as a simultaneous game, but if it were played sequentially the individual who chose second would always win.

8.2.1 Extensive Versus Normal Form

For the desertion game in which the male chooses first, Fig. 8.2 depicts the time structure in choosing actions. This representation is known as the extensive form of the game. In the normal form presentation the explicit time element is removed and the payoffs to strategies are specified, rather than payoffs to the leaf nodes of a decision tree. The male has two pure strategies: (i) C—always care; and (ii) D—always desert. The female has four pure strategies: (i) (C, C)—care if the male cares and care if he deserts; (ii) (C, D)—care if the male cares and desert if he deserts; (iii) (D, C)—desert if the male cares and care if he deserts; (iv) (D, D)—desert if the male cares and desert if he deserts. Tables 8.2 and 8.3 give the male and female payoffs for the possible combinations of strategies. From Table 8.2 we see that if the resident female strategy is (D, C) then the unique best response for the male is D. Also, from Table 8.3 we see that if the resident male strategy is D then a best response for the female is (D, C). Thus D and (D, C) are best responses to one another and are hence a Nash equilibrium. This equilibrium is the same as we previously derived by working backwards, and results in male desertion and female care.

Table 8.2 Payoffs to the male in the desertion game in which the male chooses first.

Payoff to male		Female strategy			
		(C, C)	(C, D)	(D, C)	(D, D)
Male strategy	C	6	6	3	3
	D	5	1	5	1

Table 8.3 Payoffs to the female in the desertion game in which the male chooses first.

Payoff to female		Male strategy	
		C	D
Female strategy	(C, C)	6	4
	(C, D)	6	3
	(D, C)	7	4
	(D, D)	7	3

8.3 Credible Threats and Strategic Commitment

In the previous section we analysed the scenario in which the male chooses first and found that the male strategy D and the female strategy (D, C) are in Nash equilibrium. However, when the male strategy is D it can be seen from Table 8.3 that (C, C) does equally well as (D, C), and so could potentially increase in frequency due to drift. So from this it is not clear whether the Nash equilibrium is evolutionarily stable. From Tables 8.2 and 8.3 it can also be seen that the male strategy C and the female strategy (D, D) are in Nash equilibrium. This equilibrium is maintained by the female's threat to desert regardless of the action of the male: given this threat the male should care and the female can then do no better than to desert, which is achieved by her following the strategy (D, D). So is this a credible threat and which Nash equilibrium do we expect to evolve? We analyse these questions by introducing the possibility of errors in decision-making.

We begin by focusing on the Nash equilibrium at which the male strategy is C and the female strategy is (D, D). Note that when the resident male strategy is to always care the female strategy (D, C) does equally well as (D, D). This is because all males are caring and the two female strategies then take the same action (desert). But what if a male occasionally makes a mistake and does not care? This could be because the male deserts, but it could also be because he is killed by a predator and effectively deserts. When this happens a (D, C) female will care and get a payoff of 4 while the (D, D) female will carry out her threat to desert and will get a payoff of 3. In this sense the female's threat to desert whatever the male does is not credible, as it would not evolve in the presence of errors. We conclude that this Nash equilibrium is not stable once errors are introduced.

Now consider the Nash equilibrium at which the male strategy is D and the female strategy is (D, C). When the resident male strategy is to always desert the female strategy (C, C) does equally well as (D, C) because all males are deserting and the two female strategies then take the same action (care). If a male occasionally makes a mistake and cares (although this seems a less likely mistake than deserting when he should care) then (D, C) does better than (C, C). Thus this Nash equilibrium is stabilized by errors.

Errors mean that every branch of the game tree for the extensive form representation of the game (Fig. 8.2) is explored with positive probability. It then matters what the female does in every situation and she should take the best action every time (i.e. to desert if the male cares and care if the male deserts), and the male should do the best given the female strategy (i.e. desert). In general the pair of strategies found by backwards induction is known as the subgame perfect equilibrium. It is this equilibrium that is stabilized by small errors and is then known as a limit ESS (Selten, 1983).

In two-player sequential games threats by the player who chooses last are not credible unless it is in the interest of that player to carry out the threat. In the game of Fig. 8.2 the female's threat to desert if the male deserts is not credible because if the male has deserted she does best to care. She can, however, make a credible threat if she is able to reduce her payoff for caring on her own, as in Fig. 8.3. This reduction is a handicap that commits her to deserting if the male deserts (this kind of commitment is discussed in Nesse, 2001). As a result, at the new subgame perfect Nash equilibrium (at which the male now cares and the female deserts) her payoff is 7 whereas it was previously 4. In non-game situations handicapping oneself is always suboptimal. In

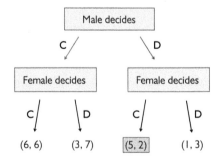

Fig. 8.3 A modified decision tree for the care versus desertion game when the male chooses first. The only change from Fig. 8.2 is that the female's payoff for caring when the male deserts has been reduced from 4 to 2. This now makes it optimal for the female to desert if the male has deserted.

contrast in games it can be highly advantageous, provided opponents are aware of, or adapted to, the handicap.

Barta et al. (2002) model the care and desertion decisions of parents over a breeding season. During this season, each population member has a sequence of reproductive bouts, where each bout is with a different partner. In each bout, the male first chooses whether to care or desert and the female then makes her choice in the light of the male's decision. The model also incorporates feeding options and allows the care/desert decisions of an individual to be affected by its energy reserves. Analysis of this model predicts that females should strategically reduce their own energy reserves prior to breeding. As a consequence of this handicap the female is unable to care alone, forcing the male to care and allowing the female to desert. Other tentative instances of handicapping have been suggested. Smith and Härdling (2000) suggest that a female bird might lay a clutch that is larger than she can care for on her own. This results in nest failure if the male does not also care, forcing him to care.

The care–desertion game has the choice between just two actions. In many games there is a continuum of actions, e.g. Engqvist and Taborsky (2016) model sperm competition allowing a continuum of possible investments in sperm. Box 8.1 analyses

---

Box 8.1 Stackelberg equilibrium for continuous actions

Let $W(x,y)$ be the payoff to an individual choosing action x when the partner takes action y in a two-player game with continuous actions. Assume that for every y there is a unique best response $\hat{b}(y)$ satisfying $\frac{\partial W}{\partial x}(\hat{b}(y),y) = 0$. Let x^* satisfying $\hat{b}(x^*) = x^*$ be the unique symmetric Nash equilibrium. Now suppose that choice is sequential: Player 1 chooses an action first and then Player 2 chooses an action in light of Player 1's choice. Thus the strategy of Player 1 is an action x, and the strategy for Player 2 is a function $b(\cdot)$ specifying the choice of action $b(u)$ for every action u by Player 1. Assume that there is some variation due to errors in the action of Player 1, so that every action is occasionally chosen. It is then reasonable to assume that $b(u) = \hat{b}(u)$ for all u. The payoff to Player 1 of choosing action x is then $V(x) = W(x,\hat{b}(x))$. At the Stackelberg equilibrium Player 1's action x_s maximizes this payoff (so that $V'(x_s) = 0$) and the action of Player 2 is $y_s = \hat{b}(x_s)$. By differentiation we have

$$V'(x) = \frac{\partial W}{\partial x}(x,\hat{b}(x)) + \frac{\partial W}{\partial y}(x,\hat{b}(x))\hat{b}'(x).$$

Evaluating this at $x = x^*$ and using that $\hat{b}(x^*) = x^*$ and $\frac{\partial W}{\partial x}(x^*,x^*) = 0$ we have

$$V'(x^*) = \frac{\partial W}{\partial y}(x^*,x^*)\hat{b}'(x^*).$$

The equation shows that $V'(x^*)$ is typically not zero, which means that the maximum of $V(x)$ is not at $x = x^*$. As $V(x^*) = W(x^*,x^*)$ is the payoff at the symmetric Nash equilibrium, it follows that Player 1 gets a higher payoff at the Stackelberg equilibrium than at the symmetric Nash equilibrium.

sequential choice in the case where there is a continuum of actions and best responses are unique.

8.4 Negotiation between Partners

If two individuals make several investment actions during the time they interact, they may have opportunities to observe and respond to each other. In Section 3.4 we considered the conflict of interest between two parents over care of their common young. We assumed that parents do not observe one another; rather, their efforts are fixed, genetically determined quantities. At evolutionary stability, the effort of each parent should then be the best given that of the other (Houston and Davies, 1985). Although we argued that this is unrealistic, we also note that the assumption is logically consistent: if all individuals of a sex are exactly the same and hence expend the same effort at evolutionary stability, a parent gains no information by observing the partner. In reality, parents differ in many ways, so that it is worth each observing the other, and adjusting their own effort as a consequence. One way of implementing this is, instead of thinking of strategies as efforts, to model strategies as rules for negotiation. This perspective can be applied to several situations, such as joint predator inspection or the vigilance of group members. With the approach, at evolutionary stability each individual employs a negotiation strategy that is the best given the strategies of others. We now consider such evolutionarily stable negotiation rules.

We specify a negotiation strategy by a response function, giving the current effort as a function of the partner's previous effort and the individual's own quality. The individual qualities are randomly drawn at the start of a generation. For concreteness, the interaction is split into a number of rounds. In round t one individual (Player 1) goes first and makes an investment u_{1t}, followed by the investment u_{2t} by the partner (Player 2). With response functions $\rho_1(q,u)$ and $\rho_2(q,u)$, we have the investment dynamics

$$u_{1t} = \rho_1(q_1, u_{2,t-1}) = m_1 + k_1 q_1 - l_1 u_{2,t-1} \qquad (8.1)$$
$$u_{2t} = \rho_2(q_2, u_{1t}) = m_2 + k_2 q_2 - l_2 u_{1t},$$

where q_1 and q_2 are the qualities of the two individuals, with $0 \leq q_i \leq 1$. The starting investment u_{11} by Player 1 also needs to be specified as a component of a strategy. We assume linear dependencies of the qualities and efforts to get a case where we can find expressions for a Nash equilibrium. This analysis was first developed by McNamara et al. (1999b). An illustration of how response functions work appears in Fig. 8.4a. In the negotiation process, the efforts quickly approach stationary values, and the negotiation outcomes $\tilde{u}_1$ and $\tilde{u}_2$ (indicated by dots in Fig. 8.4a) depend on the qualities of the individuals. Apart from the initial investment, the strategies have three components, $x_1 = (m_1, k_1, l_1)$ for Player 1 and $x_2 = (m_2, k_2, l_2)$ for Player 2. We wish to determine the best response x_1 to a resident strategy x_2. Alternatively, as in Section 6.1, we might study adaptive dynamics in the three-dimensional strategy space. For

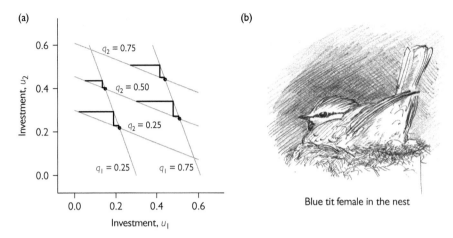

(a)

(b)

Blue tit female in the nest

Fig. 8.4 (a) Illustration of negotiation trajectories and their dependence on the qualities of interacting individuals. Individuals know their own quality q_i but not the partner's q_j and use negotiation functions as in eq (8.1). Light grey lines show the response u_2 for player 2 to u_1 by player 1, labelled with q_2, and the response u_1 for player 1 to u_2 by player 2, labelled with q_1. Four negotiation trajectories are shown in dark grey and correspond to different combinations of q_1 and q_2. Parameters: $m_1 = m_2 = 0.146, k_1 = k_2 = 0.618, l_1 = l_2 = 0.382$. (b) In the parental care of blue tits, the female incubates the eggs, but both parents feed the offspring. Experiments have shown that parents partially compensate each other's feeding efforts (Iserbyt et al., 2019). Drawing of blue tit by Jos Zwarts. Published under the Creative Commons Attribution-Share Alike 4.0 International license (CC BY-SA 4.0).

this we would use invasion fitness $\lambda(x_1, x_2)$ of the mutant into the resident population, but it is sufficient to work with a strong fitness proxy. We assume the payoffs depend on the average investments per round, $\bar{u}_1$ and $\bar{u}_2$, of the partners over a specified total number of rounds. As an approximation, in the analysis we replace these averages by the limiting negotiation-outcome values $\tilde{u}_1$ and $\tilde{u}_2$.

As detailed in Box 8.2, the payoff $W_i(\tilde{u}_i, \tilde{u}_j)$ is assumed to be a quadratic function of the negotiated investments (eqs (8.2, 8.3)), which in turn depend on the strategies of Players 1 and 2 (eq (8.4)). With this payoff we can find best responses (eq (8.6)) and a Nash equilibrium (eq (8.7)) in the strategy space. The response function illustrated in Fig. 8.4a is such a Nash equilibrium.

It is of interest to consider the 'myopic' strategy of using a response function that is an immediate best response to the partner's recent investment. From the analysis in Box 8.1 this strategy cannot be evolutionarily stable, since its use is analogous to playing second in a Stackelberg game, and so can be exploited by a partner. We can also compare a Nash equilibrium negotiation function with one where individuals are myopic. Note that if both partners use the myopic strategy, then the limiting negotiation–outcome efforts are those predicted by the analysis of Houston and Davies (1985). Using reasoning similar to that in Box 8.1, McNamara et al. (1999b) showed that the Nash equilibrium negotiated investments are smaller than one would expect from the Houston and Davies (1985) analysis (see also McNamara et al., 2003a). This is illustrated in Fig. 8.5a. This is analogous to the results in Section 7.5, where individuals attempt to gain an advantage by behaving as if they are lower quality (paying a bigger cost for a given contribution) than is actually the case. In Section 8.5 we show that there is a similar effect of reduced investments if the interaction instead is modelled as learning in combination with evolution of perceived costs of investment.

There is a role asymmetry in the negotiation dynamics in eq (8.1), because of the ordering of the moves by Players 1 and 2. Just as was done in Section 6.2, eq (6.7), we can investigate if eq (8.6) has mutual best response solutions in role-asymmetric strategies, apart from the symmetric solutions given by eq (8.7). It turns out that the symmetric solution is the only one (Exercise 8.5). We can extend the approach to cases, including parental care, where payoff functions differ between roles (Exercise 8.6).

The analysis in Box 8.2 makes simplifying assumptions, but individual-based simulations (Fig. 8.5b) show that there is robustness both with regard to the use of limiting negotiation–outcome values $\tilde{u}_1$ and $\tilde{u}_2$ instead of average investments $\bar{u}_1$ and $\bar{u}_2$, and to the adding of small random errors to the investments in the negotiation process. From Fig. 8.5b we see that population averages fluctuate around model predictions (due to a combination of selection, mutation, and genetic drift), most strongly for the strategy parameter l, less for k, and least for m. The explanation is that the negotiated investments (cf. eq (8.4)) are least sensitive to variation in l, more sensitive to k, and most sensitive to m, which influences the strength of stabilizing selection on the different strategy parameters.

For assumptions like those in Box 8.2, McNamara et al. (1999b) proved that the negotiated solution is a Nash equilibrium in a much larger strategy space, where players can use non-linear response functions and take into account the entire previous history of investments. However, if one introduces quite realistic modifications, such as more general non-linear payoff functions and random errors in the investments, a complete analysis as in McNamara et al. (1999b) becomes very difficult and is perhaps infeasible. This is a reason to view negotiation strategies as mechanisms that can be advantageous, but for which we might not be able to carry out a general small-worlds analysis. Nevertheless, locally functions are approximately quadratic, so we might expect the analysis based on quadratic functions to work well provided the range of qualities q is not too large.

Taylor and Day (2004) examined the model by McNamara et al. (1999b) in the special case of no variation in individual quality q. An analysis similar to Box 8.2 then finds that the model becomes degenerate and has a continuum of Nash equilibria. Using simulations of populations with large amounts of genetic variation in the strategy parameters, Taylor and Day (2004) showed that evolution tended to favour higher investments than predicted by the analysis with (non-genetic) variation in q. As the simulation in Fig. 8.5b illustrates, with both genetic variation in m, k, l and substantial variation in q, there is no noticeable effect of genetic variation increasing

Box 8.2 Evolutionary analysis of negotiation strategies

The payoff to player i $(i = 1,2)$ with quality q_i in an interaction with player j with quality q_j is

$$W_i(\tilde{u}_i, \tilde{u}_j) = B(\tilde{u}_i + \tilde{u}_j) - C(\tilde{u}_i, q_i) \tag{8.2}$$

with

$$B(u_{sum}) = b_0 + b_1 u_{sum} - \frac{1}{2}b_2 u_{sum}^2, \; C(u,q) = c_1 u(1-q) + \frac{1}{2}c_2 u^2. \tag{8.3}$$

With response functions from eq (8.1), the negotiated outcomes depend on the qualities and the strategies as follows (Exercise 8.2)

$$\tilde{u}_i = \frac{1}{1 - l_i l_j}(m_i - l_i m_j + k_i q_i - l_i k_j q_j), \tag{8.4}$$

with $(i,j) = (1,2)$ or $(i,j) = (2,1)$.

To find a best response $x_1 = (x_{11}, x_{12}, x_{13}) = (m_1, k_1, l_1)$ to x_2 we differentiate with respect to x_{1n} and set the derivative equal to zero, giving

$$\frac{\partial W_1}{\partial \tilde{u}_1}\frac{\partial \tilde{u}_1}{\partial x_{1n}} + \frac{\partial W_1}{\partial \tilde{u}_2}\frac{\partial \tilde{u}_2}{\partial x_{1n}} = \left(\frac{\partial W_1}{\partial \tilde{u}_1} - l_2\frac{\partial W_1}{\partial \tilde{u}_2}\right)\frac{\partial \tilde{u}_1}{\partial x_{1n}} = 0 \tag{8.5}$$

for $n = 1,2,3$ and all values of q_1, q_2 (see Exercise 8.3 for the left-hand equality). From the right-hand equality, the expression in brackets should be zero for all q_1, q_2, and it is linear in q_1, q_2. The coefficient of q_2 in the expression should then be zero, which gives the best response $\hat{l}_1$ to l_2, and the requirements that the coefficient of q_1 and the constant term be zero give the best responses $\hat{k}_1$ and $\hat{m}_1$, as follows (Exercise 8.4):

$$\hat{l}_1 = \frac{1 - l_2}{1 - l_2 + c_2/b_2}, \; \hat{k}_1 = \frac{c_1}{c_2}(1 - \hat{l}_1), \; \hat{m}_1 = \frac{b_1}{b_2}\hat{l}_1 - \hat{k}_1. \tag{8.6}$$

From the first of these we find the l-component of a Nash equilibrium from the quadratic equation for $\hat{l}_1 = l_2$ and requiring $\hat{l}_1 < 1$, with solution

$$l^* = \frac{1}{2}\left(2 + \frac{c_2}{b_2} - \sqrt{\left(2 + \frac{c_2}{b_2}\right)^2 - 4}\right), \tag{8.7}$$

and eq (8.6) then gives the other components k^* and m^*.

investments, but simulations with very limited variation in q and substantial genetic variation verify that the effect identified by Taylor and Day (2004) is real.

However, from the perspective of game theory, we should also ask which kinds of strategies might be adaptive in a situation when variation in investments by partners is mainly caused by genetic variation in strategy parameters. The analysis in Box 8.2 does not address that problem, and it is not obvious how to go about it. This could be another reason to regard negotiation strategies as possible behavioural mechanisms whose evolution in the face of genetic and other variation can be studied using simulation, as in Fig. 8.5b.

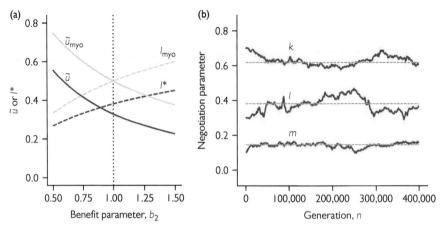

Fig. 8.5 (a) Comparison of evolutionarily stable negotiated strategies (dark grey) and the corresponding myopic responses (light grey). The dark grey solid line gives the investment from eq (8.4) for $q_1 = q_2 = 0.5$ and $b_1 = 2$, different values of b_2, $c_1 = c_2 = 1$ in eq (8.3). The light grey solid line is the corresponding, higher, myopic investment. The dashed dark grey line shows the response parameter l^* from eq (8.7). The corresponding myopic parameter (light grey dashed line) has a higher value, illustrating a higher sensitivity to the partner's investment for the myopic response. The dotted vertical line ($b_2 = 1$) gives the case in Fig. 8.4a. (b) Individual-based simulation of the evolution of the parameters m, k, and l in eq (8.1), with random errors with SD of 0.02 added to the investments. The dark grey curves show the population means, with shading giving ± 1 SD. Individual qualities q_i are randomly assigned from the interval $[0, 1]$ and there are $T = 50$ rounds of interaction, with the average investments $\bar{u}_1$ and $\bar{u}_2$ used to compute payoffs ($b_1 = 2$, $b_0 = b_2 = c_1 = c_2 = 1$ in eq (8.3)). The simulated population consists of $N = 16,000$ individuals with one diploid (unlinked) locus for each of m, k, and l, with mutation rate 0.0001 per allele and mutational increments with SD of 0.02. The dashed horizontal lines show the predictions from Box 8.2.

The issue of how partners respond to each other's parental efforts has been investigated experimentally. A main conclusion from the negotiation model is that individuals partially compensate for variation in their partner's effort: the strategy component l evolves to satisfy $0 < l < 1$. Overall, experiments tend to support partial compensation (Harrison et al., 2009), including short-term experimental manipulation of efforts (Iserbyt et al., 2019). However, experimental and modelling work (Hinde, 2006; Johnstone and Hinde, 2006) has also shown that short-term variation in partner effort can result in parallel (instead of compensatory) responses. If parents differ in their information about offspring need, an increased effort by the partner could indicate higher offspring need, with higher effort as a potentially adaptive response. Further, experiments using blue tits (Bründl et al., 2019) that manipulated a female's information about the number of eggs in a clutch (influencing her feeding effort), without the male directly gaining this information (because the female incubates the eggs; Fig. 8.4b), did not find partial compensation in male efforts. Another idea about short-term variation in efforts is that parents might alternate in their efforts, such that an individual reduces its activity right after having fed the young, and then resumes with a high effort after the partner has in turn fed the young. An effect of this type has both empirical and theoretical support (Johnstone et al., 2014; Johnstone and Savage, 2019), although observed tendencies towards alternation may be small (e.g. Iserbyt et al., 2019).

Most likely our negotiation model is overly simplistic for most situations in nature. It can still have considerable value in suggesting what to observe or manipulate in experiments, and as a basis for further modelling of parental care and other phenomena. For instance, in analysing the predator inspection game in Section 4.1 we followed Parker and Milinski (1997) and implemented a strategy as a specification of the distance to approach a suspicious object, and at equilibrium the approach distance of each individual was the best given the approach distance of its partner. But this only makes sense if actions are simultaneous. We could instead implement strategies as some form of rule for negotiating with a partner about the distance to approach. Predicted distances might then differ from those in the analysis by Parker and Milinski (1997). In general, for game-theory models of interaction over time to be useful, they need to be closely related to empirical observations, in order to investigate which behavioural mechanisms are likely to be important. As we have emphasized, the technical difficulties of analysing such models pose a significant challenge.

8.5 Evolution of Cognitive Bias

As an alternative to modelling investment in a joint project with rules for negotiation, as in the previous section, we can assume that individuals learn about their partners and the situation solely from costs and benefits of investment, interpreted as rewards. We use the model from Section 5.3, modified such that two individuals stay together over T rounds of interaction. They use actor–critic learning, as described in Box 5.2. The interaction is a repeated public goods game with learning, which was investigated by Leimar and McNamara (2019) in the more general case of groups of two or more individuals. As a starting point, let the payoffs/rewards per round be given by eqs (5.8, 5.9). Note that the cost in eq (5.9) depends on individual quality q_i: low-quality individuals pay a higher cost of investment. Individuals differ in quality but are unaware of its value. During the rounds of interaction they learn to adjust their investments based on (positive and negative) rewards. An example of the learning process appears in Fig. 8.6a. In this case there are many rounds of investment in a generation and the resulting evolutionary equilibrium learning rates are rather small. As we noted in Chapter 5, slow learning sometimes approaches an ESS of a one-shot game where individual qualities are known, and this is the case here (the derivation of the ESS is left as Exercise 8.7). This outcome of learning can help us understand why a cognitive bias might evolve, essentially following the reasoning in Box 8.1.

Let us introduce the possibility that individuals can evolve to perceive their costs of investment as different from the fitness cost, while the fitness cost is given by eq (5.9). Thus, the fitness payoff that determines the evolutionary change of genetically

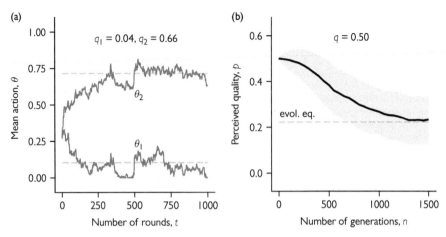

Fig. 8.6 Learning in an investment game where two individuals stay together over many repeated rounds. Individuals are unaware of their own qualities q_i, which are randomly drawn at the start of a generation, uniformly distributed on $[0, 1]$. Learning follows Box 5.2. The learning rates α_w, α_θ, and the initial estimated values w_{i0} and mean actions θ_{i0} are approximately at an evolutionary equilibrium. (a) The learning trajectories θ_{1t} and θ_{2t} of two individuals with different qualities q_i, over 1000 rounds. The dashed lines show the one-shot Nash equilibrium investments for these qualities (without any cognitive bias). (b) The evolution of cognitive bias, starting from no bias. The dark grey line shows the population mean of perceived quality and the light-grey shading indicates ± 1 SD. The dashed line shows a model prediction, which assumes that learning approaches a one-shot Nash equilibrium for the investments. There is one diploid locus coding for the bias $p_i - q_i$, where p_i is an individual's perceived quality. Parameters: number of rounds $T = 5000$; payoff, $b_0 = 1, b_1 = 2, b_2 = 0.5, c_1 = c_2 = 1.5$; mean evolved values, $\alpha_w = 0.280, \alpha_\theta = 0.006, w_{i0} = 1.76, \theta_{i0} = 0.28, \sigma = 0.08$; bias evolution: mutation rate 0.01, mutational increment standard deviation 0.04.

determined traits is the mean of $B(u_i + u_j) - C(u_i, q_i)$ over the rounds in a generation, but the rewards that drive learning can differ from this. First, the benefit

$$B(u_i + u_j) = b_0 + b_1(u_i + u_j) - \tfrac{1}{2}b_2(u_i + u_j)^2, \qquad (8.8)$$

which is from eq (5.8), represents both the fitness payoff and the perceived reward. Second, we assume that individuals have a perceived quality p_i that may differ from q_i, and that they perceive the cost

$$C(u_i, p_i) = c_1 u_i(1 - p_i) + \tfrac{1}{2}c_2 u_i^2, \qquad (8.9)$$

which for $p_i \neq q_i$ differs from the fitness cost in eq (5.9). We also assume that the difference $p_i - q_i$ is genetically determined. We refer to this difference as a cognitive bias. It turns out that zero bias is not evolutionarily stable; instead there is evolution towards a negative bias (the derivation is left as Exercise 8.8). This means that individuals act as if they underestimate their true quality. It is not necessary that individuals possess an estimate of their perceived quality p_i, it is enough that they perceive a corresponding cost of investment. For the quadratic reward functions in eqs (8.8, 8.9) one can, with some effort, work out analytically a prediction for the evolutionary equilibrium bias in the limit of slow learning (see Leimar and McNamara, 2019), and individual-based simulation is in accordance with the prediction (Fig. 8.6b).

The explanation for the effect is similar to the analysis in Box 8.1. Changing p_i influences both an individual's own investments and, through learning, the partner's investments. By acting as if its quality is lower than the true quality an individual can, because of the nature of the learning process, induce its partner to invest more. An important property of learning is that it tends to be myopic, in the sense of responding to immediate or very short-term rewards, while being blind to the longer-term strategic consequences of behaviour. The evolution of cognitive bias can then be interpreted as, to some extent, compensating for the myopic nature of learning. This kind of explanation for the evolution of cognitive bias, as an evolutionary consequence of learning being myopic, is general and might apply to many social interactions.

8.5.1 Common Interest

The cognitive bias implies that partners invest less than they would at a one-shot Nash equilibrium, just as we found for the negotiation model in Section 8.4. A crucial feature of the interaction is that partners have a common interest, in that they share the benefit of the investments (see Leimar and Hammerstein, 2010, for an overview of the

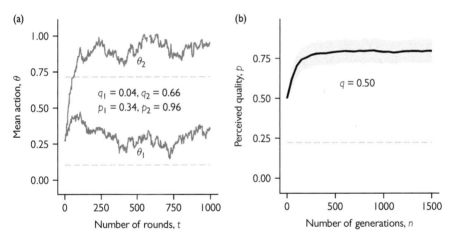

Fig. 8.7 Illustration of the effect of increasing the degree of common interest in the investment game. The difference from the situation in Fig. 8.6 is that the fitness payoff is changed: an individual's mean payoff over the rounds is multiplied by a survival factor that is a sigmoid function of the sum of the accumulated payoffs of both individuals. (a) The learning trajectories of two individuals with different q_i and a bias of $p_i - q_i = 0.30$, over 1000 rounds. The dashed lines show the one-shot Nash equilibrium investments for these qualities when there is no cognitive bias (from Fig. 8.6a). (b) The evolution of cognitive bias, starting from no bias and reaching a bias of around $p - q = 0.30$. The dashed line shows the evolutionary equilibrium without common interest (from Fig. 8.6b).

role of common interest in evolution). This is a basic explanation for why individuals are influenced by a partner's investment. The strength of common interest can also vary. For instance, in parental care, if there is lifelong monogamy parents effectively share both costs and benefits of raising offspring. Figure 8.7 illustrates the effect of a stronger common interest on the evolution of cognitive bias. The result is that a positive rather than a negative bias evolves (Fig. 8.7b), leading to higher investments than at the one-shot Nash equilibrium for the game with payoffs based on the true qualities (Fig. 8.7a). In this example the strength of common interest is increased by assuming that the survival of the pair depends on their joint payoffs from the public goods game. In general, the strength of common interest could be an important explanation for cooperation.

8.6 Social Dominance

Dominance hierarchies are common in group-living animals. They occur in small and moderately sized groups, but less often in large and unstable groups. They are formed through pairwise interactions involving threat display and fighting. They were first described for chickens by Schjelderup-Ebbe (1922), who coined the term 'peck-order' for the hierarchy (Fig. 8.8a). The adaptive reason for the fitting of individuals into dominance hierarchies is thought to be efficiency in settling conflicts over resources. By establishing dominance relationships, group members can avoid fighting every time a resource becomes available, and instead rely on the outcome of previous interactions. An individual's dominance position is typically influenced by traits like size, strength, and aggressiveness.

One suggestion is that individuals can signal their dominance position using 'badges of dominance', which could be traits like the size of the dark patch on the throat and chest of male house sparrows (Fig. 8.8b), with a larger patch indicating higher aggressiveness and fighting ability (Rohwer, 1975). If the signal has a fitness cost outside the context of social dominance, modelling indicates that the signalling can be an ESS (Johnstone and Norris, 1993). However, the empirical evidence for such badges of status, at least for house sparrows, is not very strong (Sánchez-Tójar et al., 2018).

There are game-theory models of social dominance (e.g., van Doorn et al., 2003), but it has been difficult for game theory to include elements such as individual recognition and variation in fighting ability that are likely to be important for social hierarchies in nature. It could well be that a small-worlds approach (Chapter 5) is too challenging for modellers to achieve, because it relies on each individual having representations of Bayesian prior and posterior distributions of the characteristics of other group members. Social dominance might then be a phenomenon where a large-worlds approach, for instance based on learning as described in Chapter 5, dramatically improves our ability to deliver biologically relevant predictions from game-theory models.

(a)   (b)

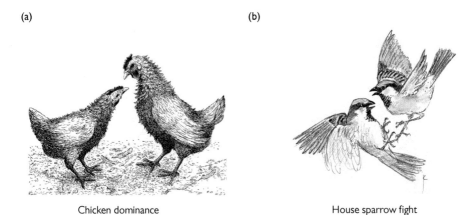

Chicken dominance   House sparrow fight

Fig. 8.8 (a) Dominance interaction between chickens. Reused from Guhl (1956). Reproduced with permission. Copyright © 1956 Scientific American, a division of Nature America, Inc. All rights reserved. (b) Illustration of a male house sparrow fight over dominance, by Jos Zwarts. Published under the Creative Commons Attribution-Share Alike 4.0 International license (CC BY-SA 4.0).

Table 8.4 Fitness payoff (left) and perceived reward (right) of aggressive (A) and submissive (S) actions in the dominance game, when individual i with quality q_i meets j with q_j. C is the cost parameter from eq (5.16) and V is the expected fitness value from a resource that may become available. The parameter v_i is the perceived reward of performing A. Also see Section 5.4 for explanation.

Fitness payoff		Act by j		Perceived reward		Act by j	
		A	S			A	S
Act by i	A	$-Ce^{-q_i+q_j}$	V	Act by i	A	$v_i - Ce^{-q_i+q_j}$	v_i
	S	0	$\frac{1}{2}V$		S	0	0

8.6.1 The Dominance Game with Individual Recognition

We extend the dominance game introduced in Section 5.4 to a situation where individuals recognize each other. The game has payoffs both in terms of fitness and perceived rewards, summarized in Table 8.4. This corresponds to nature, where dominance interactions occur also without an immediate availability of fitness-providing resources.

For individuals to be able to learn about other group members, we introduce the identity of another member of the group as an observation or state in actor–critic learning (Box 8.3). An important difference, compared with the case in Box 5.3, is that the preference intercept in the probability of action A for an individual i, in eq (8.10), is now split into two parts, θ_{iit} and θ_{ijt}.

Box 8.3 Actor–critic learning with individual recognition

Randomly selected pairs of individuals in a group play a game with two actions. The individuals, i and j, make observations, ξ_{ij} for i and ξ_{ji} for j, and they observe the identity of their opponent. These observations act as states. Individual i chooses its first action with probability

$$p_{ijt}(\xi_{ij}) = \frac{1}{1 + \exp\left(-(\theta_{iit} + \theta_{ijt} + \gamma_{it}\xi_{ij})\right)}. \tag{8.10}$$

Individual i also has an estimated value,

$$\hat{w}_{ijt} = w_{iit} + w_{ijt} + g_{it}\xi_{ij}, \tag{8.11}$$

at the start of the interaction. From this and the reward R_{it}, we get

$$\delta_{ijt} = R_{it} - \hat{w}_{ijt}. \tag{8.12}$$

This is used to update the learning parameters w_{iit}, w_{ijt}, and g_{it}:

$$w_{ik,t+1} = w_{ikt} + \alpha_w \delta_{ijt} \tag{8.13}$$
$$g_{i,t+1} = g_{it} + \alpha_w \delta_{ijt}\xi_{ij}$$

with $k = i$ or $k = j$. For the actions we have the eligibilities

$$\zeta_{ikt} = \frac{\partial \log \Pr(\text{action})}{\partial \theta_{ikt}} = \begin{cases} 1 - p_{ijt} & \text{if first action} \\ -p_{ijt} & \text{if second action} \end{cases} \tag{8.14}$$

$$\eta_{ijt} = \frac{\partial \log \Pr(\text{action})}{\partial \gamma_{it}} = \begin{cases} (1 - p_{ijt})\xi_{ij} & \text{if first action} \\ -p_{ijt}\xi_{ij} & \text{if second action} \end{cases}$$

with $k = i$ or $k = j$. The updates of θ_{iit}, θ_{ijt} and γ_{it} are then

$$\theta_{ik,t+1} = \theta_{ikt} + \alpha_\theta \delta_{ijt}\zeta_{ikt} \tag{8.15}$$
$$\gamma_{i,t+1} = \gamma_{it} + \alpha_\theta \delta_{ijt}\eta_{ijt}$$

with $k = i$ or $k = j$. In addition, there can be 'forgetting' of the characteristics of other group members, which we implement as

$$\theta_{ikt} = m\theta_{ikt} \tag{8.16}$$
$$w_{ikt} = mw_{ikt}$$

for each $k \neq i$, where $0 \leq m \leq 1$ is a memory factor.

The first of these influences the probability for i of choosing action A in any interaction, thus generalizing over opponents, whereas the second is specific to interactions with group member j, and is only updated from interactions with j. The same applies to the parameters w_{iit} and w_{ijt} for the estimated value in eq (8.11).

The starting values w_{ii0} and θ_{ii0} are genetically determined, but for the partner-specific parameters we assume that $w_{ij0} = 0$ and $\theta_{ij0} = 0$. In this way an individual has innate action preferences that can be modified by learning, both in a generalized way (θ_{iit}) and specifically for another individual j (θ_{ijt}). The starting values g_{i0} and γ_{i0} of the parameters giving the slopes of the responses to the relative-quality observation $\xi_{ij} = b_1 q_i - b_2 q_j + \epsilon_i$ (defined in eq (5.15)), as well as the learning rates α_w and α_θ and the perceived reward v of performing action A, are genetically determined and can evolve.

An example of how dominance hierarchies are formed in groups of size $g = 4$, with learning parameters at an evolutionary equilibrium, is shown in Fig. 8.9. There are on average 200 rounds per individual, and thus around 67 rounds per pair. Towards the end of the interactions in a generation, the hierarchy has arranged itself so that, in a pair, the individual with higher quality is likely to use the aggressive action A and the other to use the submissive action S (Fig. 8.9a). There are thus rather few fights (AA interactions) towards the end. Initially, most interactions are AA (Fig. 8.9b), but within the first 10 interactions per pair there is a substantial decline, with only a few per cent of interactions being AA after $t = 40$ rounds per pair.

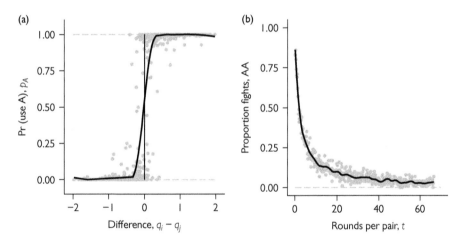

Fig. 8.9 Learning in dominance interactions with individual recognition. There are 100 groups with four members each. In each of 400 time steps in each group, two random group members interact, choosing actions A or S, giving 400/6 = 67 rounds per pair. Learning follows Box 8.3 with learning parameters approximately at an evolutionary equilibrium. (a) The probability of using A for an individual i interacting with j, with qualities q_i and q_j, for the last few rounds per pair. Each data point corresponds to a pair (6 pairs per group in 100 groups). The dark grey line is a non-linear local (LOESS) regression fit. (b) The proportion of interactions that are AA in each time step, plotted against the expected number of rounds per pair. Parameters: distribution and observation of fighting ability, $\mu_q = 0$, $\sigma_q = 0.5$, $b_1 = b_2 = 0.707$, $\sigma = 0.5$; mean evolved values, $\alpha_w = 0.054$, $\alpha_\theta = 16.2$, $w_{ii0} = -0.01$, $g_{i0} = 0.03$, $\theta_{ii0} = 3.4$, $\gamma_{i0} = 2.0$, $v = 0.15$; payoff, $V = 0.25$, $C = 0.2$.

8.6.2 Rapid Learning for the Actor

A striking aspect of the evolved learning parameters in the example in Fig. 8.9 is that the learning rate α_θ of the actor is very high. The reason is that a high learning rate speeds up the formation of a social hierarchy. This suggests that evolution of traits influencing learning processes should be taken seriously by modellers, because such evolution can be crucial for an explanation of a phenomenon. For social dominance, it might be that special neural processing of social interactions and positions (Kumaran et al., 2016; Qu and Dreher, 2018; Zhou et al., 2018) plays a role in rapid learning. Because social dominance can be of great importance and might change during the lives of social animals, it is reasonable to expect special cognitive adaptations for it.

In the example in Fig. 8.9, 50% of the variation in the observation ξ_{ij} is due to variation in $q_i - q_j$. The effect of the observations is that the emerging hierarchy correlates more strongly with the qualities q_i of groups members (cf. Fig. 8.9a) and the speed of formation increases (cf. Fig. 8.9b). For a case where the correlation between ξ_{ij} and $q_i - q_j$ is very low (not shown), the relations depicted in Fig. 8.9 would be rather similar, but with a less sharp switch for p_A and a slightly slower decrease of the proportion AA, starting from a value nearer to 1. For a correlation close to 1, on the other hand, there is an even sharper switch in p_A and a more rapid decrease in the proportion AA, starting from a lower value.

8.6.3 Effect of Group Size

Individual recognition is cognitively more demanding in larger groups. In addition, an individual i will have a smaller proportion of its interactions with a particular other group member j in a larger group, thus having fewer opportunities to learn about that individual. We might then expect that, as group size becomes larger, it will be adaptive for individuals to base their decisions mainly on the observations ξ_{ij} and the overall experience, as captured by the parameters θ_{iit} and w_{iit} in Box 8.3. They might thus approach the learning strategy illustrated in Fig. 5.8 in Section 5.4.

The effect of group size is illustrated by the example in Fig. 8.10, with 16 members in a group. Compared with Fig. 8.9, dominance relations are less sharp, and there are initially fewer AA interactions. The proportion AA also declines more slowly, and does not approach zero over the time of interaction in the group (although the number of interactions per individual is 200, the same as in Fig. 8.9). Overall, the influence of group size on dominance behaviour is rather strong. Examining the effect of the memory factor m in equation (8.16) further shows that it needs to be very close to 1 for individual recognition to play a role in social dominance in larger groups. It thus seems that very many interactions per individual and high cognitive specialization is needed for individual quality and position in a hierarchy to be highly correlated in larger groups.

So-called bystander effects (Dugatkin, 2001), where individuals observe and learn from interactions between other group members, can speed up the formation of dominance hierarchies in larger groups. The effect could be included in a learning model.

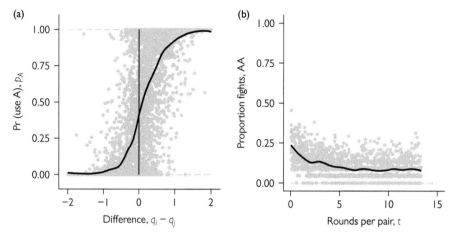

Fig. 8.10 Learning in dominance interactions in a larger group. There are 25 groups with 16 members each. In each of 1600 time steps in each group, two random group members interact, choosing actions A or S, giving $1600/120 = 13$ rounds per pair. Learning follows Box 8.3 with learning parameters approximately at an evolutionary equilibrium. (a) The probability of using A for an individual i interacting with j, with qualities q_i and q_j, for the last few rounds per pair. Each data point corresponds to a pair (120 pairs per group in 25 groups). The dark grey line is a non-linear local (LOESS) regression fit. (b) The proportion of interactions that are AA in each time step, plotted against the expected number of rounds per pair. Parameters: distribution and observation of fighting ability, $\mu_q = 0$, $\sigma_q = 0.5$, $b_1 = b_2 = 0.707$, $\sigma = 0.5$; mean evolved values, $\alpha_w = 0.175$, $\alpha_\theta = 15.9$, $w_{ii0} = -0.005$, $g_{i0} = 0.124$, $\theta_{ii0} = 1.06$, $\gamma_{i0} = 7.37$, $v = 0.094$; payoff, $V = 0.25$, $C = 0.2$.

8.7 Assessment in Contests

As we have discussed (Section 3.5), one of the first questions addressed using game theory in biology was why conflicts between animals of the same species usually are 'limited wars', not causing serious injury (Maynard Smith and Price, 1973). An important part of the answer was given by Parker (1974), who argued that the 'conventional', non-dangerous aspects of fighting behaviour serve the function of assessment by the contestants. In this way, a weaker or less motivated individual can abandon a contest before risking serious injury from fighting. Figure 8.11 shows examples of aggressive behaviour in cichlid fish that allow this kind of assessment, as has been investigated experimentally (Enquist and Jakobsson, 1986; Enquist et al., 1990). The general idea that certain behaviours, like mouth wrestling in cichlid fish, allow a weaker individual to give up uninjured was put forward by ethologists quite some time ago (e.g. Baerends and Baerends-Van Roon, 1950). In game theory, assessment in relation to intraspecific aggression has mostly been studied for conflicts between two individuals, without taking potential interactions with other individuals into account.

Tail beating　　　　　　　　　　　　　　Mouth wrestling

Fig. 8.11 Examples of aggressive behaviour in males of the South American goldeneye cichlid *Nannacara anomala*. Tail beating (left) and mouth wrestling (right). These behaviours, in particular mouth wrestling, allow individuals to assess their relative size and strength with rather high accuracy. Drawings by Bibbi Mayrhofer. Reproduced with permission from fig. 2 of Jakobsson (1987).

For these contests individual recognition is not important. We can, however, still investigate assessment using actor–critic learning in repeated plays of the game from Section 3.12, which is the same modelling approach as in the previous section. Figure 8.12 shows an example. The actor–critic learning process typically causes the individual with lower q_i in a pair to shift from initially being aggressive to becoming submissive (Fig. 8.12a), whereas the stronger individual keeps on using A, in this way obtaining any resources that become available. The number of rounds until this shift occurs is higher for smaller differences between q_1 and q_2 (Fig. 8.12b), and for small differences in q it can happen that the stronger individual becomes submissive first (not shown), thus accidentally losing the interaction. The reason for a decrease in the preference for A for the individual with lower q is that early in the interaction its reward from using A tends to be lower than estimated, making its TD error δ in eq (8.12) negative, which lowers the preference for the action A, causing the individual to use the action S more often. If then, occasionally, the individual uses A, its δ becomes negative again and pushes the preference for A to even lower values, as seen in Fig. 8.12.

8.7.1 The Sequential Assessment Game

Actor–critic learning might seem like an overly complex representation, involving several parameters (Box 8.3), of a process that could be implemented as a decision in each round whether to continue interacting or to give up. We could, for instance, modify the model in Fig. 8.12 by assuming that the contest ends when the first submissive behaviour is shown. Giving up then terminates the interaction, leaving the winner to use the contested resource.

The sequential assessment game (SAG), first developed by Enquist and Leimar (1983), is a small-worlds model that uses this kind of representation of a contest.

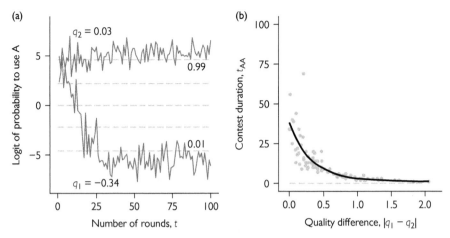

Fig. 8.12 Learning and decision-making in contests between two individuals. Pairs interact over 200 rounds during a lifetime, choosing actions A or S in each round. Learning follows Box 8.3 with learning parameters approximately at an evolutionary equilibrium for this situation. (a) The learning trajectories for an example of two interacting individuals with different qualities q_1 and q_2. The graph shows the logit of the probability of using A, which is the same as the preference for the action, as given in eq (8.10). The dashed lines indicate different probabilities p_A, ranging over 0.01, 0.10, 0.50, 0.90, and 0.99. At first both use A but after around 20 rounds, the individual with lower quality becomes submissive. (b) Contest duration, measured as the number of AA rounds out of the total of 200 in a generation, plotted against the quality difference, for 100 randomly generated pairs of contestants. Each data point corresponds to a pair and the dark grey line is a non-linear local (LOESS) regression fit. Parameters: distribution and observation of fighting ability, $\mu_q = 0$, $\sigma_q = 0.5$, $b_1 = b_2 = 0.707$, $\sigma = 0.5$; mean evolved values, $\alpha_w = 0.028$, $\alpha_\theta = 15.7$, $w_{ii0} = -0.01$, $g_{i0} = 0.02$, $\theta_{ii0} = 3.8$, $\gamma_{i0} = 1.5$, $\nu = 0.16$; payoff, $V = 0.25$, $C = 0.2$.

Figure 8.13 illustrates the SAG for a situation that is broadly similar to the one in Fig. 8.12. The fitness cost per round for i against j is $Ce^{-q_i+q_j}$ and the value of winning is V. In each round t of the game, each contestant observes a sample $\xi_{ij}(t) = q_i - q_j + \epsilon_i(t)$ of the relative fighting ability q_{diff}, with $q_{\text{diff}} = q_1 - q_2$ for Player 1 and $q_{\text{diff}} = q_2 - q_1$ for Player 2. The players form estimates of q_{diff} as sample averages, given by

$$\bar{\xi}_i(t) = \frac{1}{t}\sum_{\tau=1}^{t}\xi_{ij}(\tau), \tag{8.17}$$

and a player continues the contest until its estimate crosses a 'switching line' $S_i(t)$, i.e. until $\bar{\xi}_i(t) \leq S_i(t)$.

An evolutionarily stable switching line $S^*(t)$, together with an example of trajectories of estimates, appear in Fig. 8.13a. In this example Player 1, who has a lower fighting ability, gives up in round $t = 23$, when $\bar{\xi}_1(t)$ crosses $S^*(t)$.

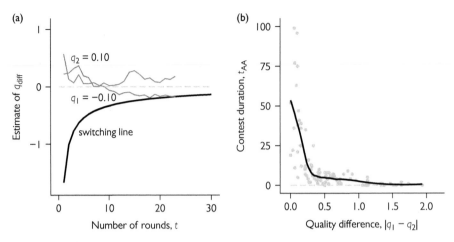

Fig. 8.13 Illustration of the sequential assessment game. (a) The light grey lines show the trajectories of the estimates $\bar{\xi}_1(t)$ and $\bar{\xi}_2(t)$ of relative fighting ability q_{diff}, labelled with the qualities (fighting abilities) q_1 and q_2. The dark grey line is an ESS switching line $S^*(t)$, defined by being a best response to itself. An individual gives up when its estimate goes below the switching line, which happens for $t = 23$ in the example. (b) Contest duration, measured as the number of rounds until one individual gives up, plotted against the absolute value of the quality difference, for 100 randomly generated contests. Each data point corresponds to a pair and the dark grey line is a non-linear local (LOESS) regression fit. Parameters: distribution and observation of fighting ability, $\mu_q = 0, \sigma_q = 0.5, b_1 = b_2 = 1.0, \sigma = 1/\sqrt{2}$; payoff, $V = 1.0$, $C = 0.004$.

Enquist and Leimar (1983) showed for the SAG that an individual's estimate $\bar{\xi}_i(t)$, together with the observation that the opponent did not give up in previous rounds, is a sufficient statistic for inferring the conditional distribution of the stochastic process representing the interaction. They also worked out a procedure to compute numerically a best response switching line, which is an extension of the analysis of the assessment game in Section 3.12. An evolutionarily stable switching line is its own best response, and can thus be determined numerically. This ESS is a Nash equilibrium of the game and, because it properly accounts for the Bayesian updating of information about fighting ability, it can be referred to as a Bayesian Nash equilibrium.

The SAG has been extended to handle such things as role asymmetries, variation in the value of winning, and lethal injuries (Leimar and Enquist, 1984; Enquist and Leimar, 1987, 1990). Parameters of the model have also been fitted to quantitative data from staged contests (Enquist et al., 1990; Leimar et al., 1991), resulting in a relatively good description of the interactions. Even so, the difficulties in developing and analysing small-worlds models make it unlikely that the approach is workable for phenomena like social dominance (Section 8.6). Another issue is that animal contests need not be settled through mutual assessment (Taylor and Elwood, 2003; Arnott and Elwood, 2009), as assumed for the SAG, but that different characteristics of an individual and its opponent might be observed and used in decision-making. Thus, individuals might base decisions on one type of information that relates to their own fighting ability and a different type of information that relates to their opponent's fighting ability. However, the technical difficulties of implementing this in a small-worlds model could be insurmountable. For game theory to handle such issues, large-worlds models are probably needed.

8.7.2 Decision-making as a Neural Random Walk

Decision-making in the SAG shows some similarity to a much-studied idea in neuroscience, namely that there is a neural accumulation of evidence for or against a decision and the decision is made when the accumulation passes a threshold, which is illustrated in Fig. 8.14.

The similarity suggests that models from neuroscience could be integrated into game theory. The study of neural processes of decision-making has a long history (e.g. Vickers, 1970) and is related to signal detection theory (Gold and Shadlen, 2007). Despite the very large volume of theoretical and experimental work in the field, the detailed neural implementation of accumulation of information and a subsequent decision is not yet conclusively established. The most widely supported prediction of these models is that decisions take longer when the differences to be discriminated

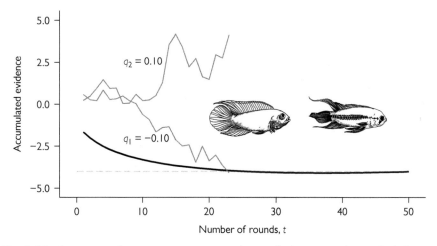

Fig. 8.14 The sequential assessment game as random-walk decision-making. The light grey lines show the same trajectories as in Fig. 8.13a, but represented as sums of observations, $\sum_{\tau=1}^{t} \xi_{ij}(\tau)$, instead of as averages. The dark grey line is the corresponding ESS switching line. Typically in neural random-walk models a decision is made when a process crosses a constant barrier, such as the one given by the dashed grey line. The inset illustrates how *N. anomala* contests end, with the loser (on the right) dramatically changing its appearance and fleeing from the winner. This switch in behaviour inhibits further aggression, both from the loser towards the winner and vice versa. Drawing by Bibbi Mayrhofer. Reproduced with permission from fig. 2 of Jakobsson (1987).

are smaller, and this is generally in agreement with the SAG (Fig. 8.13b), as well as with learning models of contests (Fig. 8.12b). In our view, an integration of basic mechanisms of learning and decision-making from neuroscience (e.g. Bogacz et al., 2006; Trimmer et al., 2013) is a promising direction for game theory in biology.

8.8 Outlook: Games with Interaction over Time

For game theory both in biology and in economics, interaction over time is a much explored and challenging topic. The greatest effort has been spent on cooperation, in particular on the repeated Prisoner's Dilemma (RPD) game. Axelrod and Hamilton (1981) argued that 'tit-for-tat' (TFT) and similar strategies implementing immediate reciprocity in the RPD game should be seen as strong predictions for cooperation in nature, with wide application ranging from interspecific mutualism to social behaviour. These predictions have not fared well when encountering observation. To give one specific illustration, several experiments have investigated whether social animals learn to cooperate in laboratory set-ups of the RPD, and if they do, which kind of strategies they seem to use. Experiments using captive blue jays (Clements and Stephens, 1995; Stephens et al., 2002; Stevens and Stephens, 2004), which are highly social birds, showed that while the birds readily learned to cooperate when rewards directly favoured mutual cooperation, any initial cooperation tended to decay when rewards corresponded to the RPD game. It was also found that, while the birds learned to cooperate with a manipulated partner ('stooge') playing TFT, they did not themselves follow TFT or some similar strategy.

Further, using a laboratory strain of Norway rats, which are social rodents, Wood et al. (2016) showed that pairs of rats learned to cooperate to a fairly high degree (around 50%) when rewards corresponded to the RPD, but they did not follow TFT or similar strategies. There is at present no definitive explanation for this tendency of rats to help each other. However, Rutte and Taborsky (2007) found that female Norway rats increased somewhat their tendency to help another unfamiliar rat after being exposed to a period of receiving help, and Schweinfurth et al. (2019) found something similar for male rats, although males only increased their tendency to help other males that previously had been helpful to them. From these studies it appears that rats can be helpful and that this tendency is at least to a certain degree modified by recent social experience, although their behaviour seems not to correspond to TFT or similar strategies. It might be that groups of Norway rats have substantial common interest and respond to various social cues by modifying their tendency to help each other to acquire food.

Game-theory models like the ones presented in this chapter could well be valuable in exploring these questions, but the modelling needs input from observation, both in terms of which mechanisms and cues guide social behaviour, and in terms of the short- and long-term fitness consequences of social interactions. The difficulty in explaining real behaviour that analyses of the RPD have faced is not a consequence of a weakness in the game-theory analysis. In fact, several proposed Nash equilibrium strategies for the RPD game (although not TFT) are robust to the introduction of errors in choosing actions. The problem lies elsewhere, for instance in that the PD game itself is not a good representation of most real social interactions (as we mentioned in Section 3.2), or in that animals discount the future strongly and mostly learn from and respond to immediate rewards (e.g. Clements and Stephens, 1995).

8.8.1 Examples of Fruitful Areas of Application

To maintain close contact with observation, it is helpful to focus modelling of interaction over time on phenomena like parental care or social dominance, which are important features of the lives of animals and for which we can expect evolution to have tuned the behaviour. These include many topics of interest to game theory, for instance reputation and signalling (Silk et al., 2000), cooperative breeding and territoriality (Quiñones et al., 2016), and market effects in interspecific mutualism (Quiñones et al., 2020).

As we have seen, learning and similar mechanisms for adjusting behaviour based on rewards can cause individuals to approach a one-shot Nash equilibrium of a game. This in turn sets the scene for the evolution of cognitive bias (Section 8.5). Because the underlying reason is quite general, essentially following the analysis in Box 8.1, one can expect the evolution of cognitive bias to occur in many situations where an individual's actions influence the behaviour of social partners. The evolution of cognitive bias is one potentially general example of how evolution can tune the parameters of behavioural mechanisms (see Marshall et al., 2013, for a discussion of this).

The strength of common interest can have a profound influence on the evolutionary tuning of behavioural mechanisms, with particular relevance for conflict and cooperation in social groups (compare Figs 8.6 and 8.7). The general idea of common interest has been discussed for quite some time (e.g. Connor, 1995; Kokko et al., 2001; Clutton-Brock, 2009; Leimar and Hammerstein, 2010), but there are few game-theory models addressing it. As an illustration, we pointed out that assessment in contests (Section 8.7) allows individuals to resolve a conflict at lower cost, and this contains an element of common interest. There is a further strengthening of common interest if one takes into account that aggressive behaviours that enable assessment (Fig. 8.11) might also permit contestants to maintain attention to other stimuli apart from the opponent, for instance an approaching predator, to a greater extent than they could during escalated fighting. There is experimental evidence that contestants are considerably more vulnerable to predator attacks during escalated fighting than during aggressive displays (Jakobsson et al., 1995). The additional common interest might thus cause contestants to be more motivated to perform displays rather than escalated fighting when they estimate predation risk to be higher.

8.8.2 Behavioural Mechanisms in Large and Small Worlds

Small-worlds models (Chapter 5) take into account all kinds of information available to a decision maker. An important advantage of these models is that they fully represent how selection operates in a given situation. When considering which behavioural mechanisms to implement in game theory, a possible approach is to prefer mechanisms that can be shown to be optimal in very simple environments or under simplifying assumptions. It is then reasonable to think that these mechanisms will perform rather well also in a broader set of circumstances, which is a reason to implement them also in large-worlds models. Ideally, of course, it would be preferable to be able to work out such reasoning in greater detail.

We proceeded in this way in our analysis of negotiation rules (Section 8.4). Under simplifying assumptions, the equilibrium strategy derived in Box 8.2 is its own best response in a larger space of strategies where individuals can use all information available to them to make decisions. We argued that the mechanism has a certain robustness (Fig. 8.5), suggesting that it is adaptive under more general assumptions. We did not, however, prove that there is such robustness. Arguing in a similar way for the sequential assessment game (Section 8.7), the correspondence between an ESS for this small-worlds model and a neural random walk (Figs 8.13, 8.14) could be a reason to make such neural random walks an element of more complex models of contests, for instance with different kinds of display behaviours.

Behavioural mechanisms have long been a cornerstone of the study of animal behaviour (e.g. Ridley, 1995; McNamara and Houston, 2009; Rubenstein and Alcock, 2018), including also the modelling of animal decisions in ecology (e.g. Eliassen et al., 2016; Budaev et al., 2019), and they are addressed by one of Tinbergen's four fundamental questions about animal behaviour (Tinbergen, 1963). However, game theory needs specific instances of such mechanisms that map observations and states to action choices and allow learning to occur. We suggest that animal psychology and neuroscience can provide inspiration for such mechanisms.

As an illustration, it can be useful for game theory to incorporate motivation into behavioural mechanisms. As used in psychology, motivation refers broadly to something giving direction to behaviour, thus providing a weighing of different activities or goals against each other. This includes weighing positive aspects (rewards) inherent in an activity against negative aspects (costs), for instance the value of gaining food in foraging against the risk of predation, or the value of a high dominance position against the costs of fighting with strong opponents. Cues perceived by an individual can change its motivation and thus guide its behaviour. An example from neuroscience of such a mechanism is opponent actor learning (Collins and Frank, 2014; Möller and Bogacz, 2019). This mechanism implements a competition between 'Go' and 'No-Go' pathways in the brain, regulating whether or not to go ahead with a certain kind of activity. It is thought to be based on dopamine signalling. For game-theory modelling, the mechanism would be an elaboration of the actor–critic learning formulation (e.g. Box 5.1 or Box 8.3), with separate representation of positive and negative components of action preferences that in turn can be guided by cues influencing motivation.

8.8.3 Evolutionary Stability for Large-worlds Models

A basic idea for large-worlds models of interaction over time is to let behavioural mechanisms play the role of strategies in game-theory models. As mentioned above, it is then important which mechanisms one considers. For instance, reinforcement learning (Chapter 5) has its origin in optimization theory for Markov decision processes (Sutton and Barto, 2018). It is natural to ask if these procedures will perform optimally at least in certain environments. The question has been explored (Dayan et al., 2000; Courville et al., 2006; Daw, 2014), giving rise to the general suggestion (also mentioned above) that evolution should favour organisms that approximate optimal behaviour using efficient mechanisms. In addition to such general arguments for implementing certain behavioural mechanisms in large-worlds models, one can also make parallels with evolutionary modelling in general.

A common approach for instance in life-history theory is to identify a number of potentially important traits from empirically based insights, and then to study the evolution of these traits using modelling. Studying component traits of behavioural mechanisms in evolutionary analyses of interaction over time is a version of this general approach, with the possible distinction that it is particularly challenging to identify the traits and mechanisms.

Evolution can act by tuning the component traits of a mechanism, such as the parameters of a negotiation rule (Figs 8.4, 8.5) or learning rates and the amount of cognitive bias in the investment game (Fig. 8.6). This corresponds to a constrained strategy space, given by variation in the component traits of the behavioural mechanism. Evolutionary stability amounts to finding stable evolutionary equilibria for these traits in some particular environment or environments that one wants to consider. This can then give valuable insight into which trait values one can expect to find in nature.

8.9 Exercises

Ex. 8.1. Consider a two-player game with possibly different payoffs to the two contestant. Suppose that there are unique best responses for each player. Player 1 has payoff W_1^* at a Nash equilibrium in the simultaneous choice version of the game. Show that Player 1 has a payoff that is at least W_1^* at a Stackelberg equilibrium.

Ex. 8.2. Suppose that a parent, e.g. the female, of quality q_1 is paired with a male of quality q_2. The two negotiate their parental efforts using response rules in eq (8.1). Assuming that $|l_1 l_2| < 1$, derive eq (8.4) for the limiting efforts $\tilde{u}_1$ and $\tilde{u}_2$ by inserting the limiting efforts in eq (8.1).

Ex. 8.3. Derive the expression for the local payoff derivative in eq (8.5). For the derivation it is helpful to use eq (8.4) to obtain the following results: $\partial \tilde{u}_2 / \partial m_1 = -l_2 \partial \tilde{u}_1 / \partial m_1$, $\partial \tilde{u}_2 / \partial k_1 = -l_2 \partial \tilde{u}_1 / \partial k_1$ and, which is more work to verify, $\partial \tilde{u}_2 / \partial l_1 = -l_2 \partial \tilde{u}_1 / \partial l_1$. These can then be used for the derivatives in eq (8.5).

Ex. 8.4. Derive the best responses $\hat{l}_1, \hat{k}_1, \hat{m}_1$ to l_2, k_2, m_2 in eq (8.6). You might use the approach suggested in Box 8.2.

Ex. 8.5. Use a graphical approach to check that the role symmetric solution to the best response mapping in the left-hand part of eq (8.6) is the only Nash equilibrium. Plot the iterated local best response function $\hat{l}_1(\hat{l}_2(l_1))$ against l_1, for simplicity assuming that $b_2 = c_2$, and refer to Section 6.2.

Ex. 8.6. Female and male parents negotiate their efforts. Payoff are $W_f(u_f, u_m; q_f) = B(u_f + u_m) - ((1 - q_f)u_f + \alpha_f u_f^2)$ and $W_m(u_f, u_m; q_m) = B(u_f + u_m) - ((1 - q_m)u_m + \alpha_m u_m^2)$, where $B(z) = 2z - z^2$. Adapt the analysis of Box 8.2 to show that the evolutionarily stable response slopes satisfy $l_f = \frac{1 - l_m}{1 - l_m + \alpha_f}$ and $l_m = \frac{1 - l_f}{1 - l_f + \alpha_m}$.

Ex. 8.7. Work out the Nash equilibrium for the one-shot public goods game where each of a pair of individuals knows both their qualities, q_i and q_j, where $i, j = 1, 2$ or $i, j = 2, 1$. The payoffs are $W_i(u_i, u_j, q_i) = B(u_i + u_j) - C(u_i, q_i)$ where B and C are specified in eqs (5.8, 5.14). The Nash equilibrium investments $u_i^*(q_i, q_j)$ can be found by solving the two equations $\partial W_i / \partial u_i = 0$, $i = 1, 2$, i.e. $B'(u_i + u_j) = \partial C(u_i, q_i)/\partial u_i$. Solve these linear equations and show that the solution $u_i^*(q_i, q_j)$ depends on the qualities in a linear fashion. Also note that $u_i^*(q_i, q_j)$ increases with increasing q_i but decreases with increasing q_j.

Ex. 8.8. Show that, in the limit of slow learning, zero cognitive bias is not an evolutionary equilibrium to the investment interaction described in Section 8.5. First, note that if the players have perceived qualities p_i and p_j, with $i, j = 1, 2$ or $i, j = 2, 1$, learning will approach a Nash equilibrium of a one-shot game with payoffs given by eqs (8.8, 8.9). As in Exercise 8.7, one finds that the equilibrium investments $u_i^*(p_i, p_j)$ satisfy $B'(u_i + u_j) = \partial C(u_i, p_i)/\partial u_i$. Note that for this equilibrium, a partner's investment $u_j^*(p_j, p_i)$ does not just depend on the partner's perceived quality p_j but, through learning, also depends on the focal individual's own perceived quality p_i. Second, to examine the evolution of cognitive bias, approximate the fitness payoff as $W_i(u_i^*(p_i, p_j), u_j^*(p_j, p_i), q_i) = B(u_i^*(p_i, p_j) + u_j^*(p_j, p_i)) - C(u_i^*(p_i, p_j), q_i)$. To see if zero bias, i.e. $p_i = q_i$ and $p_j = q_j$, is an evolutionary equilibrium, examine the derivative of this fitness payoff with respect to p_i at zero bias. Using the condition that $u_i^*(p_i, p_j)$ is a Nash equilibrium, the derivative can be written as $[\partial C(u_i^*, p_i)/\partial u_i - \partial C(u_i^*, q_i)/\partial u_i]\partial u_i^*/\partial p_i + B'(u_i^* + u_j^*)\partial u_j^*/\partial p_i$. For $p_i = q_i$ the term in square brackets is zero, and the last term is negative (Exercise 8.7), so individual i can increase its fitness payoff by decreasing p_i from q_i.

9

Games Embedded in Life

9.1 Self-consistency

In specifying the structure of a game, such as the Hawk–Dove game of Section 3.5, the payoff to a contestant is specified for each action of that contestant and each action of any opponent. In many of the games presented so far we have followed this procedure, specifying payoffs in advance; in economic jargon, we have taken payoffs to be exogenous. There are two principal reasons why it might not be reasonable to do this. We can loosely summarize these reasons as (i) the game needs to be embedded in an ecological setting, and (ii) the game needs to be embedded in the life of the contestants. We illustrate these points with two cases, which we later go on to analyse in more detail.

Care or desert. Consider a breeding season in which individuals pair up and produce young. Once the young are born each parent can decide whether to care for the young or to desert. One advantage of desertion is that there is the possibility of remating and producing another brood that year, whereas there is no time for this if an individual cares. In this scenario we might be interested in the pattern of care, for example under what circumstances do we predict that at evolutionary stability females care and males desert. Consider a focal breeding pair in the population. The payoff to the male if he deserts and the female cares reflects his probability of remating once he deserts. But this probability depends on the numbers of males and females who also desert and are seeking new partners. In other words the payoff to the male depends on the resident population strategy. Thus the payoff cannot be specified in advance, but emerges once one has found the ESS.

Territory owner versus intruder. In Section 6.2 we introduced a version of the Hawk–Dove game in which two individuals contest a territory. In this contest one individual is the current owner of the territory and the other is an intruder. In this scenario we are interested in the levels of aggression of the owner and the intruder at evolutionary stability. As the analysis of Section 6.2 shows, the value V of the territory is central to predictions. But this value will depend on how long an owner can maintain ownership. This will depend on the level of aggression of the owner and any intruders in the future. It will also depend on the frequency of future challenges. This frequency is affected by the population size, which depends on the mortality rate, and hence

Game Theory in Biology: Concepts and Frontiers. John M. McNamara and Olof Leimar,
Oxford University Press (2020). © John M. McNamara and Olof Leimar (2020).
DOI: 10.1093/oso/9780198815778.003.0009

depends on the level of aggression of residents. Thus V cannot be specified in advance as it depends on the future behaviour of the owner and the resident strategy.

The above examples show that considering a game in isolation may be problematic in that payoffs cannot be specified in advance, but may emerge from a solution of the game. In economic terms, payoffs are endogenous. This highlights a problem of consistency: the solution of the game must be in terms of the payoffs that are generated by the solution itself.

One approach to the problem is to construct models that include the whole of the lives of all population members, considering not only the behaviour of each individual but the combined effect of all individuals on ecological variables such as population density and the food supply. Having found the ESS over a lifetime, one can, if one wishes, derive game payoffs for local situations that are consistent with this ESS in the sense that these payoffs predict the same ESS. However, payoffs emerge rather than being specified in advance (Section 9.4). Exogenous payoffs may predict a range of possible evolutionary outcomes as these payoffs vary. Endogenous payoffs are more likely to be restricted and correspond to a restricted class of solutions.

This more holistic approach comes at a cost. Predictions may be harder to obtain and harder to understand. Furthermore, testing predictions poses a significant challenge to empiricists as much more must be measured. Thus it is not a good idea to abandon approaches that isolate interactions; rather, isolating interactions from the rest of an animal's life has merits as a theoretical and empirical tool provided one is aware of the limitations. In this chapter we highlight some of these limitations, illustrating that placing a game within a life and within an environment generated by the population can give new insights.

9.2 The Shadow of the Future, and the Past

The two examples presented above illustrate that it is not always possible to isolate a focal game from future circumstances. In many situations it is not possible to isolate an interaction from the past either. The reason is that current circumstances depend on past events, so that the states and actions of interacting individuals depend on the past. When either is the case, we need to abandon the simple notion of a game with given payoffs and analyse the whole life, of which the game is a part, in a holistic manner. In this way we revert to first principles and can work directly with lifetime reproductive success (LRS). We define the LRS of an individual as the total number of surviving offspring (discounted by relatedness) produced over its lifetime. As we discuss in Section 10.6, when all surviving offspring are the same, LRS is a fitness proxy in the sense defined in Section 2.5. We deal exclusively with such situations in this chapter. Situations in which surviving offspring can differ in state, and hence differ in their ability to survive and reproduce, are analysed in Chapter 10.

Assuming sufficient flexibility in possible strategies, an organism maximizes LRS if and only if it maximizes its future LRS in every possible situation (see Houston and McNamara, 1999; and Section 10.6). Consider a two-player game in which a mutant individual that is in a given state contests a resource with a member of the resident population; the contest might, for example, be over a food item or a mating opportunity. The mutant gains an immediate contribution to reproductive success from the game and then gains additional reproductive success in the future. If the mutant is to maximize its LRS it has to maximize the sum of these two components of LRS. In Fig. 9.1 the combined actions of the mutant and the opponent are represented as determining the number of offspring produced at that time by the mutant and the state of the mutant immediately after the game. Those offspring that survive to maturity are the mutant's immediate reproductive success and the future reproductive success of the mutant depends on its state after the game. The factors that determine these quantities can be summarized as follows.

The ecological background. The resident strategy determines the size of the population and the distribution of resources in the environment. For example, in the desertion game of Section 9.3 the availability of mates after desertion depends on this strategy. In the foraging model of Section 9.6 the availability of food and risk of predation in a local habitat depends on the number of residents that use this habitat. The frequency of future interactions between the mutant and residents may also be affected by the resident strategy. For example, in the models of territory owners and intruders of Section 9.5 the availability of vacant territories is a function of the mortality rate of residents, which can depend on the level of aggression of residents and intruders when competing for a territory.

   In the schematic representation in Fig. 9.1 the ecological background affects the current state of the opponent, the number of the mutant's offspring that reach maturity, and the mutant's expected future LRS after the game.

The resident strategy in the past. The current action of the resident opponent is determined by its state and its strategy. Its state is affected by its behaviour in the past

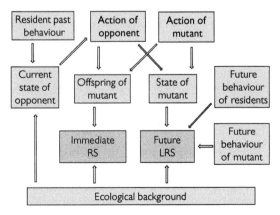

Fig. 9.1 Factors influencing the best action of a mutant individual (in a given state) in a contest with a resident over a resource.

and the ecological background, so that past behaviour affects the choice of current action. This consideration is particularly relevant in state-dependent dynamic games. For example, Houston and McNamara (1988) model the strategy of a small bird that maximizes the probability that it will survive a winter. During winter food is uncertain and some food items are contested by other population members. A strategy specifies whether to play Hawk or Dove, depending on the current energy reserves and time of day; the disadvantage of playing Hawk is that a fight may result in death. Suppose that the resident strategy specifies that near dusk an individual plays Hawk if its reserves are insufficient to survive the night. Then the resident strategy in the early part of the day will determine the likelihood that reserves are below this critical level near dusk and hence the probability a resident will play Hawk at this time. Section 9.6 presents a state-dependent dynamic game in which there are no two-player interactions. Instead previous behaviour affects current reserves and hence the foraging habitat chosen by individuals, and so affects the current food available in these habitats. One can then regard this as the resident strategy determining the ecological background.

Future behaviour of the mutant and residents. Even in a non-game-theoretic context, the expected future LRS of an organism depends on what it does in the future. When the mutant faces a sequence of interactions with others, the mutant's future prospects depend on its own future behaviour and that of residents in those interactions. In Section 9.4 we consider a population in which each male meets, and attempts to mate with, a sequence of females. The LRS of a male is proportional to the number of matings over its lifetime. If a mutant male meets a female that is contested by a resident male, the mutant must choose its level of aggression. High aggression increases the chances of mating with the female, but is more likely to result in death. The best choice of action depends in part on how valuable it is to be alive after the contest, i.e. the future LRS if one survives. This depends on how many females are liable to be contested in the future and the behaviour of both the mutant and residents in those contests. Similar considerations about future behaviour are relevant to the models of territory owners and intruders of Section 9.5 and the foraging model of Section 9.6.

The philosophy behind the approach to modelling in this chapter is that when modelling one should specify the consequences of actions, and not the costs or benefits (Houston and McNamara, 1988, 1991, 2006). An action affects (i) the number of offspring produced at that time, (ii) whether or not the individual survives, (iii) the state of the individual after the action, and (iv) the state of others. By these means one achieves a model that is consistent.

9.3 Resident Strategy Affects Future Opportunities

Parents often face a trade-off between providing care and seeking additional matings. In modelling this trade-off it is important to account for these additional opportunities in a self-consistent manner. We illustrate this in a simple model in which each parent can care or desert that is based on the model of Webb et al. (1999).

9.3.1 Care for Young or Desert

We model behaviour during the annual breeding season in a population in which there are equal numbers of adult males and females. At the beginning of the season all adults pair up and produce young. There are no extra-pair copulations (EPCs), so that a male has full paternity of his female partner's young. Each pair member then decides whether to care for their young or desert. This decision is made independently of the decision of the other parent (simultaneous choice). The number of young that survive is V_0, V_1, or V_2 depending on whether zero, one, or two parents care. We assume that $V_0 < V_1 < V_2$. If a parent cares there is no time for another mating that season. A deserting parent seeks an additional mating, producing an extra V_+ surviving offspring on average during that season. Population members attempt to maximize the expected number of surviving offspring produced over the season.

If we simply analyse this model by assuming a value for V_+, then at evolutionary stability biparental care, uniparental care by the male, uniparental care by the female and biparental desertion are all possible, depending on the values assumed for the other V's. Furthermore, by the results of Selten (1980) there can be no mixed-strategy ESSs (Section 6.2). However, assuming a given value for V_+ is dubious in most biological scenarios. For example, Model 2 of Maynard Smith (1977) assumes a fixed value for V_+ for males. For this model at one ESS all males desert and all females care. But if all females are caring, there are no females available for the deserting males to mate with. In other words, assuming a given value for V_+ typically produces a model that lacks consistency (Webb et al., 1999). A consistent model must be explicit about where the extra matings come from, and must hence consider the whole breeding season. When this is done V_+ will typically depend on the care/desert decisions of other population members.

Suppose that in the resident population a proportion x of females and y of males desert. Let $r_f(x,y)$ be the probability that a deserting female remates and $r_m(x,y)$ be the probability a deserting male remates in this population. Then consistency demands that $xr_f(x,y) = yr_m(x,y)$ since the number of matings by all deserting females must equal that by all deserting males (Webb et al., 1999). A mutant male within this resident population has local payoff (measured in surviving offspring)

$$W_C^m(x,y) = (1-x)V_2 + xV_1$$

if he cares, and since he has the same remating probability as resident males, has payoff

$$W_D^m(x,y) = (1-x)V_1 + xV_0 + r_m(x,y)V_2$$

if he deserts. Here we have assumed that the expected number of surviving offspring from the second brood is V_2 since there is no time for further matings and both parents should thus care. Analogous formulae hold for mutant females.

Figure 9.2 illustrates the direction of selection acting on mutants of both sexes. In this figure we have taken $r_f(x,y) = \frac{y}{0.15+x+y}$ and $r_m(x,y) = \frac{x}{0.15+x+y}$. For these functions the probability of finding a new partner increases as the number of deserting

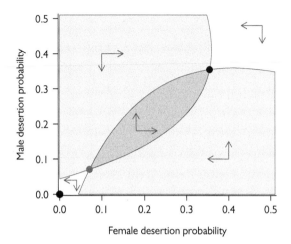

Fig. 9.2 Selection on the probability of desertion for members of a breeding pair that have produced young, depending on the resident strategy. Arrows indicate the direction of selection on each trait. Solid circles indicate singular points. Care parameters: $V_0 = 0$, $V_1 = 1$, $V_2 = 1.25$.

members of the opposite sex increases and decreases as the number of the deserting members of the same sex increases. The probability also increases with total numbers deserting for fixed ratio x/y. This latter property captures the idea that partners will be difficult to find when rare, even if there is no competition from members of the same sex. For the case illustrated, biparental care by all population members is an ESS. There is also a mixed-strategy ESS at which both males and females care with probability $x^* = y^* = 0.3545$.

The existence of a mixed ESS when feedback is allowed is in contrast to the case of fixed V_+ where no mixed ESS can occur. To understand why, consider the Nash equilibrium of Fig. 9.2 at which both sexes have probability $x^* = y^* = 0.3545$ of deserting. Since this is a Nash equilibrium resident males have equal payoffs from caring and deserting. Thus a single mutant male that, say, always deserted, would have the same payoff as a resident male that deserted, regardless of whether V_+ were fixed or feedback were allowed. Now suppose that the number of mutants started to increase by drift. Then for fixed V_+ there would be no selection against mutants since mutant males never affect the payoff of other mutant males. Thus the increase by drift could continue, and the Nash equilibrium strategy cannot be an ESS. In contrast, when there is ecological feedback, as numbers of mutant males increase, the proportion of all males that desert increases. This reduces the chances that a deserting male remates, so that resident males, who have a lower desertion probability than mutant males, do better than mutants. Thus the mutant is selected against. A formal proof of convergence stability of the mixed strategy that specifies that both sexes desert with probability $x^* = y^* = 0.3545$ could be based on the analysis of Box 6.2.

The existence of mixed strategy Nash equilibria is, perhaps, rather academic. We have argued that in nature there are usually underlying state variables, so that when simple models predict a mixed-strategy equilibrium, organisms will instead evolve pure contingent strategies where their decision is based on the value of the state variable. In the case of parental desertion this is liable to be the case. For example, a male may be more likely to desert if he observes a neighbouring female that is receptive. Furthermore, the care/desertion decisions of the two parents might often not be independent, so that one parent may desert in order to prevent the other parent from deserting (Section 8.2). Our simple analysis based on no state variable and independent decisions might then have poor predictive power.

In our model there are at most two broods per breeding season. McNamara et al. (2000) analyse care/desertion decisions over a breeding season when there is time for more breeding attempts. They allow ecological feedback, with the probability of remating dependent on the numbers of individuals of both sexes that are currently single. They show that this feedback can result in a pattern of stable oscillations between different forms of care over the breeding season.

In our simple model we have excluded the possibility of EPCs. When males can obtain EPCs the paternity of other males will be reduced. As Queller (1997) has shown, it is then important to be consistent in models of parental effort, ensuring that every offspring has exactly one mother and one father (the 'Fisher condition'; see also Houston and McNamara, 2006; Kokko and Jennions, 2008; Jennions and Fromhage, 2017). For the analysis of a general care/desert model that allows EPCs and corrects the mistakes of others, see Fromhage et al. (2007).

9.4 Dependence on Future Actions

Suppose that males in a population must compete with each other for access to females. How aggressive do we expect two males to be towards one another if they contest for the same female? We might expect that as the level of aggression of a male increases his probability of winning the contest and his risk of death or serious injury both increase. His optimal level of aggression will then depend on his future prospects if he comes out of the current encounter unscathed. However, future prospects depend on the number of future females that are not contested and the level of aggression of the male and his opponents in future contests. To analyse this situation we develop a simple model based on that of Houston and McNamara (1991).

9.4.1 Model of Repeated Contests

We suppose that each male in the population encounters a sequence of receptive females over his lifetime. If a male encounters a female that is not contested by another male he mates with her. A proportion θ of the females that a male encounters are contested by another randomly selected male. When this happens they play the Hawk–Dove game against one another. The winner of this contest mates with the female. If the two males both play Hawk, so that a fight occurs, the loser dies with probability z. If a male survives a contest he may still die from other causes; after each encounter with a female the probability he goes on to encounter at least one more female is r. A strategy is specified by a probability p; a male following strategy p plays Hawk with this probability in each contest. The payoff to a strategy is the expected lifetime number of matings of a male following the strategy. We assume that this payoff is proportional to mean LRS. Then, assuming that all offspring are the same, so that mean LRS is a fitness proxy (Section 10.6), our payoff is also a fitness proxy.

Consider a mutant male with strategy p' that has just encountered his first female in a population with resident strategy p. On encounter with this female the probability that he mates with her and the probability he survives until the next encounter are functions $B(p',p)$ and $S(p',p)$ of his strategy and the resident strategy (and the environmental parameters). Exercise 9.2 spells out the exact dependence. Given he survives until the next encounter, the future number of matings of the male is exactly the same as when he encountered his first female. Thus the expected lifetime number of matings of the male, $W(p',p)$, satisfies $W(p',p) = B(p',p) + S(p',p) W(p',p)$, so that

$$W(p',p) = \frac{B(p',p)}{1 - S(p',p)}. \tag{9.1}$$

This is the payoff function for this game.

First suppose that the probability of death in a Hawk–Hawk fight, z, is small or that the probability of surviving until the next encounter, r, is small. It can be shown that there is then a single convergence-stable ESS (a CSS) (cf. Houston and McNamara, 1991). When θ is low there are enough uncontested females so that it is not worth the risk of always fighting. Consequently, the probability of playing Hawk is less than 1 at the CSS. In contrast when θ is close to 1 it is worth fighting for every female, so that the CSS is to play Hawk with probability 1.

When the probability of death in a fight and the probability of surviving between encounters are larger, there can be two CSSs for some values of θ (cf. Houston and McNamara, 1991). Figure 9.3 illustrates the best response function and the direction of change of the resident trait value under adaptive dynamics in such a case. As can be seen, at one CSS the probability of playing Hawk is less than one, while males always play Hawk at the other CSS. The existence of two CSSs can be understood as follows. When the resident probability of playing Hawk lies in the range $p < \frac{5}{8}$, it is best for a mutant to play Hawk since the risk that the opponent will fight is low. When $\frac{5}{8} < p < \frac{7}{8}$ the risk is higher. Furthermore, some opponents will play Dove so that a mutant that plays Dove is liable to still mate with some contested females. Thus it is best to always play Dove. When the resident strategy is $p = \frac{5}{8}$ both of the actions Hawk and Dove are equally good and any strategy does as well as the resident strategy. Thus $p = \frac{5}{8}$ is a Nash equilibrium. It is also a convergence-stable point. However, when the resident probability of playing Hawk is sufficiently large ($p > \frac{7}{8}$), very few females are either uncontested or contested by an opponent that plays Dove. Thus, unless a male is aggressive he is liable to die before he can mate with another female. The best response is therefore to always play Hawk. There is therefore selection pressure to increase p and the population evolves to $p = 1$.

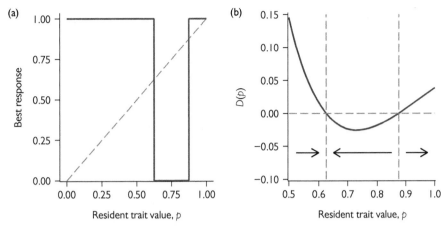

Fig. 9.3 Males encounter a sequence of females, some of which are contested. (a) The optimal probability of playing Hawk for a mutant male when the resident strategy is to play Hawk with probability p. (b) The function $D(p) = \frac{\partial W}{\partial p'}(p,p)$. Arrows show the direction of evolution under adaptive dynamics. Parameters: probability female is contested, $\theta = 0.8$; probability loser of a Hawk–Hawk fight dies, $z = \frac{4}{7}$; probability of survival between encounters, $r = \frac{8}{9}$. For these parameters there are Nash equilibria at $p = \frac{5}{8}$, $p = \frac{7}{8}$, and $p = 1.0$. The Nash equilibria at $p = \frac{5}{8}$ and $p = 1.0$ are convergence stable and are ESSs (Exercise 9.2), and are hence CSSs (Section 4.2). The Nash equilibrium at $p = \frac{7}{8}$ is neither convergence stable nor an ESS (Exercise 9.2). In both panels the vertical lines are at $p = \frac{5}{8}$ and $p = \frac{7}{8}$.

We note that in specifying the above model, the probability a female is contested is a given exogenous parameter. It would be more realistic to allow the parameter to depend on the density of mature males, which would depend on their mortality rate and hence on the resident strategy. However, for simplicity of exposition we will continue to make the simplifying assumption that θ is exogenously defined in what follows.

9.4.2 Reproductive Value and the Emergence of Costs

In Section 3.5 we analysed a single round of the Hawk–Dove game by first specifying payoffs in terms of the value, V, of the resource and the cost, C, of losing a fight. In the above model of repeated contests a male plays a sequence of rounds, each against a different opponent. Rather than specifying V and C our measure of performance for a male is his mean lifetime number of matings, and our analysis of evolutionary stability is based on this payoff, or could equivalently have been based on his mean LRS. As we now illustrate, the concept of reproductive value provides a link between the standard analysis of what happens in a single round of the Hawk–Dove game, or more generally a game that forms part of an organism's life, and our whole-life perspective.

LRS measures the total number of offspring (discounted by relatedness) produced over the lifetime of an individual that survives until their first annual census time at age 0. We can extend this concept by looking at reproductive success in just a part of the life of an individual. Specifically, define the expected future reproductive success (EFRS) of an individual as the mean number of surviving offspring (discounted by relatedness) produced by the individual in the future. Then mean LRS is the EFRS of an age 0 individual.

Assume that mean LRS is a fitness proxy and focus on some resident strategy x. Residents are typically in a range of different states during their lives. For example, they may have different energy reserves or social status. This means that they have different potentials to leave offspring in the future, and so have different EFRSs. It is usual to refer to the function that specifies how the EFRS of residents depends on their state as the reproductive value function. We let $V(s)$ denote the EFRS of a resident in state s; i.e. the reproductive value of state s. We now illustrate the use of this function in the repeated Hawk–Dove game. Although we are here concerned with a situation in which mean LRS is a fitness proxy, as we describe in Section 10.3, the idea of reproductive value is more general and can be applied more generally.

Suppose males play Hawk with probability p under the resident strategy in the repeated Hawk–Dove game. Resident males are in one of two states: they have either just encountered a female (call this state s_{enc}) or have just finished an encounter and survived any fight during that encounter (call this state s_{after}). Assume that each mating results in $2K$ surviving offspring on average. Thus, discounting for relatedness, a male leaves K descendants from a mating. (We note that K depends on the resident strategy if density dependence acts through survival of the young—see Exercise 9.3). Then the reproductive value of a resident male that encounters a random female is $V(s_{enc}) = W(p,p)K$. Similarly, if a male has just survived an encounter with a female his reproductive value is $V(s_{after}) = rW(p,p)K$ where r is the probability he survives to meet the next female.

We now ask if it is worthwhile for a resident male to deviate from the resident strategy in a single contest, given that he will then revert to the resident strategy in future contests. Suppose that our focal male has just encountered a female who is contested by another resident male. If he plays Hawk with probability p' in the current contest his expected reproductive success in the current encounter is easily shown to be $0.5[1 + p' - p]K$. The probability he survives the encounter is $(1 - 0.5p'pz)$. If he does survive his EFRS (his reproductive value) is just $V(s_{after})$ as he will behave as a resident from then on. His EFRS is thus $\tilde{W}(p',p) = 0.5(1 + p' - p)K + (1 - 0.5p'pz)V(s_{after})$. We can write this as

$$\tilde{W}(p',p) = F(p) + \frac{1}{2}[V - pC(p)]p', \tag{9.2}$$

where $F(p)$ is a term that does not depend on p', $V = K$ and $C(p) = zV(s_{after})$. This expression is in the form of the standard payoff for a single Hawk–Dove game with benefit V and cost C (Section 3.5). $V = K$ is the value of mating with the female (in terms of offspring). The cost of losing a fight, $C(p) = zV(s_{after})$, is the probability of death times the future loss as a result of death. The difference $\frac{1}{2}[V - pC(p)]$ is the advantage of playing Hawk over playing Dove in the current contest, given the male reverts to playing Hawk with probability p in future contests.

Let the resident strategy be $p = p^*$ where p^* satisfies the equation $V - p^*C(p^*) = 0$, so that $p^* = \frac{V}{C(p^*)}$. From eq (9.2) we see that the EFRS of our focal male does not depend on his probability of playing Hawk. In particular, he cannot do better than play Hawk with the resident probability p^* in the current contest given that he will play Hawk with this probability in all future contests. Standard results from the theory of Markov decision processes (Puterman, 2005) then imply that no strategy can do better than to play Hawk with probability p^* in all contests. In other words p^* is a Nash equilibrium strategy. Conversely, if $p < 1$ and $V - pC(p) \neq 0$ then there exists a $p' \neq p$ such that our focal male can increase his LRS playing Hawk with probability p' in all contests, so that p is not a Nash equilibrium strategy. It can similarly be shown that if $V > C(1)$ then the strategy $p = 1$ is the unique best response to itself and hence an ESS.

Figure 9.4 shows the zeros of the function $V - pC(p)$ for the case illustrated in Fig. 9.3. One zero is at the ESS at $p = \frac{5}{8}$. In other words, the Nash equilibrium in the current model at $p = \frac{5}{8}$ is consistent with the analysis of Section 3.5 for a single round of the Hawk–Dove game when we take the cost of losing a fight to be the resultant loss in future reproductive success. A similar conclusion holds for the Nash equilibrium at $p = 1$; since $V > C(1)$ the analysis of a single round predicts a Nash equilibrium at $p = 1$. Finally, we note that our current model also predicted a Nash equilibrium at $p = \frac{7}{8}$. Again we have $V - pC(p) = 0$ for this p. Note, however, that it can be shown that $p = \frac{7}{8}$ is not an ESS in our current model. In contrast the analysis of a single round with fixed costs and benefits shows that any Nash equilibrium is also an ESS.

The moral of this example is that the costs of a single encounter should not be assumed in advance. Instead any competitive encounter should be placed in the

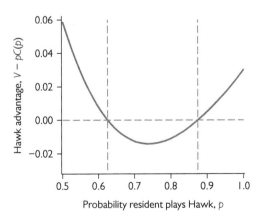

Fig. 9.4 The advantage of playing Hawk over playing Dove, $V - pC(p)$, for the case illustrated in Fig. 9.3. For simplicity we have taken $K = 1$ so that a mating results in two surviving offspring on average.

context of the rest of the life of the individual within a given resident population. To do so, the model should specify the consequences (in terms of matings and survival) of actions. This allows one to deal directly with LRS as the payoff to an individual. The situation is then analysed in terms of this measure. Having done so, it is possible, if one wishes, to reinterpret the analysis in terms of what happens in a single competitive encounter in terms of suitably defined benefits and costs, but here the cost of fighting emerges from the analysis rather than being specified in advance.

We note, however, that even though we have specified consequences in the above model it may still not be fully consistent. That is because we have not specified how density dependence acts in the population. When including density dependence the manner in which it acts can be crucial to predictions (cf. Mylius and Diekmann, 1995). If in the above model density dependence acts through survival of juveniles, then our model is consistent for every value of the survival parameter r (provided the population is viable). Suppose instead that the number of surviving offspring per mating is density independent and is fixed, and density dependence acts through adjustment of r. Then for each resident strategy the population is only at a constant size for one value of r, so that r emerges from the model and cannot be taken as an input parameter.

9.4.3 Assessment and Plastic Adjustment

The above analysis assumes that a strategy is determined by the probability that an individual plays Hawk in a contest. At a CSS population members adopt a Nash equilibrium strategy appropriate to the actual values of parameters such as the frequency with which females are contested (θ). In practice we might expect parameters to vary over time, either within a generation or across generations. We would then expect population members to follow some learning rule that bases their current level of aggression on their previous experience. From this new perspective it is the link between experience and behaviour that is under selection. Given a specification of how an environment varies over time one could, in principle, seek an evolutionarily stable learning rule. Such a rule would do best given other population members also adopted the rule. Here we have more limited aims. We consider a plausible, but not necessarily optimal, adjustment rule and examine the consequences of following the rule. We consider a situation in which there are two alternative CSSs in the original model, so that which CSS is reached depends on the evolutionary history. We illustrate that when there is plasticity, the adjustment of behaviour over the lifetimes of individuals can lead the population to become locked into one of two alternative quasi-stable demographic states corresponding to the CSSs, where the state reached depends on recent experience.

We consider the case illustrated in Fig. 9.5. In this case z and r are held constant and we assume behaviour is adapted to these values. We assume that the probability that a female is contested, θ, varies over time as a result of external factors. Population members form estimates of the current values of θ and the probability, p, that an opponent will play Hawk in a contest. Each individual then adjusts its behaviour to be the best given that these estimates represent true values. Although the behaviour of individuals reflects the risk of mortality, for simplicity there are no actual births or deaths in our simulation. This might roughly capture a population with births and deaths where the estimates of a parent were passed on to its offspring epigenetically.

In Fig. 9.5a θ first increases from 0.96 to 1 over the first 5000 encounters of a male with females. θ then decreases from 1 to 0.9 over the next 12,000 encounters. Initially 40% of males play Hawk in an encounter with another male. As θ increases this percentage increases and then suddenly jumps to around 100% when $\theta \approx 0.988$. The percentage remains at around 100% until θ decreases to below $\theta \approx 0.927$ when there is a transitory drop in the percentage to zero, after which the population stabilizes with around 35% of population members playing Hawk. It can therefore be seen that the percentage playing Hawk depends on both the current value of θ and recent history, so that there is hysteresis.

These effects can be understood from Fig. 9.5b. For $\theta > 0.9958$ $p^* = 1$ is the unique ESS and this is also a CSS. For $0.9 \leq \theta < 0.9958$ there are three ESSs $p_1^* < p_2^* < p_3^* = 1$. Of these p_1^* and p_3^* are CSSs and p_2^* is not convergence stable. Initially, the population roughly tracks the CSS at p_1^* as θ increases, although there is always considerable between-individual variation in estimates for θ and p. As the critical value of $\theta = 0.9958$ is approached some population members estimate θ to be above this value and switch to playing Hawk all the time. This increases the estimate of p

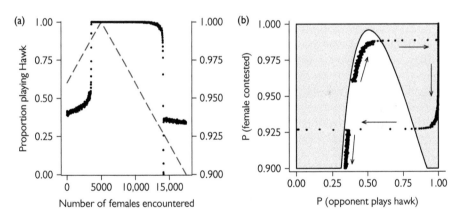

Fig. 9.5 Plastic adjustment in the repeated Hawk–Dove game. In (a) the dashed line shows the dependence of θ on the number of encounters of each male with a female (right-hand scale). In both (a) and (b) the dots indicate the proportion of the population playing Hawk in a competitive encounter (averaged over the last 10 encounters); in (a) this is given as a function of the encounter number, in (b) the combination of θ and this proportion is plotted, with arrows indicating the direction of time. The shaded region in (b) shows those combinations of θ and p values for which it is optimal to play Hawk. Other parameters $r = 0.96, z = 0.325$. Estimates of θ are updated according to $\hat{\theta}' = (1 - \beta)\hat{\theta} + \beta Y$ where $Y = 0$ if a female is uncontested and $Y = 1$ if the female is contested. Estimates of p are similar. The simulation is based on 1000 individuals that use the learning parameter $\beta = 0.02$.

for other population members and all switch to playing Hawk, so that the population rapidly switches to the CSS at $p_3^* = 1$. The population then tracks this value until the boundary at p_2^* is approached. The estimates of some population members crosses this boundary and they consequently play Dove. This then pulls down the estimate for p for others and results in almost all population members playing Dove for a short while, after which the population again tracks the CSS at p_1^*.

9.5 Territorial Defence and the Desperado Effect

Many animals need to first obtain a territory before they can reproduce. If territories are in short supply we can then expect competition and possibly fighting over their possession. But how aggressive do we expect an animal to be when contesting a territory with another individual? We begin by discussing the stylized case of contests modelled as Hawk–Dove games. As we saw, when there is a role asymmetry (Section 6.2), at evolutionary stability the probability of using Hawk may differ between individuals who possess territories and those who do not (floaters) and are intruding onto another's territory. In particular, the strategy Bourgeois in which owners play Hawk and intruders play Dove is an ESS when the value of possessing a territory V is less than the cost of losing a fight C. At this ESS there is respect for prior ownership. A second 'paradoxical' ESS at which owners play Dove and intruders play Hawk also exists. At this ESS territories change hands at every intrusion.

The Bourgeois and the paradoxical strategy are only ESSs when $V < C$. However, as Grafen (1987) pointed out, V and C are liable to depend on the resident strategy. V will be higher when ownership is respected, as individuals tend to keep territories longer. C is the loss in EFRS as a result of losing a fight, which will depend on whether the individual is an owner or floater and how easy it is to obtain a territory in the future. Grafen concludes that if territories last for some time, reproduction requires a territory, and vacant territories are rare, then respect for ownership means that V is high and the EFRS of floaters, and hence their C value, is low. Thus we expect $V > C$ for floaters, so they should play Hawk and risk death in a fight since there is little to lose—what Grafen calls the desperado effect. Thus Bourgeois cannot be an ESS.

The analysis of Grafen (1987) allows V and C to emerge. It nevertheless takes the rate at which floaters find free territories, which we here call v, as a given parameter. However, v will depend on the numbers of floaters, which depends on the mortality rate of floaters, and hence on how often floaters fight for territories. Thus v depends on the resident strategy and should also emerge from a more holistic model of what is happening at the population level. Eshel and Sansone (1995) present such a model. They show that if v and certain other parameters are held fixed then there are no mixed-strategy ESSs. In contrast, when v emerges there may exist a mixed-strategy ESS in which owners always play Hawk and intruders play Hawk with a probability that is strictly between 0 and 1. At this ESS intruders are indifferent between playing Hawk and Dove. To see why ecological feedback stabilizes the proportion of intruders playing Hawk, consider first what would happen if this proportion were to increase. There would be more intruder deaths resulting in a smaller number of intruders and hence more empty territories and a larger v. It would then be best for an intruder to play Dove at each already occupied territory and wait for a vacant territory. Conversely if the proportion of intruders fighting were reduced, there would be a bigger intruder population and hence fewer available territories and a smaller v, so that it would be best for an intruder to always fight if the territory were already occupied. In an analysis of a similar model, Kokko et al. (2006) noted that this type of negative feedback does not operate on the level of aggression of owners.

9.5.1 Owner–Intruder Contests with Assessment and Learning

Assessment of fighting ability is likely to be important in most animal contests (Section 3.5), including those between territory owners and intruders. We can allow for this feature by using the assessment game (Sections 3.12, 5.4, 8.6) as a building block, modelling owner–intruder contests as a sequence of assessment games, in a similar way as in Section 8.7. We assume that a contest continues as long as both individuals use the aggressive behaviour A and that it ends at the first use of submissive behaviour S by one or both of them, or if one or both are killed. An individual using S leaves the territory and becomes a floater, which means that aggressive behaviour is needed to succeed in competing for territories. There are N females and N males in the population and K territories.

A season lasting one generation is divided into discrete time periods of unit duration. The sequence of events in each time period is: (i) a male owning a territory has an expected reproductive success V, with $0 < V \leq 1$, whereas floaters have no reproductive success; (ii) floater males search for territories, finding a random territory with probability v; (iii) if there is an owner at the territory, or if more than one floater arrives, there is a territorial contest, with a risk of mortality and with the winner becoming the owner; and (iv) males suffer mortality, with rate d_i for male i, outside the context of contests. For simplicity, we assume that there is at most one contest per territory and time period, so any additional arriving floaters must continue searching. Males that are present on a territory at the end of a time period are owners in the next time period. Note also that in this model V is an owner's expected reproductive success per time period. Reproduction is implemented such that the owner has a probability V of mating with a random female, which produces one offspring The sequence continues until $2N$ offspring have been produced, N females and N males, with diploid genetics. Males differ in quality (fighting ability) q. At the start of a generation, all males are floaters and a male's quality is random and drawn from a normal distribution with mean μ_q and standard deviation σ_q. In a round of a territorial contest, in addition to the observation of relative quality, $\xi_{ij}(t) = b_1 q_i - b_2 q_j + \epsilon_i(t)$, as given in eq (5.15), a male only distinguishes whether he is an intruder or an owner. Table 9.1 gives the perceived rewards of the actions A and S for these states.

Males use these rewards in actor–critic learning, as detailed in Box 9.1. The perceived values v_{ik} in Table 9.1 are genetically determined and can evolve. The starting values ($t = 0$) of a number of other learning parameters from Box 9.1 are also genetically determined ($w_{i00}, w_{i10}, g_{i0}, \theta_{i00}, \theta_{i10}, \gamma_{i0}$), thus allowing the aggressive

Table 9.1 Perceived reward of aggressive (A) and submissive (S) actions in the assessment game, when individual i with quality q_i meets j with q_j, where individual i is either an intruder ($k = 0$) or an owner ($k = 1$). The parameter v_{ik} is the perceived reward of performing A as either an intruder or an owner ($k = 0, 1$).

Perceived reward R_{ik}		Act by j	
		A	S
Act by i	A	$v_{ik} - e^{-q_i + q_j}$	v_{ik}
	S	0	0

Box 9.1 Territorial contests with learning

In each round t of the contest, with t_0 being the first round, the two individuals i and j play the assessment game with two actions, A and S. The individuals make observations, ξ_{ij} for i and ξ_{ji} for j. Each individual also knows whether it is an intruder ($k = 0$) or an owner ($k = 1$), but they do not observe the identity or ownership status of their opponent. Individual i chooses its first action A with probability

$$p_{ikt}(\xi_{ij}) = \frac{1}{1 + \exp\left(-(\theta_{it} + \theta_{ikt} + \gamma_{it}\xi_{ij})\right)}, \tag{9.3}$$

and has an estimated value

$$\hat{w}_{ikt} = w_{it} + w_{ikt} + g_{it}\xi_{ij}, \tag{9.4}$$

at the start of the interaction. From this and the reward R_{ikt}, we get

$$\delta_{ikt} = R_{ikt} - \hat{w}_{ikt}. \tag{9.5}$$

This is used to update the learning parameters w_{it}, w_{ikt}, and g_{it}:

$$w_{i,t+1} = w_{it} + \alpha_w \delta_{ikt} \tag{9.6}$$
$$w_{ik,t+1} = w_{ikt} + \alpha_w \delta_{ikt}$$
$$g_{i,t+1} = g_{it} + \alpha_w \delta_{ikt}\xi_{ij}$$

with $k = 0$ or $k = 1$. For the actions we have the eligibilities

$$\zeta_{i\cdot} = \frac{\partial \log \Pr(\text{action})}{\partial \theta_{i\cdot}} = \begin{cases} 1 - p_{ikt} & \text{if first action} \\ -p_{ikt} & \text{if second action} \end{cases} \tag{9.7}$$

$$\eta_{ikt} = \frac{\partial \log \Pr(\text{action})}{\partial \gamma_{it}} = \begin{cases} (1 - p_{ikt})\xi_{ij} & \text{if first action} \\ -p_{ikt}\xi_{ij} & \text{if second action} \end{cases}$$

with $k = 0$ or $k = 1$ and $\zeta_{i\cdot}$ denotes ζ_{it} or ζ_{ikt}, and the updates

$$\theta_{i,t+1} = \theta_{it} + \alpha_\theta \delta_{ikt}\zeta_{it} \tag{9.8}$$
$$\theta_{ik,t+1} = \theta_{ikt} + \alpha_\theta \delta_{ikt}\zeta_{ikt}$$
$$\gamma_{i,t+1} = \gamma_{it} + \alpha_\theta \delta_{ijt}\eta_{ikt}.$$

If an individual is involved in more than one contest during its lifetime, we assume that it keeps updating the learning parameters, continuing from any previous learning from rounds before t_0. The value of k, either 0 or 1, determines if the parameters w_{ikt} and θ_{ikt} are updated as above in a given contest or are left unchanged.

behaviour of owners and intruders to evolve. The initial preferences θ_{ik0} for the action A determine how aggressive a male is as intruder ($k = 0$) and owner ($k = 1$) at the start of a generation. The individual-specific learning parameters θ_{it} and w_{it} are assumed to be zero at the start of a generation and allow a male to adjust his behaviour to his own quality q_i, either as intruder or as owner.

Concerning fitness payoffs, in addition of the benefits for territory owners of having a chance to mate, there are mortality costs both in and outside of contests. A male suffers a rate of mortality per round of AA interaction of

$$d_{iAA} = \varphi e^{-q_i + q_j} \tag{9.9}$$

and also accumulates damage increments $e^{-q_i + q_j}$ per round of AA interaction to a summed damage D_i from injuries or exhaustion. This results in per period rates of mortality

$$d_i = \mu + cD_i \tag{9.10}$$

outside of contests. The parameters V, v, φ, μ, and c, can be chosen to represent different ways in which owner–intruder interactions depend on the population processes that give rise to differences in reproductive success.

Figure 9.6 shows two examples of this. For the case in panels (a) and (b) of this figure, contests with escalated AA rounds are dangerous, as given in eq (9.9) with $\varphi = 0.01$. Floaters also suffer noticeable mortality while searching, with $\mu = 0.01$ and $v = 0.1$, needing on average 10 periods of searching to locate a territory. A consequence is that owners win only slightly more than half of their contests when they meet a matched intruder (Fig. 9.6a). Also, a male on average experiences only around 0.8 contests, but these tend to be dangerous, lasting on average 12 rounds and for nearly matched opponents lasting considerably longer (Fig. 9.6b), the latter often ending with one male dying. Because of the low number of contests per male, there is little opportunity for stronger individuals to accumulate as owners. This, together with the mortality risks for a floater to locate an additional territory, is the reason for the limited owner advantage.

For the contrasting case in panels (c) and (d) of the figure, escalated rounds are not very dangerous in terms of direct mortality ($\varphi = 0.001$) but accumulated contest damage somewhat increases the rate of mortality outside of contests ($\mu = 0.002$, $c = 0.002$). Floaters readily find territories ($v = 0.5$) resulting in on average five contests per male and, as a consequence, there is a pronounced owner advantage, with owners winning 99% of all contests, the only exception being when intruders have much higher fighting ability (Fig. 9.6c). Contests are typically short, lasting on average about four rounds, and sometimes the intruder uses submissive behaviour already in the first round of a contest. However, when intruders have considerably higher fighting ability contests are sometimes long (Fig. 9.6d). Altogether, one reason why owners typically win is that there are enough contests per male for stronger males to accumulate as owners and another is that an intruder readily can find another territory.

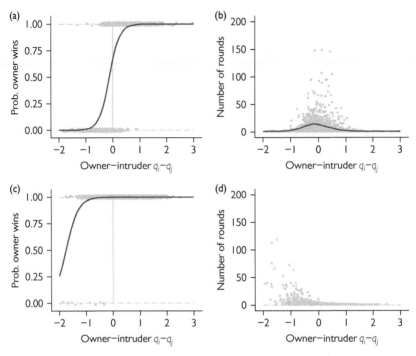

Fig. 9.6 Outcomes of owner–intruder interactions for two cases of population processes. There are $N = 2000$ each of males and females and $K = 1000$ territories. Learning follows Box 9.1 with perceived rewards from Table 9.1. The learning parameters are approximately at an evolutionary equilibrium. Panels (a) and (b) show the probability of the owner winning and the duration of contests plotted against the owner–intruder quality difference, $q_i - q_j$, for a case with substantial direct mortality costs both in and outside of contests and no accumulation of damage ($\varphi = 0.01$, $\nu = 0.1$, $\mu = 0.02$, $c = 0$). The points show individual contests. The dark line in (a) shows a logistic regression fit to the points and the line in (b) shows a kernel regression (using method npreg in R). Panels (c) and (d) illustrate the same for a case where direct mortality rates are low, floaters readily find territories and with accumulation of fighting damage ($\varphi = 0.001$, $\nu = 0.5$, $\mu = 0.002$, $c = 0.002$). Other parameters: payoff, $V = 0.1$; distribution and observation of fighting ability, $\mu_q = 0$, $\sigma_q = 0.5$, $b_1 = b_2 = 1.0$, $\sigma = 0.2$; learning rates, $\alpha_w = 0.10$, $\alpha_\theta = 0.50$; approximate evolved values panels (a, b), $w_{i00} = 3.50$, $w_{i10} = 3.80$, $g_{i0} = 0.10$, $\theta_{i00} = 3.60$, $\theta_{i10} = 5.00$, $\gamma_{i0} = 3.80$, $v_0 = 1.60$, $v_1 = 1.50$; panels (c, d), $w_{i00} = -0.12$, $w_{i10} = 0.60$, $g_{i0} = 1.10$, $\theta_{i00} = 0.60$, $\theta_{i10} = 11.0$, $\gamma_{i0} = 2.80$, $v_0 = 0.12$, $v_1 = 0.60$.

These examples illustrate how contests can be embedded into processes that determine the fitness effects of finding, maintaining, and utilizing territories. The only part of these effects that is contained strictly within a contest is the risk of fatal injury in a fight; the other effects derive from processes outside of the contest. Depending on the details of these processes, the evolutionary outcome for owner–intruder contests can range from rather dangerous fights with limited owner advantage (Fig. 9.6a, b), in which intruders might resemble desperadoes, to a very pronounced owner advantage (Fig. 9.6c, d), resembling somewhat the Bourgeois ESS for the Hawk–Dove game.

Concerning real owner–intruder interactions, a number of papers have surveyed both field observations and experiments (e.g. Leimar and Enquist, 1984; Kokko et al., 2006; Sherratt and Mesterton-Gibbons, 2015), finding that owner advantages in winning contests are widespread but not universal. There are fewer instances of fieldwork investigating the population processes underlying fitness effects. One example is the work by Austad (1983) on male combat in the bowl and doily spider, estimating male floater mortality and chance of finding a web with a female. To some extent this situation resembles the illustration in Fig. 9.6a, b, and the data were used by Leimar et al. (1991) to model these spider contests as a sequential assessment game with risk of fatal injury. Another example is the work by Hammerstein and Riechert (1988), investigating the processes having fitness effects on contests between female desert grass spiders over sites for webs. This type of work is challenging and in general rather little is known about the life of floaters. It could be that floaters often face fairly high rates of mortality, in a similar way as is often the case for dispersing individuals (Bonte et al., 2012).

9.6 State-dependent Ideal Free Distributions

When foraging does not incur any risk from predation or other dangers we might expect a foraging animal to choose the food source that maximizes its net rate of energetic gain (Stephens and Krebs, 1986). Typically, however, the rate of intake at a food source will decrease with the number of other populations members that have also chosen this source, either because of direct competition over food items, or because others deplete the source. Thus the choice of food source of each population member can be expected to depend on the choice made by others. If animals in the population are well informed about the current rate of intake at each possible source and can switch freely between sources at no cost, we might expect individuals to distribute themselves across sources so that no individual can increase its intake rate by switching to another source. Such a distribution is referred to as an ideal free distribution (Fretwell and Lucas, 1970).

The situation is more complex if different food sources have different predation risks since we can no longer use energy gain as our performance criterion, but must find some way of incorporating both energy gain and predation risk within the same currency. Reproductive value, that is the function that measures future reproductive success, provides this common currency (McNamara and Houston, 1986). Since the immediate effect of gaining food is to change the energy reserves of the animal, the dependence of reproductive value on reserves is central to understanding the food–predation trade-off. However, since current reproductive value depends on future actions, the best current actions depend on future actions at all possible future levels of reserves of the animal. We illustrate these concepts in a specific setting that is motivated by the model of McNamara and Houston (1990).

9.6.1 Survival Over the Winter

A winter starts at time $t = 0$ and ends at time $t = T$. At time $t = 0$ there is some given distribution of energy reserves over members of a large population. At each of the times $t = 0, 1, \ldots, T - 1$ each population member must choose whether to forage in habitat 1 or habitat 2. Having chosen a habitat an animal forages in that habitat for one unit of time, before making a choice again. An animal foraging in habitat i gains a food item of energetic value e_i with probability p_i and gains no food with probability $1 - p_i$. Energy gained from food is stored as fat reserves. Here p_i is a function of the current number of other animals that have also chosen to forage in habitat i. (More realistically we might allow p_i to depend on both the current and past numbers of animals in the habitat, but for simplicity we ignore this complication here.) Similarly, the probability that the focal animal is not killed by a predator, r_i, depends on the number of other animals present, but otherwise what happens to one animal in the time unit is independent of what happens to other animals. p_i and r_i may also explicitly depend on the time in winter. During unit time each animal expends 1 unit of energy reserves. An animal dies of starvation if its reserves ever drop to 0. If an animal survives until the end of winter at time T its reproductive value is a function $R(s_T)$ of its reserves s_T at this time.

For this model a strategy specifies how the choice of habitat depends on current energy reserves and time in winter. Each population member is attempting to maximize its EFRS (measured at the start of winter). This scenario is a game since the best action of any individual depends on food availability and predation risk, but these quantities depend on the strategy adopted by the resident population. An ESS for the game can be found numerically by using the following two approaches.

Best response of a mutant. The procedure to find the best response of a mutant to a given resident strategy has two phases, as we now describe.

Forward iteration. We first follow the resident population forwards in time as follows. We have assumed that the distribution of reserves over resident members is given at time $t = 0$. Since the strategy specifies how the choice of habitat at time $t = 0$ depends on reserves, this allows us to determine the numbers choosing each habitat at this time. These numbers then determine the probability that food is found and predation risk in each habitat. This allows us to compute the numbers surviving until time $t = 1$ and the distribution of reserves over those that do survive. We can then use exactly the same procedure to calculate the numbers surviving until time $t = 2$ and the distribution of reserves over survivors, and so on. In this way, we find the probability that an individual finds food and the probability of avoiding predation in each habitat at each time in winter. These probabilities apply to both resident population members and any mutant (provided the population is large) and set the ecological background. This then allows us to find the optimal strategy of a mutant as follows.

Backward iteration. Consider a single mutant within this resident population. We can find the optimal strategy of the mutant by the technique of dynamic programming

(Houston and McNamara, 1999). The logic behind the technique is as follows. The reproductive value at time T is known; it is zero if the animal is dead and specified by the function R otherwise. Consider a mutant individual at time $T - 1$. There are three possible outcomes for mutant at time T: (i) it has died due to predation or starvation, (ii) it is alive but has found no food, and (iii) it is alive and has found food. The probabilities of these outcomes depend on the action of the mutant at time $T - 1$. An action at time $T - 1$ thus determines the expected reproductive value of the mutant at time T. The optimal action at time $T - 1$ maximizes this expected reproductive value. Furthermore, the expected reproductive value under this optimal action gives the reproductive value at time $T - 1$ under the assumption that future behaviour over the winter is optimal. This procedure is carried out for every level of reserves at time $T - 1$. The resulting reproductive value at time $T - 1$ can then be used to calculate the optimal action at time $T - 2$ and the reproductive value at this time, at every level of reserves, again assuming future behaviour is optimal. In this way we can work backwards to time $t = 0$, finding the optimal action at each level of reserves at each time. Box 9.2 gives the mathematical details of the scheme we have just described.

Computing an ESS. If the resident strategy is an ESS then the best mutant response will coincide with the resident strategy. One method of attempting to find such as strategy is to iterate the best response to obtain a sequence of strategies. One starts with some arbitrary strategy, finds the best response to that, then finds the best response to this

---

Box 9.2 Working backwards using dynamic programming

Let $V(s,t)$ denote the reproductive value (expected number of future offspring) of an animal with reserves s at time of winter t, given that behaviour from time t until the end of winter is optimal (i.e. maximizes expected future offspring number). For simplicity it will be assumed that the energy contents e_1 and e_2 of items are integers and we restrict attention to non-negative integer levels of reserves.

We are given that

$$V(s, T) = R(s) \text{ for } s \geq 1 \tag{9.11}$$

and have $V(0, T) = 0$.

We express V at time t in terms of V at time $t + 1$ as follows. For t in the range $t = 0, 1, \ldots, T - 1$ and for $s \geq 1$ set

$$H_i(s,t) = r_i(t)\left[(1 - p_i(t))V(s - 1, t + 1) + p_i(t)V(s + e_i - 1, t + 1)\right] \tag{9.12}$$

for each $i = 1, 2$, where action i is to forage for unit time in habitat i. Here $r_i(t)$ and $p_i(t)$ are the probabilities of survival and of finding food in habitat i at time t. Then

$$V(s,t) = \max\{H_1(s,t), H_2(s,t)\}. \tag{9.13}$$

Furthermore, the optimal choice of habitat at this time is the choice that achieves this maximum. Finally we have $V(0, t) = 0$.

These equations allow us to start at time T and work backwards over time until time $t = 0$.

best response, and so on. If this process leads to a strategy that is the unique best response to itself then this strategy is an ESS. The method does not always work, since the sequence of strategies may oscillate between two or more strategies. If this occurs then some form of damping may force convergence (McNamara et al., 1997). Even with damping there may be problems, due to the fact that reserves have usually been computed on a discrete grid, although if oscillations are very small it is usually reasonable to assume an approximate ESS has been computed. The technical fix of allowing errors in decision-making may also help (McNamara et al., 1997).

To illustrate the above concepts we consider an example in which Habitat 1 is safe from predators while Habitat 2 is risky. The probability of finding food in Habitat 1 is a decreasing function of the number of other current users. In Habitat 2, neither the probability of finding food nor the predation risk depend on the number of other animals present. We assume that there is a maximum capacity of 100 units of energy that can be stored as fat.

For this scenario we would expect an animal that is trying to survive the winter should take more risks in terms of predation when reserves are low (McNamara, 1990). Thus a best response strategy will be of threshold form; i.e. there will be a function $d(t)$ such that the best response is to choose Habitat 1 at time t when reserves are above $d(t)$; otherwise choose Habitat 2. Figure 9.7a shows an example in which we have chosen two different resident strategies, each of threshold form. In this example all population members have initial energy reserves of $s = 100$. Thus initially, all population members use Habitat 1 (Fig. 9.7b). This leads to a low mean net gain in this habitat (Fig. 9.7c), so that reserves tend to decrease until many population members are using Habitat 2. The proportion of the surviving population using a habitat at a given time then becomes fairly stable over time (not shown), although the absolute numbers decrease due to mortality, leading to an increase in mean net gain in Habitat 1 (Fig. 9.7c). The best response to the resident strategy depends on food availability: the greater the mean net gain in Habitat 1 the lower the switching threshold (Fig. 9.7d). Figure 9.7d also shows an approximate ESS threshold; it is approximate in the sense that successive iterates of the best response map oscillate around a value that never deviates from this threshold by more than 1 unit.

When all that is important is mean energy gain, at the ideal free distribution individuals in every habitat are doing equally well. This is no longer true when there are state-dependent effects. For instance, at the ESS in this example, all those individuals on Habitat 2 have lower reserves and hence a small probability of survival than individuals in Habitat 1. Moreover, all the mortality (from starvation and predation) occurs in Habitat 2 (McNamara and Houston, 1990).

9.6.2 Further Dynamic Games

A number of other dynamic game models have been developed that have a similar structure to the state-dependent ideal free model we develop above. For example, Houston and McNamara (1987) consider the daily routine of male birds that are

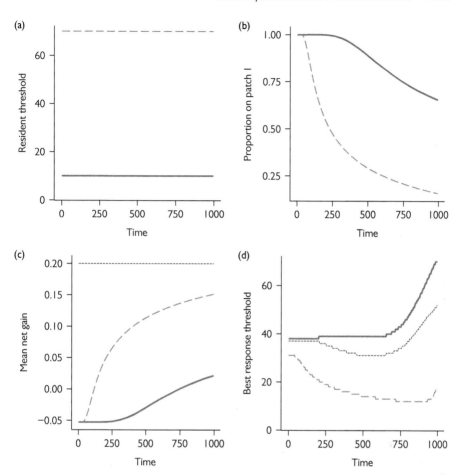

Fig. 9.7 The state-dependent ideal free distribution. Best response strategies for two resident populations are shown. (a) Resident thresholds. (b) The proportions of the original resident populations that are foraging in Habitat 1. (c) Mean net gain rates in Habitat 1, $p_1(t)e_1 - 1$. The constant mean net gain rate in Habitat 2 is also shown (fine dashed line). (d) Best response thresholds. The fine dashed line shows an approximate ESS threshold. In panels (b)–(d) the solid (dashed) curves refer to the resident strategy in (a) indicated by the solid (dashed) curve. Terrminal reward, $R(s) = s$. Parameters: $p_1(\rho) = 4.5/(15 + 4\rho)$ where ρ is the proportion of the original population choosing Habitat 1, $p_2 = 0.3$, $r_1 = 1.0$, $r_2 = 0.9975$, $T = 1000$.

singing to attract a mate, when there is competition between males. Lucas and Howard (1995) and Lucas et al. (1996) model the calling strategies of male anurans that are in competition to attract females. Houston and McNamara (1988) model the daily foraging routine of foraging organisms that compete pairwise for food. Alonzo (2002) and Alonzo et al. (2003) consider coupled state-dependent dynamic games between predators and prey. All of these model are solved by the combination of forwards and backwards iteration outlined above. All are limited in that factors such as food availability are deterministic functions of time of day, allowing strategies to depend on time of day and energy reserves alone. Models would have to be extended to incorporate a real-time response if there were stochastic changes at the population level.

Rands et al. (2003) consider a foraging game in which there is a food versus predation trade-off, but in contrast to the above games, they are concerned with a game between just two individuals, with the focus of the paper on the interaction between them. The importance of this work is that it highlights the spontaneous emergence of leaders and followers. In their model, at any decision epoch each individual can decide whether to forage or rest. If an individual rests its reserves decrease. If it forages food is found stochastically, as in the above model of survival over winter. In terms of predation risk, resting is the safest option, followed by foraging together, with foraging alone incurring the greatest risk. In this model players can die of either predation or starvation (if reserves hit zero) and the objective of each is to maximize their long-term survival probability.

In order to find an ESS strategy Rands et al. (2003) assume that each individual bases its decision on whether to forage on its own reserves and those of its partner. At the ESS, an individual rests if reserves are very high and forages if reserves are very low. For intermediate levels of reserves it forages if the reserves of its partner also specify that the partner should forage, otherwise it too rests. As a consequence, for long periods there are two distinct roles: one individual has low reserves and initiates foraging, the other has high reserves and acts as a follower. The difference in reserves is maintained because the individual with high reserves forages when the low-reserve individual is foraging. In this way the high-reserve individual builds up its reserves to a very high level, at which time it ceases to forage. As a consequence the low-reserve individual is forced to give up foraging while its reserves are still low because of the dangers of foraging alone. Thus the low-reserve individual remains with fairly low reserves. Although the disparity in roles is maintained for extended periods of time, there are occasional switches in roles.

9.7 Is it Worth it?

In this chapter we have explored models in which a game, or a round of a game, occupied just a part of the life of an individual. We have then asked whether it is reasonable to consider the game in isolation, or whether it is necessary to also consider other aspects of the lives of contestants and how the behaviour of population members affects the environment experienced by these contestants. The answer partly depends on the theoretical and empirical questions that are the focus of the modelling. For example, the focus might be on understanding a local empirical phenomenon in a particular situation in a given population. Alternatively, the focus might be on explaining a pattern across populations or species.

We can illustrate approaches to these two different issues by considering a situation where population members are involved in a sequence of contests over resources during their life, as in the model of Section 9.4. If the focus were on a single contest, then it might be possible to estimate the loss of future reproductive success as a result of fighting from empirical data. This would then specify the cost of a fight, and assuming the value of the resource could also be specified, one could model a single contest in isolation, comparing model prediction with data.

There might, however, be a problem with this approach if a strategy required individuals to be flexible. In the model of Section 9.4, a male faced the same situation each time a female was contested, so that an ESS specified the same behaviour in each contest. Instead, each contestant might face a range of situations in its contests, for example due to variation in its own energy reserves, or variation in the size or aggressiveness of opponents. If strategies were necessarily limited in their flexibility, then at evolutionary stability the resident strategy might perform reasonable well on average, but its performance in any one contest might not be the best according to a cost–benefit analysis that considered this contest in isolation (cf. Section 11.2).

In the situation envisaged in Section 9.4, the approach based on isolating a single contest would not reveal that there could be also be an alternative ESS for the same model parameters. Thus if the focus were on patterns across populations, it might be best to take a more holistic view. In general, in order to understand patterns it might be necessary to construct models that include the whole of the lives of all population members, following them forward in time recording births, deaths, and state changes. This is certainly how the idea of invasion fitness in Metz et al. (1992) is formulated. Models might also account for the combined effect of population members on ecological variables such as population density and the food supply. The importance of such an approach, incorporating environmental feedbacks, has been advocated by many, including Levins (1970), Metz et al. (1992), Mylius and Diekmann (1995), Geritz et al. (1997), Dieckmann and Law (1996), Dieckmann and Metz (2006), Meszéna et al. (2006), Fronhofer et al. (2011), Argasinski and Broom (2013, 2018), and Krivan et al. (2018). The formulation of game theory in terms of invasion fitness (Section 2.3) does take the effect of the resident strategy on the environment it creates into account. However, even in that approach there are assumptions about a separation of timescales: it is assumed that a given resident population reaches demographic stability on a timescale that is rapid compared with evolutionary change. Adaptive dynamics (Section 4.2) also makes assumptions about timescales: in its simplest formulation it is assumed that one mutation goes to fixation before the next arises.

In this book we use the concept of invasion fitness, but one can even dispense with the concept of fitness by just following a population forwards in time, as in an evolutionary simulation. This is essentially the approach to game theory promoted in Argasinski and Broom (2013, 2018). Their work formulates equations that specify the rate of change of numbers of individuals that follow each of the possible strategies. In contrast, in the approach we advocate, one first proves that a certain measure is a fitness proxy (LRS in the case of our analyses above), and then analyse evolutionary stability using this proxy. The two approaches will yield the same ESSs if a suitable proxy can be found, and given suitable assumptions about relative timescales.

We believe, however, that in dealing with the maximization of a suitable currency, our approach is usually better at providing understanding.

In general, complex models come at a cost because more must be specified and predictions might be harder to understand. The level of complexity chosen should depend on the questions posed. As all models have their limitations, it is important to understand these limitations and how making a more complex model has the potential to change predictions.

9.8 Exercises

Ex. 9.1. Consider the care versus desert example illustrated in Fig. 9.2. Find all Nash equilibrium strategies (x^*, y^*) for which $x^* = y^*$.

Ex. 9.2. Consider the game of Section 9.4 in which males attempt to mate with a sequence of females over their lifetime. In this game some females are contested by two males and each must play Hawk or Dove. Let the resident strategy be to play Hawk with probability p. Consider a mutant male with strategy p'. (i) Show that on encountering a female he mates with her with probability $B(p', p) = 1 - 0.5\theta(1 + p - p')$. Show that the probability the male survives until the next encounter with a female is $S(p', p) = r(1 - 0.5\theta p' pz)$. Hence use eq (9.1) to write down $W(p', p)$. (ii) Show that $D(p) = \frac{\partial W}{\partial p'}(p, p)$ has the same sign as $\Delta = \theta zrp^2 - zr(2 - \theta)p + 2(1 - r)$. Consider the special case in which $\theta = 0.8$, $r = \frac{8}{9}$, and $z = \frac{4}{7}$. Show that the equation $\Delta(p) = 0$ has roots at $p_1^* = \frac{5}{8}$ and $p_2^* = \frac{7}{8}$, and hence identify which of these two points is convergence stable under adaptive dynamics. (iii) Show that the Nash equilibrium at $p^* = \frac{5}{8}$ is an ESS and that the equilibrium at $p^* = \frac{7}{8}$ is not an ESS. Note that the Nash equilibrium strategy $p^* = 1$ is a unique best response to itself and hence an ESS.

Ex. 9.3. Consider again the game of Section 9.4 in which males attempt to mate with a sequence of females over their lifetime. Assume that each mating by a male results in one male and one female offspring, and that the male survives to maturity and encounters his first female (potential mating) with probability $\frac{1}{2}$. The survival probability between matings, r, is density dependent. Let $\theta = 1$ and $z = 1$. (i) Suppose that the resident population plays Hawk with probability p. Use the fact that this resident population is of stable size to find the relationship between r and p. Hence show that the population is only viable if $p \leq \frac{1}{\sqrt{2}}$. (ii) Show there is a unique convergence-stable Nash equilibrium at $p = (\sqrt{17} - 3)/2$.

Ex. 9.4. An individual must gain a single food item, and has two opportunities to do so on successive days. If it gains an item on day 1 it stops foraging; otherwise if it is still alive and has failed to gain an item it forages again on day 2. A strategy specifies the foraging efforts (u_1, u_2) on these two days. If it expends effort u ($0 \leq u \leq 1$) on day k ($k = 1, 2$) then it immediately dies with probability u, lives and gains an item with probability $(1 - u)up_k$, and lives but fails to gain an item with probability $(1 - u)(1 - up_k)$. We assume that $p_1 = 0.5$. The reproductive value of the individual after the two days is 1 if it is alive and has an item; otherwise it is 0. An optimal strategy (u_1^*, u_2^*) maximizes its expected reproductive value.

(i) Show that $u_2^* = 0.5$. Assuming that behaviour on day 2 is optimal, show that if the individual survives day 1 without obtaining an item its reproductive value is $V_2 = p_2/4$. Let $H_1(u) = (1 - u)up_1 + (1 - u)(1 - up_1)V_2$. Note that this function is maximized at $u = u_1^*$. Show that $u_1^* = \frac{1}{2}\frac{(4 - 3p_2)}{(4 - p_2)}$.

(ii) Now suppose every member of a large population faces the same challenge at the same time, with what happens to one individual independent of what happens to another. Suppose that $p_2 = 1 - \frac{\rho_2}{2}$ where ρ_2 is the proportion of the original population that forages on day 2. By noting that $\rho_2 = (1 - u_1^*)(1 - \frac{u_1^*}{2})$, and employing a suitable numerical technique, show that at evolutionary stability we have $u_1^* = 0.2893$.

10

Structured Populations and Games over Generations

As we have pointed out in previous chapters, individuals often differ in state variables. In many circumstances the state and actions of an individual in one generation affect the state of its offspring. When this is the case, the mean number of surviving offspring produced by an individual (mean LRS) is not a fitness proxy, since the ability of offspring to leave offspring themselves depends on their state and hence depends on the state and action of a parent. This chapter is concerned with game theory in this intergenerational context.

Aspects of quality such as health are states that may be passed on to future generations, but to what extent this occurs depends on behaviour. For example, a high-quality female may have more resources to expend on offspring production and care. She will then tend to produce higher quality offspring than a low-quality female if both produce the same number of offspring. However, if instead she produces more offspring these could potentially be of the same quality as those of a low-quality female. Whether or not an individual has a territory can be another aspect of an individual's state. In the red squirrel, a mother may bequeath her territory to one of her offspring (Price and Boutin, 1993). In group-living animals dominance status can be an important state affecting survival and reproduction. This state sometimes tends to be inherited; for example, in some hyenas high-dominance females behave so as to ensure that their daughters have high status (e.g. Holekamp and Smale, 1993). Many environments are composed of different local habitats linked by the dispersal of juveniles, with individuals maturing and reproducing in a single habitat. The location of an individual is then a state whose change over generations is determined by the dispersal strategy. Other states need not be directly inherited from a parent but are important in influencing the lives of offspring; the sex of an individual is a major example. This state can interact with aspects of quality. So for example, in deciding whether to produce daughters or sons a female may take into account how her phenotypic quality would be passed on to female and male offspring. A reason for this is that the advantages of high quality might differ between the sexes.

When population members differ in state we use a projection matrix to analyse the change over time of the cohort of individuals following a rare mutant strategy (Section 10.1). The elements of the projection matrix specify how the current state of a cohort member determines the number of descendants in each state in the

Game Theory in Biology: Concepts and Frontiers. John M. McNamara and Olof Leimar,
Oxford University Press (2020). © John M. McNamara and Olof Leimar (2020).
DOI: 10.1093/oso/9780198815778.003.00010

following time period, for instance the next year. This then specifies how the cohort changes from one year to the next. Over time the distribution of cohort members over the different states settles down to a steady-state distribution. The per-capita annual growth in cohort numbers also settles down to a limiting value that is the largest eigenvalue of the projection matrix. This growth rate is the invasion fitness for the mutant strategy.

The analysis of projection matrices requires some mathematics and can seem technical. In this chapter we mostly restrict ourselves to the simplest case, where the states needed to describe the population dynamics are also recognized by individuals when they make decisions. This is a small-worlds approach and, in the same way as we have discussed in previous chapters, is often an idealization of real behavioural and developmental mechanisms.

We illustrate projection matrices and invasion fitness with a simple model based on the trade-off between producing many low-quality offspring or fewer high-quality offspring (Section 10.2). In this example, it can be the case that if individuals carrying a rare mutation start in poor states, then the mutant cohort may initially decline in number, but make up for this later as the states of cohort members improve in subsequent generations. This emphasizes that invasion fitness is a measure of the growth rate over many generations.

The quality versus number example also illustrates that in determining the best response of a mutant strategy to a given resident strategy it is not possible to ask what is the best action in one state without consideration of what actions are taken in other states. This is because the future states of the individual and its descendants depend on the current action, and the value of being in each future state depends on the action taken in that state.

Despite the above remark, in determining whether a given resident strategy is evolutionarily stable one can, nevertheless, consider the different possible states in isolation (Section 10.3). In order to do so one first finds the reproductive value function for the resident strategy. This function specifies how the contribution of a resident individual to future generations depends on its current state. Reproductive value then acts as a currency to evaluate actions. If in every state, the action taken under the resident strategy maximizes the mean reproductive value of descendants left at the next annual census time, then the resident strategy is a Nash equilibrium strategy. Conversely, if in some state s there is an action u under which the descendants have a greater reproductive value than under the resident action, then the strategy that agrees with the resident strategy at all states except s and takes action u in state s can invade the resident population. These results provide a way of testing whether a given strategy is evolutionarily stable and provide the basis of a computational method for finding an ESS.

Having developed the technical machinery that can deal with evolutionary stability in structured populations we are at last in a position to justify the fitness proxies used in the sex-allocation problems of Sections 3.8 and 3.11. We do so in Section 10.4, where we also consider the question of whether high-quality females should produce daughters or sons when offspring tend to inherit their mother's quality.

Section 10.5 is similarly concerned with a population in which the sex of individuals is a key state variable. However, rather than allowing the allocation to sons and daughters to evolve when mating is at random, in this section we assume a fixed allocation and consider the evolution of female mating preference. Males are characterized by an inherited trait, such as tail length in the male widowbird. We consider the co-evolution of this trait and the strength of preference of females for the trait when there are costs of being choosy. Fisher (1930) hypothesized a process in which an initial preference of females for, say, long-tailed males is amplified, leading to a further increase in tail length. This occurs because by mating with a long-tailed male a female produces sons that also have long tails and so are preferred by other females. We show how the evolutionarily stable outcomes of this co-evolutionary process can be characterized in game-theoretical terms by defining a function that specifies how the 'value' of mating with a male depends on his tail length. A key ingredient in the analysis is that the projection matrices take into account that the female preference gene is carried but not expressed in sons.

A population may be state structured but all newborn offspring are in the same state. When this is so, the mean number of surviving offspring produced over the lifetime (mean LRS) acts as a fitness proxy, and so can be used to establish whether a resident strategy is evolutionarily stable (Section 10.6). This currency is often easier to deal with than invasion fitness itself. We use the currency to analyse a model in which individuals only differ in age. In this model we are concerned with the scheduling of reproduction over the lifetime of an organism when there is competition for resources with older individuals outcompeting younger ones. As we show, for some parameter values there are two alternative ESSs. At one population members put maximum effort into reproduction in their first breeding season leading to their death, while at the other ESS population members delay reproduction until they are more competitive.

When there are many local habitats linked by dispersal the location of an individual can be a state variable. Dispersal then changes this state variable. In Section 10.7 we illustrate the use of projection matrices in this setting.

Finally, in Section 10.8 we briefly discuss how one can deal with situations where the states recognized by individuals for decision-making differ from the states needed to describe the population dynamics. We mention how the concept of reproductive value can still be helpful in these situations, and we also go into possible fitness proxies.

10.1 Invasion Fitness for Structured Populations

For convenience of presentation we restrict attention to a population in which there is an annual cycle. The population is censused at the same time each year; by time t we will mean the annual census time in year t. At an annual census each individual is in one of the J states $s_1, s_2, \ldots, s_J$. At this time each individual chooses an action u from the set A of possible actions, where this choice can depend on the current state of the individual. A strategy specifies this choice, and will be described by a vector $\mathbf{x} = (u_1, u_2, \ldots, u_J)$ where u_j is the action taken in state s_j. Population size is regulated by density-dependent factors, and we assume that for each resident strategy $\mathbf{x}$ the population settles down to a unique demographic equilibrium at which the population size and the proportion of population members in any state is the same each annual census time.

Suppose that a resident $\mathbf{x}$ population is at its demographic equilibrium. Consider a possibly mutant individual in this resident population. Let this individual be in state s_j and take action u at time t. Let $d_{ij}(u, \mathbf{x})$ denote the mean number of descendants left by the individual that are alive at time $t + 1$ and are in state s_i at this time. Here the term 'descendant' can refer to the focal individual itself if it is still alive at time $t + 1$, any offspring of the individual that were produced between t and $t + 1$ that are alive at $t + 1$, any grandchildren that are descended from offspring produced between t and $t + 1$ that are alive at $t + 1$, etc. In counting descendants we are really counting the spread of genes. Thus if the genes of interest are at an autosomal locus in a diploid species, each offspring counts as one half of a descendant, each grandchild as one quarter of a descendant, etc.

Let $\mathbf{x}' = (u_1', u_2', \ldots, u_J')$ be a rare mutant strategy within the resident $\mathbf{x}$ population. Then a mutant in state s_j leaves an average of $a_{ij}(\mathbf{x}', \mathbf{x}) = d_{ij}(u_j', \mathbf{x})$ descendants in state s_i the following year. The $J \times J$ matrix $\mathbf{A}(\mathbf{x}', \mathbf{x})$ that has (i, j) element $a_{ij}(\mathbf{x}', \mathbf{x})$ is called the projection matrix for the mutant strategy. To consider the growth rate of the cohort of mutants, let $n_j(t)$ be the total number of mutants that are in state s_j at time t. Then numbers of mutants in state s_i at time $t + 1$ are given approximately by

$$n_i(t+1) = \sum_j a_{ij}(\mathbf{x}', \mathbf{x}) n_j(t). \tag{10.1}$$

The approximation assumes that mutants do not significantly interact with each other, that mutant numbers are large enough so that we can average over demographic stochasticity, and that mutant numbers are still small compared with the population as a whole so that the background biological environment can still be taken to be the resident demographic equilibrium. We can also write eq (10.1) in matrix notation as $\mathbf{n}(t+1) = \mathbf{A}(\mathbf{x}', \mathbf{x})\mathbf{n}(t)$ where the column vector $\mathbf{n}(t) = (n_1(t), n_2(t), \ldots, n_J(t))^T$ specifies the number of mutants in each state at time t. Let $N(t) = n_1(t) + n_2(t) + \ldots + n_J(t)$ be the total number of mutants at time t and let $\rho_j(t) = n_j(t)/N(t)$ be the proportion of these mutants that are in state s_j. Then eq (10.1) can be rewritten as

$$\frac{N(t+1)}{N(t)} \rho_i(t+1) = \sum_j a_{ij}(\mathbf{x}', \mathbf{x}) \rho_j(t). \tag{10.2}$$

Given appropriate assumptions about the projection matrix $\mathbf{A}(\mathbf{x}', \mathbf{x})$ (e.g. that it is primitive, see Caswell, 2001; Seneta, 2006), as time increases the proportions $\rho_j(t)$ and the annual per-capita growth rate $N(t+1)/N(t)$ tend to limiting values. Equation (10.2) then gives

$$\lambda(\mathbf{x}',\mathbf{x})\rho_i(\mathbf{x}',\mathbf{x}) = \sum_j a_{ij}(\mathbf{x}',\mathbf{x})\rho_j(\mathbf{x}',\mathbf{x}), \tag{10.3}$$

where $\rho_j(\mathbf{x}',\mathbf{x}) = \lim_{t\to\infty}\rho_j(t)$ and $\lambda(\mathbf{x}',\mathbf{x}) = \lim_{t\to\infty}(N(t+1)/N(t))$. We may write this equation in vector notation as $\lambda(\mathbf{x}',\mathbf{x})\rho(\mathbf{x}',\mathbf{x}) = \mathbf{A}(\mathbf{x}',\mathbf{x})\rho(\mathbf{x}',\mathbf{x})$. Equation (10.3) is known as an eigenvalue equation; $\lambda(\mathbf{x}',\mathbf{x})$ is an eigenvalue and $\rho(\mathbf{x}',\mathbf{x})$ a column (right) eigenvector of the matrix $\mathbf{A}(\mathbf{x}',\mathbf{x})$. $\mathbf{A}(\mathbf{x}',\mathbf{x})$ may have more than one eigenvalue, but given suitable assumptions, $N(t+1)/N(t)$ converges to the largest eigenvalue. $\rho(\mathbf{x}',\mathbf{x})$ is then the unique eigenvector corresponding to this eigenvalue that also satisfies the normalization condition

$$\sum_j \rho_j(\mathbf{x}',\mathbf{x}) = 1. \tag{10.4}$$

We can consider a mutation that is exactly the same as the resident strategy $\mathbf{x}$. For this 'mutant' we must have $\lambda(\mathbf{x},\mathbf{x}) = 1$ since the resident strategy is at a density-dependent equilibrium. By eq (10.3) the stable distribution for the mutant satisfies

$$\rho(\mathbf{x},\mathbf{x}) = \mathbf{A}(\mathbf{x},\mathbf{x})\rho(\mathbf{x},\mathbf{x}) \tag{10.5}$$

(as well as the normalization condition (10.4)). The distribution $\rho(\mathbf{x},\mathbf{x})$ is both the stable distribution of this rare 'mutant' and the stable demographic distribution of the resident strategy as a whole.

Since $\lambda(\mathbf{x}',\mathbf{x}) = \lim_{t\to\infty} N(t+1)/N(t)$ the number of mutants in the cohort has asymptotic growth rate

$$N(t) \sim K\lambda(\mathbf{x}',\mathbf{x})^t, \tag{10.6}$$

where K depends on the initial number of cohort members in each state. Thus if $\lambda(\mathbf{x}',\mathbf{x}) < 1$ mutant numbers will eventually tend to zero and the resident strategy cannot be invaded by this mutant, whereas if $\lambda(\mathbf{x}',\mathbf{x}) > 1$ the mutant will invade: mutant numbers will grow until they become sufficiently common to change the population composition. Thus the Nash equilibrium condition giving a necessary condition for the strategy $\mathbf{x}^*$ to be evolutionarily stable becomes

$$\lambda(\mathbf{x},\mathbf{x}^*) \leq \lambda(\mathbf{x}^*,\mathbf{x}^*) = 1 \text{ for all } \mathbf{x}. \tag{10.7}$$

In other words, for structured populations the invasion fitness of a mutant strategy is the largest eigenvalue of the projection matrix for this mutant.

In order to better appreciate the assumptions made we might, as we did in Section 6.1, consider a potentially multidimensional general strategy $\mathbf{x} = (x_1, x_2, \ldots, x_L)$. The components x_k represent traits and the number of traits can differ from the number of demographic states, i.e. $L \neq J$ in general. For a rare mutant following strategy $\mathbf{x}'$ that is in demographic state s_j, the number of descendants in state s_i the following year would then be $a_{ij}(\mathbf{x}',\mathbf{x})$, where each element of this matrix can depend on any component, or several different components, of $\mathbf{x}'$. Most of the results in this chapter hold also for this more general case, but some criteria for evolutionary stability can differ (see Section 10.8).

10.2 Offspring Quality versus Number

We illustrate the above ideas with an example in which individuals vary in reproductive quality, where the idea of quality is motivated by examples such as adult size, social status, and territory quality. We do not, however, attempt to build a realistic model of any situation here. Instead we present a toy model in which there is density dependence but otherwise no frequency dependence, and consider the trade-off between the number of surviving offspring and their quality when quality is to some extent inherited (cf. McNamara and Houston, 1992b). The idea is to bring out aspects of invasion fitness in the simplest possible setting.

We assume a large asexual population with discrete, non-overlapping generations, with a generation time of 1 year. The population is censused annually at the start of a year. At this time individuals are of one of two qualities, High or Low, and each must decide its clutch size: whether to produce one, two, or three offspring. Surviving offspring are mature by the autumn, and are either High or Low quality at this time. Table 10.1 specifies the mean number of mature offspring in the autumn that are of each quality, and how this depends on parental quality and clutch size. Between autumn and the next annual census time mortality acts to limit total population size in a density-dependent manner, with each population member subject to the same mortality risk regardless of their quality or strategy.

It can be seen from Table 10.1 that for a given clutch size a High-quality parent has more surviving offspring than a Low-quality parent, and a greater proportion of these are High quality. Furthermore, regardless of the quality of the parent, the number of surviving offspring increases with clutch size while the proportion of these offspring that are High quality decreases.

Suppose the resident strategy is $\mathbf{x} = (3,3)$, so that each resident maximizes the number of their offspring that are present in the autumn. Let $K_{3,3}$ be the probability of overwinter survival at the demographic equilibrium for this resident population. The projection matrix for the resident strategy can then be expressed as $\mathbf{A}(\mathbf{x},\mathbf{x}) = K_{3,3}\mathbf{G}_{3,3}$, where $\mathbf{G}_{3,3}$ is the projection matrix before the action of density dependence overwinter:

Table 10.1 Mean number of High- and Low-quality offspring that survive to maturity in autumn as a function of parental quality and clutch size.

Parent High		Offspring		Parent Low		Offspring	
		High	Low			High	Low
Clutch size, q_H	1	0.9	0.1	Clutch size, q_L	1	0.3	0.6
	2	1.2	0.3		2	0.2	0.8
	3	1.0	0.7		3	0.1	1.0

$$\mathbf{G}_{3,3} = \begin{bmatrix} 1.0 & 0.1 \\ 0.7 & 1.0 \end{bmatrix}. \tag{10.8}$$

A direct calculation (Box 10.1) shows that $\mathbf{G}_{3,3}$ has (largest) eigenvalue $g_{3,3} = 1.2646$. Since $\mathbf{A}(\mathbf{x},\mathbf{x})$ has eigenvalue 1, it follows that $K_{3,3} = g_{3,3}^{-1} = 0.7908$.

Now consider the mutant strategy $\mathbf{x}' = (2,1)$ within this resident $\mathbf{x} = (3,3)$ population. The projection matrix for the mutant strategy is $\mathbf{A}(\mathbf{x}',\mathbf{x}) = K_{3,3}\mathbf{G}_{2,1}$ where

$$\mathbf{G}_{2,1} = \begin{bmatrix} 1.2 & 0.3 \\ 0.3 & 0.6 \end{bmatrix}. \tag{10.9}$$

Since the largest eigenvalue of $\mathbf{G}_{2,1}$ is $g_{2,1} = 1.3243$ we have $\lambda(\mathbf{x}',\mathbf{x}) = K_{3,3}g_{2,1} = g_{2,1}/g_{3,3} = 1.3243/1.2646 = 1.047$. Since this value exceeds 1 the mutant can invade.

Note that in determining whether $\mathbf{x}' = (2,1)$ can invade the resident $\mathbf{x} = (3,3)$ population, the value of $K_{3,3}$ is irrelevant. We could have seen this by noting that for any two strategies we have $\lambda((q'_H,q'_L),(q_H,q_L)) = g_{q'_H,q'_L}/g_{q_H,q_L}$. This example, with density dependence acting in the same manner on all population members, is therefore an optimization problem rather than a game. Table 10.2 gives the values of g_{q_H,q_L} for every strategy (q_H,q_L). The unique ESS is the strategy $(2,1)$ that achieves the maximum in this table.

The ability of $(2,1)$ to invade $(3,3)$ can be understood by comparing the proportion of residents that are High quality at the demographic equilibrium (i.e. ρ_H) with the

---

Box 10.1 Eigenvalues and eigenvectors for 2 × 2 matrices

Let

$$\mathbf{A} = \begin{bmatrix} a_{11} & a_{12} \\ a_{21} & a_{22} \end{bmatrix}. \tag{10.10}$$

be a 2 × 2 matrix. Then any eigenvalue λ of $\mathbf{A}$ satisfies the quadratic equation $(\lambda - a_{11})(\lambda - a_{22}) - a_{12}a_{21} = 0$, so that the largest eigenvalue is

$$\lambda = \frac{1}{2}\left(B + \sqrt{B^2 - 4C}\right), \tag{10.11}$$

where $B = a_{11} + a_{22}$ is the trace and $C = a_{11}a_{22} - a_{12}a_{21}$ is the determinate of $\mathbf{A}$. Let ρ be the stable proportion of individuals in state 1 (so that the stable proportion in state 2 is $1 - \rho$). Then the first row of the eigenvector equation for the stable proportions (cf. eq (10.3)) gives

$$\rho = \frac{a_{12}}{\lambda + a_{12} - a_{11}}. \tag{10.12}$$

Similarly the ratio of reproductive values (introduced later in Section 10.3) satisfies

$$\frac{v_2}{v_1} = \frac{\lambda - a_{11}}{a_{21}}. \tag{10.13}$$

We can then use eq (10.45) of the Exercises to find the absolute values of v_1 and v_2, although for most purposes only the relative values are important.

Table 10.2 Growth rates g_{q_H, q_L} before the action of density dependence for each of the nine possible strategies.

Growth rate		q_L		
		1	2	3
q_H	1	0.9791	1.0000	1.0618
	2	1.3243	1.3162	1.3000
	3	1.3000	1.2873	1.2646

proportion of mutants that are High quality when the mutant cohort settles down to stable growth. All individuals following the resident strategy $(3,3)$ maximize their mean number of surviving offspring, but these offspring tend to be Low quality, and by having large clutches themselves ensure the next generation is also mainly Low quality, and so on. Consequently, at the demographic equilibrium only 27.43% of the population is High quality (this can be calculated using the results in Box 10.1). In contrast, the cohort of $(2,1)$ mutants produce less total offspring but more High-quality ones. As a result 70.71% of cohort members are high quality under stable growth. High-quality individuals tend to leave more offspring than Low-quality individuals for the same clutch size, so that the mutant strategy does (just) better than the resident, even though for any given quality residents produce more offspring. We can quantify this argument by noting that in general we can write invasion fitness as

$$\lambda(\mathbf{x}', \mathbf{x}) = \sum_j a_j(\mathbf{x}', \mathbf{x}) \rho_j(\mathbf{x}', \mathbf{x}), \tag{10.14}$$

where $a_j(\mathbf{x}', \mathbf{x}) = \sum_i a_{ij}(\mathbf{x}', \mathbf{x})$ is the total number of descendants left by a mutant individual in state s_j (Exercise 10.2). Applying this formula we have

$$\lambda((3,3),(3,3)) = K_{3,3}\,[0.2743 \times (1.0 + 0.7) + 0.7257 \times (0.1 + 1.0)] = 1$$
$$\lambda((2,1),(3,3)) = K_{3,3}\,[0.7071 \times (1.2 + 0.3) + 0.2929 \times (0.3 + 0.6)] = 1.047.$$

Figure 10.1 illustrates the growth in numbers of the cohort of $(2,1)$ mutants. In this figure we have started with 1000 mutants in generation 0, with 27.4% of these mutants of High quality; so that it is as if 1000 residents were instantaneously changed to mutants. Initially mutant numbers decline, but the proportion that are of High quality increases. By generation 3 this proportion becomes sufficiently large so that mutant numbers start to increase, but it takes until generation 8 until the number of mutants is above initial values. By that time around 70.7% of mutants are of High quality and the per-capita growth rate in mutant numbers is close to 1.047.

Figure 10.1 illustrates that in order to see if a mutant can invade it might be necessary to count descendants several generations into the future, rather than just looking at offspring or even grandchildren. By measuring the asymptotic growth rate

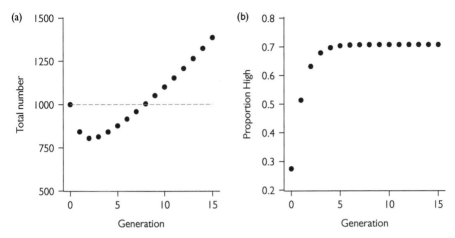

Fig. 10.1 Invasion of the mutant strategy $(2,1)$ when the resident strategy is $(3,3)$ for the quality versus offspring number example. (a) Total number of mutants. (b) Proportion of the cohort of mutants that are High quality. There are initially 1000 mutants with the proportion that are High quality equal to that in the resident population.

in mutant numbers, the largest eigenvalue of the projection matrix essentially looks infinitely far into the future. It is worth noting that if one needs to look very far into the future, processes such as new mutations occurring or environmental changes can make this kind of analysis of the invasion process approximate or even unrealistic (see, e.g., Livnat et al., 2005).

Equation (10.14) expresses the eigenvalue in terms of the mean number of descendants left in just 1 year's time. However, in this formula the mean is an average over the stable distribution of states, which is itself defined asymptotically far into the future, so there is no contradiction. Equation (10.14) highlights that a change in strategy has two effects: it changes the number of descendants left next year (the a_j) and it changes the long-term distribution of descendants over states (the ρ_j). This insight can be conceptually useful in applications, particularly when environments are positively autocorrelated so that local conditions tend to persist into the future. For example, McNamara and Dall (2011) show that natal philopatry can evolve because it is worth individuals suffering an immediate loss in reproductive success in order to ensure that their descendants are on good breeding sites in future generations. McNamara et al. (2011) show that individuals can increase the proportion of descendants that are in good local habitats by being overly 'optimistic', behaving as if the probability they are currently in a good local habitat is greater than its actual value.

This example also illustrates that the best action in one state can depend on what the individual would have done in another state. For example, it can be seen from Table 10.2 that if the action of High-quality individuals is to produce a clutch of size 1 then the best action of Low-quality individuals is to produce a clutch of size 3, whereas if High-quality individuals produce a clutch of size 2 then Low-quality individuals do best to produce a clutch of size 1.

10.3 Reproductive Value Maximization

Consider a population with resident strategy **x**. Different members of this population will be in different states and have different abilities to produce surviving offspring and to leave descendants further into the future. So for example, in the case of dominance status in female hyenas, a high-status female will leave (on average) more offspring, and even more grandchildren, than a low-status female.

To quantify this ability we define $d_j(t)$ to be the mean number of descendants (discounted by relatedness) left in t years' time by a resident that is currently in state s_j. The initial state s_j will be correlated with the states of future descendants. This correlation may be high for low t, but will eventually tend to zero as t increases; so, for example the descendants left in 1000 years' time by a high-status female hyena are hardly more likely to be high status than those left by an average female. For large t the distribution of states of descendants will approximate the steady-state distribution $\rho(\mathbf{x}, \mathbf{x})$, at which the per-capita annual growth rate in descendant numbers is 1. Thus we might expect to have

$$d_j(t) \rightarrow v_j(\mathbf{x}) \text{ as } t \rightarrow \infty, \tag{10.15}$$

where $v_j(\mathbf{x})$ represents the transitory effect of the initial state. $v_j(\mathbf{x})$ is referred to as the reproductive value of state s_j. Figure 10.2 illustrates eq (10.15) for the offspring quality versus number example of Section 10.2.

Since the descendants left in $t + 1$ years' time are the descendants left in t years' time by the descendants present next year we have $d_j(t+1) = \sum_i d_i(t) a_{ij}(\mathbf{x}, \mathbf{x})$. Taking the limit as $t \rightarrow \infty$ and making use of eq (10.15) then gives

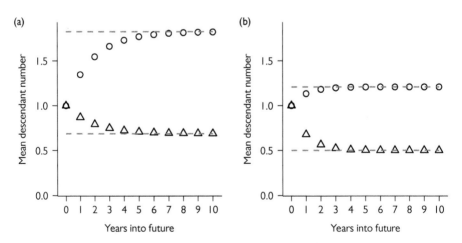

Fig. 10.2 Reproductive values for the offspring quality versus number model. Resident strategies are (a) $(3,3)$ and (b) $(2,1)$. Upper (lower) points are the mean number of descendants left by a High-quality (Low-quality) individual. The horizontal dashed lines are at the asymptotic values of these points, and are the reproductive values.

$$v_j(\mathbf{x}) = \sum_i v_i(\mathbf{x})a_{ij}(\mathbf{x},\mathbf{x}). \qquad (10.16)$$

In vector notation this becomes $\mathbf{v}(\mathbf{x}) = \mathbf{v}(\mathbf{x})\mathbf{A}(\mathbf{x},\mathbf{x})$. This analysis shows that the resident projection matrix has $\mathbf{v}(\mathbf{x})$ as its row (left) eigenvector, with again the eigenvalue being $\lambda(\mathbf{x},\mathbf{x}) = 1$.

Reproductive value acts as a common currency in that it can be used to specify the value, in terms of leaving descendants in the future, of being in a state. We can compare the expected values of descendants left next year by an $\mathbf{x}'$ mutant and an $\mathbf{x}$ resident that are both in the same state s_j. These expected values are $\sum_i v_i(\mathbf{x})a_{ij}(\mathbf{x}',\mathbf{x})$ and $\sum_i v_i(\mathbf{x})a_{ij}(\mathbf{x},\mathbf{x}) = v_j(\mathbf{x})$, respectively. Thus the resident's descendants are at least as 'valuable' as the mutants if

$$\sum_i v_i(\mathbf{x})a_{ij}(\mathbf{x}',\mathbf{x}) \le v_j(\mathbf{x}). \qquad (10.17)$$

If a resident's descendants are at least as valuable as a mutant's for all states s_j we write this as $\mathbf{v}(\mathbf{x})\mathbf{A}(\mathbf{x}',\mathbf{x}) \le \mathbf{v}(\mathbf{x})$. A central result is that

$$\mathbf{v}(\mathbf{x})\mathbf{A}(\mathbf{x}',\mathbf{x}) \le \mathbf{v}(\mathbf{x}) \implies \lambda(\mathbf{x}',\mathbf{x}) \le \lambda(\mathbf{x},\mathbf{x}) = 1. \qquad (10.18)$$

If inequality (10.17) holds for all states s_j, with the inequality strict for at least one state, we write this as $\mathbf{v}(\mathbf{x})\mathbf{A}(\mathbf{x}',\mathbf{x}) < \mathbf{v}(\mathbf{x})$. When this inequality holds we have $\lambda(\mathbf{x}',\mathbf{x}) < \lambda(\mathbf{x},\mathbf{x}) = 1$. The converse to equation (10.18) is also true

$$\mathbf{v}(\mathbf{x})\mathbf{A}(\mathbf{x}',\mathbf{x}) \ge \mathbf{v}(\mathbf{x}) \implies \lambda(\mathbf{x}',\mathbf{x}) \ge \lambda(\mathbf{x},\mathbf{x}) = 1. \qquad (10.19)$$

Conditions (10.18) and (10.19) appear to have been first explicitly stated in biological game theory by Leimar (1996). The results follow from the optimality analysis in McNamara (1991), but Leimar (1996) also gave a direct derivation using the approach of Taylor (1990). A proof is set as Exercise 10.4. The condition allows us to recognize a Nash equilibrium strategy. Specifically, in Section 10.1 we assumed that strategies are flexible in the sense that the choice of action in one state does not constrain the choice in other states (cf. Sections 10.8 and 11.2.2), and for this case we have

$$\mathbf{x}^* \text{ is a Nash equilibrium} \iff \mathbf{v}(\mathbf{x}^*)\mathbf{A}(\mathbf{x}',\mathbf{x}^*) \le \mathbf{v}(\mathbf{x}^*) \text{ for all } \mathbf{x}'. \qquad (10.20)$$

The condition says that $\mathbf{x}^*$ is a Nash equilibrium strategy if and only if, when this is the resident strategy, each individual is behaving so as to maximize the mean value of descendants left in 1 year's time, where value is quantified using the residents' own reproductive value $\mathbf{v}(\mathbf{x}^*)$.

We illustrate the above formulae using the quality versus number model of Section 10.2. Let $\mathbf{x} = (3,3)$ be the resident strategy, and consider the performance of the mutant strategy $\mathbf{x}' = (2,1)$. The reproductive value vector under the resident strategy is $(1.8229, 0.689)$ (Box 10.1, see also Fig. 10.2a). Consider first a High-quality individual. The mean value of descendants left next year from clutch sizes 3 (the resident) and 2 (the mutant) are

$$w_H(3) = K_{3,3}[1.0 \times 1.8229 + 0.7 \times 0.689] = 2.3052K_{3,3} \, (= 1.8229),$$
$$w_H(2) = K_{3,3}[1.2 \times 1.8229 + 0.3 \times 0.689] = 2.3942K_{3,3}.$$

Thus $w_H(3) < w_H(2)$ so that we can conclude from criterion (10.20) that the resident strategy is not at a Nash equilibrium. We now consider the decision of a Low-quality individual. A similar calculation reveals that $w_L(3) = 0.8713K_{3,3}$ and $w_L(1) = 0.9603K_{3,3}$, so that $w_L(3) < w_L(1)$. Thus, regardless of their state, mutants produce more valuable descendants next year than residents; i.e. $\mathbf{v}(\mathbf{x})\mathbf{A}((2,1),(3,3)) > \mathbf{v}(\mathbf{x})\mathbf{A}((3,3),(3,3)) = \mathbf{v}(\mathbf{x})$. From the results above we thus have $\lambda((2,1),(3,3)) > \lambda((3,3),(3,3)) = 1$, so that the $(2,1)$ mutant can invade the resident $(3,3)$ population.

The reproductive value vector $\mathbf{v}(\mathbf{x})$ is a row eigenvector of the resident projection matrix (for the eigenvalue equal to 1). Let $\tilde{\mathbf{v}}(\mathbf{x}) = \alpha\mathbf{v}(\mathbf{x})$ where α is a positive constant. Then $\tilde{\mathbf{v}}(\mathbf{x})$ is still a row eigenvector of the projection matrix (for the eigenvalue equal to 1) and all the above analyses of reproductive value maximization still hold with $\mathbf{v}(\mathbf{x})$ replaced by $\tilde{\mathbf{v}}(\mathbf{x})$. In applications one often works with the row eigenvector normalized to, say, $\tilde{v}_1(\mathbf{x}) = 1$ rather than use the eigenvector given by eq (10.15).

The quality versus number model of Section 10.2 is especially simple. In cases with more than two states, the eigenvalues and eigenvectors of a projection matrix must typically be calculated by some numerical scheme. Nash equilibria are also typically found by some numerical iterative procedure. For example, given a resident strategy one might use criterion (10.19) to find a mutant strategy that can invade. This mutant strategy could then be taken to be the new resident, and so on, until a strategy is found that cannot be invaded. The problem with such schemes is that there is no guarantee of convergence, although damping and the introduction of small error can help (McNamara et al., 1997).

10.4 Sex Allocation as a Game over Generations

In considering sex allocation in Sections 3.8 and 3.11 we emphasized that, when a female can determine the sex of her offspring, invasion fitness is not just the mean number of surviving offspring. This is because a son and a daughter could have different abilities to leave offspring themselves. Here we justify the fitness proxy used in those sections. We also briefly consider sex allocation when individuals differ in quality when quality is inherited, addressing the question of whether high-quality females should produce sons or daughters.

10.4.1 The Shaw–Mohler Fitness Proxy

For simplicity we again assume discrete non-overlapping generations with a generation time of 1 year. Each year there is a mating season in which each female mates with a randomly chosen male. She then gives birth to offspring, and her strategy determines the sex of these offspring.

It is convenient to choose the annual census time at the start of the mating season. At this time individuals are in one of two states, female or male, and we can refer to these as s_1 and s_2, or simply by using the subscripts F and M. Let the resident sex-allocation strategy be x and let x' be a rare mutant strategy. We assume that a mutant female that is present at an annual census time leaves $K(x)F(x')$ daughters and $K(x)M(x')$ sons at the next annual census time. The factor $K(x)$ takes density dependence into account. The proportion of females to males during mating is $s(x) = F(x)/M(x)$, so that each male mates with $s(x)$ females on average. Thus the projection matrix for the mutant strategy is

$$\mathbf{A}(\mathbf{x'},\mathbf{x}) = \frac{1}{2}K(x)\begin{bmatrix} F(x') & s(x)F(x) \\ M(x') & s(x)M(x) \end{bmatrix}. \tag{10.21}$$

Here we have taken into account the fact that (i) mutant individuals mate with residents (since mutants are rare), (ii) the offspring of such matings have a probability of $\frac{1}{2}$ of inheriting the mutant strategy, and (iii) it is always the female who determines the sex allocation of offspring.

Setting $x' = x$ we can use eq (10.5) in order to consider the demographic stability of the resident population. This equation yields $\frac{\rho_F(x,x)}{\rho_M(x,x)} = \frac{F(x)}{M(x)}$; a fact we already used to write down the projection matrix. The equation also yields $K(x)F(x) = 1$. This says that a resident female leaves one surviving daughter on average at demographic stability, which is of course required for stationarity.

Reproductive value under the resident strategy satisfies eq (10.16). From this equation we can deduce that $F(x)v_F(x) = M(x)v_M(x)$. It follows that the total reproductive value of all females equals the total reproductive value of all males. This result must hold since every individual has one mother and one father, and was used by Fisher (1930) to argue for a 1:1 sex ratio.

To analyse whether the mutant can invade we set $\alpha = F(x)v_F(x) = M(x)v_M(x)$. We can then write $\mathbf{v}(x) = \alpha\left(\frac{1}{F(x)}, \frac{1}{M(x)}\right)$, so that, using eq (10.21),

$$\frac{2}{\alpha K(x)}\mathbf{v}(x)\mathbf{A}(x',x) = \left(\frac{F(x')}{F(x)} + \frac{M(x')}{M(x)}, 2s(x)\right). \tag{10.22}$$

It follows from eqs (10.18, 10.19) that

$$w(x',x) = \frac{F(x')}{F(x)} + \frac{M(x')}{M(x)} \tag{10.23}$$

is a fitness proxy. This proxy was first derived by Shaw and Mohler (1953). If we set $W(x',x) = F(x)w(x',x)$ then this also a fitness proxy. We have

$$W(x',x) = F(x') + s(x)M(x'), \tag{10.24}$$

which is the mean number of matings obtained by the offspring of an x' female. This justifies the proxy used in Section 3.8. Alternatively, if all matings result in the same number of offspring we may take the number of grandchildren produced as a fitness proxy.

In Section 3.11 females differ in their state (external temperature) when they decide on the sex of their offspring, and their decisions depend on this state. However, this state is not inherited and the distribution of states of females is the same each year and does not depend on their sex-allocation strategy. Under these assumptions $W(x',x)$ is still a fitness proxy provided $F(x')$ and $M(x')$ are taken to be averages over the different states of x' females.

10.4.2 Quality-dependent Sex Allocation

When certain properties of a state are inherited the fitness proxy W given by eq (10.24) is no longer appropriate. Aspects of quality such as adult size, health, social status, or territory quality can be passed down from a mother to her offspring. We now outline how this transmission of quality might affect sex-allocation decisions. In particular, how might we expect the sex of offspring to depend on maternal quality?

In Section 3.11 the maternal state, in the sense of the temperature during the development of offspring, affected the differential survival of daughters and sons, but had no effect once offspring reached maturity. In this case the relative probabilities of survival to maturity influenced the Nash equilibrium sex-allocation strategy (eq (3.20), which is based on the fitness proxy in eq (10.24)). The situation becomes more intricate when maternal quality affects the quality of offspring when they reach maturity.

Assume that a population is at evolutionary stability and that a female has quality q. Let $v_f(q)$ and $v_m(q)$ denote the reproductive values of one of her daughters and sons, respectively, at independence. Then assuming that the same resources and time are required to produce independent sons and daughters, the female will be producing offspring whose sex has the higher reproductive value. Two hypothetical dependences of reproductive value on quality are illustrated in Fig. 10.3. In one case high-quality females should produce sons, in the other they should produce daughters. So the key question is how the underlying biology determines the functions v_f and v_m.

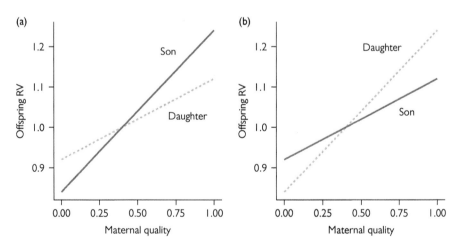

Fig. 10.3 Illustration of possible dependence of the reproductive value of offspring on maternal quality. In (a) high-quality females should produce sons, in (b) they should produce daughters.

Trivers and Willard (1973) considered a scenario in which: (i) females vary in quality with high-quality females producing high-quality offspring, (ii) quality persists until maturation and breeding, and (iii) reproductive success is more strongly dependent on quality in males than in females. They argued that in this scenario high-quality females are predicted to produce sons and poor-quality females to produce daughters. However, as Leimar (1996) pointed out, there is a flaw in this reasoning since Trivers and Willard took the number of grandchildren as their fitness proxy. Under the assumption that maternal quality affects daughter quality, the quality of the offspring of daughters will depend on the quality of their grandmother. Thus not all grandchildren are the same and just counting their number is not appropriate.

In order to specify projection matrices when quality is inherited it is first necessary to specify the range of possible values of quality in females and males. Any model then needs to additionally specify:

- How female quality affects the number of offspring she is able to produce.
- How female quality affects the probability any daughter survives and her quality if she survives.
- How female quality affects the probability any son survives and his quality if he does so.
- How male quality affects his mating success.
- Whether male quality is also inherited.

A simple model incorporating these factors might have two qualities for females and two for males. There would then be four states (female-high, female-low, male-high, male-low) and the projection matrices would be 4×4 matrices.

Leimar (1996) constructed two models: in one each sex had two qualities, in the other each sex had a continuum of qualities. In both models male quality is not passed on to offspring. The general conclusions of these models is that the factors that favour high-quality females preferring sons at evolutionary stability are: (i) a weak correlation between female quality and female fecundity, (ii) a weak correlation between female quality and daughter quality, (iii) a strong correlation between female quality and male quality, and (iv) a strong correlation between male quality and male ability to compete with other males for females. The converse properties leads to high-quality females preferring daughters.

10.4.3 Empirical Evidence

Theories of sex allocation (e.g. Charnov, 1982; Pen and Weissing, 2002) are among the most developed and successful in evolutionary ecology. In their pioneering work, Trivers and Willard (1973) initiated the evolutionary study of facultative sex ratios, i.e. when particular environmental conditions favour the production of either male or female offspring. Haplodiploid insects, where males are haploid and females are diploid, provide many examples of such facultative sex ratios. An egg-laying female directly determines the sex of offspring in these insects, through her decision of whether to fertilize an egg that is being laid (using stored sperm). Some of these insects are parasitoids, meaning that females lay eggs in or on host organisms (e.g. other insects). The offspring develop in and consume the host. Typically females are larger than males in parasitoid insect species, and therefore it should benefit a mother to lay female-destined eggs in larger hosts. This prediction has strong empirical support (Charnov, 1982; West and Sheldon, 2002).

Trivers and Willard (1973) dealt specifically with the quality or condition of the mother as the environmental factor influencing sex-allocation decisions. Much effort has been devoted to testing their hypothesis, both using observational field studies and in experiments. While there is empirical support for the hypothesis, including for mammals, there is also a fair amount of discussion and controversy (Sheldon and West, 2004). One reason for the difficulty could be that many species have chromosomal sex determination, for instance with X/Y chromosomes, potentially making it harder for mothers to influence the sex of offspring. Another reason is that many factors apart from the condition of the mother can influence whether offspring of one sex or another should be favoured, as the example above of host size illustrates. As another example, in some bird species offspring can stay on at the nest and help rearing their younger siblings, and sometimes mostly one sex, for instance females, acts as helpers. Komdeur et al. (1997) showed that Seychelles warbler parents with nests in good sites, where helpers are particularly beneficial, strongly skewed the offspring sex towards females if they were without a helper, and there are similar results for other bird species (West and Sheldon, 2002). A question related to but different from the Trivers and Willard (1973) hypothesis is how much a mother should invest in an offspring given that it is either a son or a daughter. Theoretical predictions for this related question appear to depend on modelling details (Veller et al., 2016).

As the work by Leimar (1996) shows, to predict which sex of offspring should be preferred by mothers in good condition, one needs good estimates of the reproductive value of sons compared with daughters, and such estimates are difficult to obtain. This is because reproductive value depends on effects across generations and also that in many species life histories differ between the sexes. Schindler et al. (2015) addressed this by developing models with life-histories adapted to particular species. Even so, at the present time there is still a need for further work to evaluate the Trivers and Willard (1973) sex-allocation hypothesis (e.g. Douhard, 2017).

10.5 The Fisher Runaway Process

In many species males have conspicuous traits that females find attractive. Among birds the display feathers of a male peacock and the tail of a male long-tailed widowbird are striking examples. In a classic experiment on the widowbird Andersson (1982) experimentally manipulated the tail length of males, shortening some and elongating others (Fig. 10.4). He showed that the mating success of males increased with their tail length. Since increased tail length appears to be costly (Pryke and Andersson, 2005), this experiment and subsequent work suggest that the extreme tail length of males is maintained by female preference and reflects a balance between

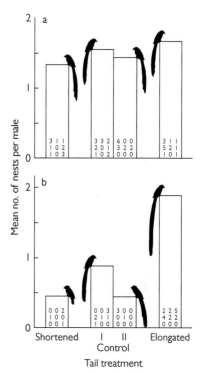

Fig. 10.4 A male long-tailed widowbird and fig. 1 from Andersson (1982). In Andersson's classic experiment the tails of some males were shortened, some were extended, with other birds acting as controls. Part (a) male success before the manipulation, (b) success after. Original photo of the long-tailed widowbird (*Euplectes progne*) male by Bernard Dupont. Published under Creative Commons Attribution-Share Alike 2.0 Generic license (CC BY-SA 2.0). Data figure reprinted by permission from Springer Nature: M. Andersson, Female choice selects for extreme tail length in a widowbird. *Nature*, 299, 818–820, Copyright (1982).

reduced survival and increased mating success. Such a balance had previously been proposed by Darwin (1871) and Fisher (1930), but raises the question of why the preference by females might arise in the first place and why it might be maintained.

Fisher (1930) suggested that if the resident female mate-choice strategy resulted in long-tailed males gaining more matings, then any mutant female should prefer to mate with a long-tailed male because their sons would inherit the long tail and so gain more matings, resulting in more grandchildren for the female. He also argued that an initial preference by females, for whatever reason, could be amplified by this feedback loop, resulting in a runaway process in which increased preference for long tails and increased tail length would co-evolve. Females might initially prefer males because they yield benefits such as good genes. It has also been suggested (e.g. Andersson, 1982; Pryke and Andersson, 2005) that the runaway process might be triggered by the fact that increased tail length increases the visibility of males in flight and hence makes it more likely that they are detected by females. Here we illustrate how mate-preference evolution could work.

A female's mate-choice genes are passed on to future generations through both her daughters and her sons, so that any analysis of the evolution of female mating preferences has to incorporate these two channels of inheritance. Most analyses of the Fisher runaway process explicitly model the genetics of traits, often using quantitative genetics (e.g. Lande, 1981; Pomiankowski et al., 1991; Iwasa et al., 1991). However, as we illustrate, the evolutionarily stable endpoints of a runaway process can be characterized in phenotypic terms. Each resident female mate-choice strategy determines a function that specifies how the payoff in terms of reproductive value of mating with a male depends on his trait value. The resident strategy is then an ESS if and only if females are behaving to maximize the payoff they obtain from mating.

For simplicity we present our analysis for the case in which the male trait is restricted to one of two values, referring to males as either type 1 or type 2. We assume discrete non-overlapping generations with one breeding season per generation and a 1:1 birth sex ratio with no differential mortality between the sexes. During the breeding season each female mates at most once. Males may mate many times.

We start by considering a male trait that is not genetically determined, but is a phenotypic trait that is only present in males and is transmitted phenotypically from father to son, e.g. via cultural inheritance. This transmission is subject to error: with probability ϵ the trait value of a son is different to that of his father. Phenotypic inheritance avoids the complication that females carry genes for male type and hence avoids the need to consider the covariance between the male-type genes and the female mate-choice genes, which is an important element of quantitative genetic analyses. After exposing the logic in this simpler case we outline how the analysis can be extended to a male trait that is genetically determined.

The genes determining a given mate-choice strategy can be present in three categories of individual: (i) carried but not expressed in a type 1 male, (ii) carried but not expressed in a type 2 male, and (iii) carried and expressed in a female. We regard a strategy as being determined by the allele at a specific haploid locus. The projection matrix for the strategy $\mathbf{x}'$ in a resident $\mathbf{x}$ population is then specified by two functions:

$a_j(\mathbf{x})$ = mean number of daughters with the $\mathbf{x}'$ allele produced by a type j $\mathbf{x}'$ male,

$\alpha_i(\mathbf{x}',\mathbf{x})$ = mean number of type i sons with the $\mathbf{x}'$ allele produced by an $\mathbf{x}'$ female.

Note that male transmission of the mutant allele depends only on the resident mate-choice strategy since mutant females are rare. In contrast, the transmission by a mutant female depends on her mate-choice strategy and that of resident females since their action determines the population proportion of males of different types (and there may also be direct competition for males). Taking account of the 1:1 offspring sex ratio and errors in transmission of male type, the projection matrix can then be written as

$$\mathbf{A}(\mathbf{x}',\mathbf{x}) = \begin{bmatrix} a_1(\mathbf{x})(1-\epsilon) & a_2(\mathbf{x})\epsilon & \alpha_1(\mathbf{x}',\mathbf{x}) \\ a_1(\mathbf{x})\epsilon & a_2(\mathbf{x})(1-\epsilon) & \alpha_2(\mathbf{x}',\mathbf{x}) \\ a_1(\mathbf{x}) & a_2(\mathbf{x}) & \alpha_1(\mathbf{x}',\mathbf{x})+\alpha_2(\mathbf{x}',\mathbf{x}) \end{bmatrix}. \tag{10.25}$$

To proceed further denote the probability that a mutant female chooses a type i male as her partner by $m_i(\mathbf{x'},\mathbf{x})$. Assume that the mean number of offspring left by a female if she chooses a type i male, N_i, depends only on the male's type. A mutant female mates with a resident male (since mutants are rare), so that she passes on her mutant allele to each offspring with probability $\frac{1}{2}$. Since each offspring is male with probability $\frac{1}{2}$ we have

$$\alpha_1(\mathbf{x'},\mathbf{x}) = \frac{1}{4}\left[m_1(\mathbf{x'},\mathbf{x})N_1(1-\epsilon) + m_2(\mathbf{x'},\mathbf{x})N_2\epsilon\right] \quad (10.26)$$

$$\alpha_2(\mathbf{x'},\mathbf{x}) = \frac{1}{4}\left[m_1(\mathbf{x'},\mathbf{x})N_1\epsilon + m_2(\mathbf{x'},\mathbf{x})N_2(1-\epsilon)\right]. \quad (10.27)$$

For a type i male, the mean number of resident females that will choose him is $m_i(\mathbf{x},\mathbf{x})/\rho_i(\mathbf{x})$, where $\rho_i(\mathbf{x})$ is the proportion of males that are type i in the resident population. Thus, using the 1:1 offspring sex ratio, we have

$$a_i(\mathbf{x}) = \frac{1}{4}\left(\frac{m_i(\mathbf{x},\mathbf{x})}{\rho_i(\mathbf{x})}\right)N_i. \quad (10.28)$$

The machinery developed in Section 10.3 gives a criterion for the invasion of a mutant strategy in terms of maximization of reproductive value. McNamara et al. (2003b) derived the invasion criterion by exploiting the fact that the $a_i(\mathbf{x})$ do not depend on $\mathbf{x'}$. This derivation is presented in Box 10.2. From the analysis in the box it can be seen that the best response, in terms of the payoff of a mutant in the resident population, maximizes $W(\mathbf{x'},\mathbf{x})$ given by eq (10.34). By eqs (10.26) and (10.27), this is equivalent to a female's mate-choice strategy $\mathbf{x'}$ maximizing

$$m_1(\mathbf{x'},\mathbf{x})r_1 + m_2(\mathbf{x'},\mathbf{x})r_2, \quad (10.29)$$

where the payoffs for choosing her partner are

$$r_1 = N_1\left[v_1(1-\epsilon) + v_2\epsilon + 1\right] \quad (10.30)$$
$$r_2 = N_2\left[v_1\epsilon + v_2(1-\epsilon) + 1\right], \quad (10.31)$$

where v_i is the reproductive value of a type i male under the resident strategy. The best response only depends on the ratio $\frac{r_2}{r_1}$ of these payoffs. Note that, according to (10.35), $W(\mathbf{x'},\mathbf{x})$ in eq (10.34) is a fitness proxy, but it might not be a strong fitness proxy (see Section 2.5). This means that $W(\mathbf{x'},\mathbf{x})$ can be used to determine Nash equilibria, but the best responses given by the proxy might differ from the best responses in terms of invasion fitness.

We illustrate the above ideas with a specific model of mate choice. Assume that during the annual breeding season each female attempts to choose her mate in a mating window that starts at time of year 0 and ends at time of year T. During the mating window she encounters a sequence of males. The female is assumed to prefer type 2 males, and if she encounters a male of this type she chooses him as her mate. If she has not encountered a type 2 male by time x she ceases to be choosy and mates with the first male she encounters after this time. Her strategy is thus specified by this switch time x. There is a cost of choosiness: the larger the value of x the more likely it is that she will not encounter any male after this time and so does not mate at all.

Box 10.2 Reproductive value maximization for the Fisher process

Let $\mathbf{v} = (v_1, v_2, 1)$ be the row vector of reproductive values under the resident mate-choice strategy. Thus $\mathbf{v}\mathbf{A}(\mathbf{x}, \mathbf{x}) = \lambda(\mathbf{x}, \mathbf{x})\mathbf{v}$ where the matrix $\mathbf{A}(\mathbf{x}, \mathbf{x})$ is given by eq (10.25) (with $\mathbf{x}'$ set equal to $\mathbf{x}$). Let ρ_i' be the stable proportion of all mutant males that are type i in the resident population. Thus the column vector $\boldsymbol{\rho}' = (\rho_1', \rho_2', 1)^T$ satisfies $\mathbf{A}(\mathbf{x}', \mathbf{x})\boldsymbol{\rho}' = \lambda(\mathbf{x}', \mathbf{x})\boldsymbol{\rho}'$. By these two eigenvector equations

$$\mathbf{v}\left[\mathbf{A}(\mathbf{x}, \mathbf{x}) - \mathbf{A}(\mathbf{x}', \mathbf{x})\right]\boldsymbol{\rho}' = \left[\lambda(\mathbf{x}, \mathbf{x}) - \lambda(\mathbf{x}', \mathbf{x})\right]\mathbf{v}\boldsymbol{\rho}'. \tag{10.32}$$

All the entries in the first and second columns of the matrix $\mathbf{A}(\mathbf{x}, \mathbf{x}) - \mathbf{A}(\mathbf{x}', \mathbf{x})$ are zero. Thus

$$\mathbf{v}\left[\mathbf{A}(\mathbf{x}, \mathbf{x}) - \mathbf{A}(\mathbf{x}', \mathbf{x})\right]\boldsymbol{\rho}' = W(\mathbf{x}, \mathbf{x}) - W(\mathbf{x}', \mathbf{x}), \tag{10.33}$$

where

$$W(\mathbf{x}', \mathbf{x}) = (v_1 + 1)\alpha_1(\mathbf{x}', \mathbf{x}) + (v_2 + 1)\alpha_2(\mathbf{x}', \mathbf{x}). \tag{10.34}$$

By eqs (10.32) and (10.33)

$$\lambda(\mathbf{x}', \mathbf{x}) \leq \lambda(\mathbf{x}, \mathbf{x}) \iff W(\mathbf{x}', \mathbf{x}) \leq W(\mathbf{x}, \mathbf{x}) \tag{10.35}$$

since $\mathbf{v}\boldsymbol{\rho}' > 0$.

Let ρ_i denote the proportion of resident males that are type i. We assume that each female encounters type 2 males as a Poisson process of rate ρ_2. Type 1 males may be less visible than type 2 males, and we assume that a female encounters males of this type as a Poisson process of rate $\kappa\rho_1$, where $\kappa \leq 1$ takes account of the reduced visibility of these males.

Given the resident strategy x there is a complication in finding the probabilities, $m_1(\mathbf{x}, \mathbf{x})$ and $m_2(\mathbf{x}, \mathbf{x})$, that a resident female mates with males of each type. This is because the mating probabilities depend on ρ_1 and ρ_2, but these proportions depend on the mating probabilities. Nevertheless, one can compute numerically the values of the mating probabilities and the proportions of the two types of male that are consistent with one another by an iterative scheme. Once $\rho_1, \rho_2, m_1(\mathbf{x}, \mathbf{x})$, and $m_2(\mathbf{x}, \mathbf{x})$ have been found one can find the coefficients of the projection matrix for the resident population. This can then be used to solve for the reproductive values v_1 and v_2. These reproductive values then determine the payoffs r_1 and r_2 (eqs (10.30) and (10.31)) and the mutant strategy $\mathbf{x}'$ that maximizes expression (10.29) can then be found.

Figure 10.5 illustrates how payoff ratios and best responses in terms of payoff depend on the resident female switch time. When $\kappa = 1$, so that males of each type are equally visible, the best response to the resident strategy of being totally non-choosy ($x = 0$) is to also be totally non-choosy. The relative payoff from mating with a type 2 male rather than a type 1 male increases as the resident x increases. This eventually results in females that follow the best response strategy being choosy. As can be seen there are three Nash equilibria at $x_1^* = 0$, $x_2^* = 0.234$, and $x_3^* = 1.503$. These Nash equilibria are also ESSs as best responses are unique. The outer ESSs are convergence stable and the middle ESS is not convergence stable (McNamara et al., 2003b). Thus

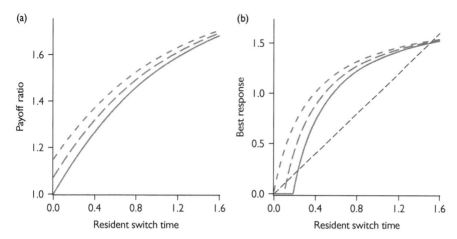

Fig. 10.5 The Fisher process with phenotypic inheritance of the male trait. The female mate-choice strategy is the time at which she switches to accepting any male. (a) The payoff ratio $\frac{r_2}{r_1}$. (b) Best responses. The 45° line is also shown. The visibility of type 1 males is $\kappa = 1$ for the solid lower curves, $\kappa = 0.9$ for the middle curves, and $\kappa = 0.8$ for the upper curves. Mating window of length $T = 2.5$. $N_1 = N_2$.

although there is a CSS at which females are choosy, it is separated from the non-choosy ESS by an invasion barrier and cannot be reached from this ESS. Decreasing the visibility of type 1 females to $\kappa = 0.8$ changes the situation. The best response to $x = 0$ is to be choosy, and there is a unique CSS at $x^* = 1.526$. It can be shown that $\rho_2 = 0.887$ at this ESS, so that preference for type 2 results in this type being more common.

One can easily extend this model to deal with more than two male types. With a large number of types that lie on a continuum, at evolutionary stability extreme male types tend to predominate unless there are factors such as biased mutation or reduced fecundity of extreme types that resist a runaway in the evolutionary process (Pomiankowski et al., 1991; McNamara et al., 2003b). In general the mating payoffs provide a common currency that takes into account fecundity differences and bias as well as the resident female mate-choice strategy. Similarly, in models with genetically determined male type they provide a common currency that reflects genetic quality and preference (Kokko et al., 2002).

Our illustrative example lends credence to the idea that the exaggerated tail of the widowbird is the result of a runaway process that was triggered by the greater visibility of long-tailed males. The model is, however, unrealistic for a variety of reasons, not least because the male type is phenotypically inherited. Genetic determination of the male type can be incorporated into a similar model by assuming that in addition to the mate-choice locus there is a separate locus determining male type. In the case of just two male types the genes determining a given mate-choice strategy **x** can be present in four categories of individual: (i) males that express the type 1 allele and that carry but do not express **x**, (ii) males that express the type 2 allele and that carry but do not express **x**, (iii) females that express **x** and carry the male type 1 allele, and (iv) females that express **x** and carry the male type 2 allele. The projection matrix for the change in numbers of carriers of a mutant mate-choice strategy is then four-dimensional.

Two models of this sort are possible. In one the male type allele is not expressed in females. In the other we allow the mate-choice strategy expressed by a female to depend on the male type allele she is carrying. The mate-choice strategy of a female can then be represented by a vector $\mathbf{x} = (x_1, x_2)$ where x_i is the mate-choice tactic she expresses if she is carrying the allele for male type i. To understand why there might be selection to have different mate-choice behaviour in the two circumstances, we can extend the idea of payoff maximization to four payoffs $r_{11}, r_{12}, r_{21}, r_{22}$, where r_{ij} is the payoff assigned to a female that carries the type j allele if she mates with a type i male. The analogue of eq (10.29) is then that under strategy $\mathbf{x}' = (x_1', x_2')$ the local payoff to a female that carries male type allele j is $m_{1j}(x_j', \mathbf{x})r_{1j} + m_{2j}(x_j', \mathbf{x})r_{2j}$, where $m_{ij}(x_j', \mathbf{x})$ is the probability the female mates with a type i male under local mate-choice strategy x_j'. A best response strategy maximizes both local payoffs. When $N_1 = N_2$ it can be shown that

$$r_{21} - r_{11} = r_{22} - r_{12} \tag{10.36}$$

(Exercise 10.5). Thus when $N_1 = N_2$ the additional payoff that a female gets from mating with a type 2 male, as opposed to a type 1 male, does not depend on the type allele she carries. However, a female that carries the type 2 allele is more likely to produce type 2 offspring than a female that carries the type 1 allele if both mate with the same male. Thus if males of type 2 are more valuable, then both sides of eq (10.36) are positive and we also have $r_{12} > r_{11}$. Equation (10.36) then implies that $\frac{r_{21}}{r_{11}} > \frac{r_{22}}{r_{12}}$. It follows that females that carry the male type 1 allele should be more choosy than females that carry the male type 2 allele.

When the mate-choice strategy of females cannot change with the male type allele they carry, we cannot employ the above analysis. When females are constrained to use the same strategy regardless of the type allele they carry, at evolutionary stability their mate-choice strategy is doing the best on average. Section 10.8 outlines approaches that can be used to find an ESS in this case.

10.6 Maximizing Lifetime Reproductive Success

Projection matrices are very powerful as they can incorporate intergenerational effects. However, in counting the number of descendants left at the next annual census time, the matrices do not differentiate between the surviving offspring of an individual and the individual itself, given that it survives. The emphasis is on the cohort of individuals following the same strategy rather than any one individual. Thus for species in which individuals live for many years the analysis using projection matrix gives no sense of how successful an individual is over its lifetime. This section examines how statements formulated in terms of projection matrices can be reformulated in terms of individual optimization.

The simplest situation to analyse is that in which age is the only state variable. Assume that the annual census time is just prior to the single breeding season in a year. Young produced in the breeding season do not themselves reproduce that year. Any that survive until the next annual census time are classified as age 0 at this time. Age then increases by 1 year at every subsequent census time. Define $w_i(\mathbf{x}',\mathbf{x})$ to be the expected future reproductive success (EFRS) of an individual of age i. By the term 'future reproductive success' we mean the number of offspring (discounted by relatedness) produced by the individual in the current and future years that manage to survive until their first annual census time at age 0. As the notation indicates, EFRS depends on the strategy adopted and the resident strategy. The EFRS of an individual at age 0, $w_0(\mathbf{x}',\mathbf{x})$, then measures the average total number of surviving offspring produced by an individual over its lifetime, and is referred to as mean lifetime reproductive success (mean LRS). Note that, since the resident strategy $\mathbf{x}$ is assumed to be at a demographic equilibrium, each age 0 resident leaves on average one surviving offspring in this state, so that $w_0(\mathbf{x},\mathbf{x}) = 1$.

In this simple setting, it is clear that if $w_0(\mathbf{x}',\mathbf{x}) < 1$ then the size of the cohort of x' mutants is declining so that $\lambda(\mathbf{x}',\mathbf{x}) < 1$. Conversely, if $w_0(\mathbf{x}',\mathbf{x}) > 1$ then the size of the cohort of x' mutants is increasing so that $\lambda(\mathbf{x}',\mathbf{x}) > 1$. Finally if $w_0(\mathbf{x}',\mathbf{x}) = 1$ then $\lambda(\mathbf{x}',\mathbf{x}) = 1$. From these relationships we have:

$$\lambda(\mathbf{x}',\mathbf{x}) \leq \lambda(\mathbf{x},\mathbf{x}) = 1 \text{ for all } \mathbf{x}' \tag{10.37}$$
$$\Longleftrightarrow w_0(\mathbf{x}',\mathbf{x}) \leq w_0(\mathbf{x},\mathbf{x}) = 1 \text{ for all } \mathbf{x}'. \tag{10.38}$$

This result was first proved by Taylor et al. (1974) for an age-structured population and extended to a state-structured population in which all offspring are in the same state by McNamara (1993). The result shows that a strategy is a Nash equilibrium for invasion fitness if and only if it is a Nash equilibrium for the payoff function w_0. The analogous result with strict inequalities also holds. These results show that mean LRS is a fitness proxy (Section 2.5).

Assuming appropriate conditions on the flexibility of strategies, the technique of dynamic programming can be used to find the strategy maximizing mean LRS (Taylor et al., 1974). One starts by considering behaviour at the maximum possible age, finding the action maximizing EFRF at this age. This then allows the action that maximizes EFRS at the penultimate age to be calculated. In this way one works backwards to find the strategy maximizing ERFS at age 0, i.e. maximizing mean LRS. To illustrate these results we look at an example of an age-structured population in which individuals must decide how to schedule reproductive effort over their lifetime.

10.6.1 Reproductive Scheduling

We model a large asexual population with an annual breeding season. The annual census time is just prior to breeding. At this time individuals are classified as being either age 0 or age 1, where by age 1 we mean this actual age or older. During the breeding season each individual decides on the effort to expend on reproduction, where efforts lie between 0 and 1. Thus a strategy is specified by the vector of efforts (e_0, e_1), where

e_0 is the effort at age 0 and e_1 is the effort when older. Breeding individuals compete for limited resources, with older individuals tending to outcompete age 0 individuals. Suppose that the resident strategy is (e_0, e_1) and that stable numbers in this population are n_0 and n_1. We assume that a mutant of age i has the potential to gain resources at rate r_i where

$$r_0 = \frac{R}{2 + n_0 e_0 + 2n_1 e_1} \quad \text{and} \quad r_1 = \frac{R}{1 + n_0 e_0 + n_1 e_1}, \quad (10.39)$$

where R is a constant that measures the richness of the environment. These formulae capture the idea that age 0 individuals get fewer resources than age 1 individuals, and furthermore they are disproportionately affected by the presence of age 1 competitors. If the mutant has strategy (e_0', e_1'), at age i its actual rate of resource gain is $r_i e_i'$, resulting in $r_i e_i' \theta$ offspring that survive until the annual census time the following year. Here $\theta < 1$ is a second environmental parameter that decreases as the harshness of winter increases. Offspring that have survived from last year are classified as age 0 at this time. The probability that an individual that expends effort e' on reproduction survives until the following annual census time is $(1 - e')\theta$. Surviving individuals are classified as age 1.

Let the resident strategy result in potential rates of resource gains r_0 and r_1. We first consider the effort of an age 1 mutant individual in this population. Let $w_1(e_1')$ be the EFRS of the mutant given it expends effort e_1'. Then $w_1(e_1') = r_1 e_1' \theta + (1 - e_1')\theta w_1(e_1')$, so that $w_1(e_1') = \frac{r_1 e_1' \theta}{1 - (1 - e_1')\theta}$. Since this is a strictly increasing function of e_1', the action that maximizes the mutant's EFRS is to expend the maximal effort $e_1' = 1$. As this result holds for all resident strategies, in looking for a Nash equilibrium strategy we can henceforth restrict attention to strategies that satisfy $e_1 = 1$. The EFRS of an age 1 individual is then $w_1^* = r_1 \theta$ where r_1 is determined by the resident strategy.

We now consider an age 0 mutant that expends effort e_0'. The EFRS of the mutant is then $w_0(e_0') = r_0 e_0' \theta + (1 - e_0')\theta w_1^* = r_1 \theta^2 + \theta(r_0 - r_1 \theta)e_0'$. Thus the best mutant effort is

$$e_0^* = 1 \quad \text{if} \quad r_0 > r_1 \theta \quad (10.40)$$
$$e_0^* = 0 \quad \text{if} \quad r_0 < r_1 \theta. \quad (10.41)$$

Motivated by the above we now suppose that the resident strategy is to expend maximal reproductive effort at age 0, i.e. $e_0 = 1, e_1 = 1$. Let n_0 be the equilibrium number of individuals of age 0 in the population (there are no age 1 residents as all age 0 residents die after reproduction). The demographic equilibrium condition is $n_0 = n_0 r_0 \theta$, so that $r_0 = \theta^{-1}$. Thus $n_0 = R\theta - 2$, by eq (10.39), so that $r_1 = \frac{R}{R\theta - 1}$. By eq (10.40), the resident strategy is the unique best response to itself if $r_0 > r_1 \theta$. Thus if $R > \frac{1}{\theta(1 - \theta)}$ then the strategy of maximum reproductive effort at age 0 (followed by death) is a strict best response to itself for the payoff w_0. By eq (10.37) the strategy is also a strict Nash equilibrium for invasion fitness, and is hence an ESS. We refer to this strategy as ESS0. Note that we also require that $R\theta > 2$ to ensure population viability.

Now suppose that the resident strategy is to expend zero effort at age 0 and then maximal reproductive effort at age 1; i.e. $e_0 = 0, e_1 = 1$. The demographic equilibrium

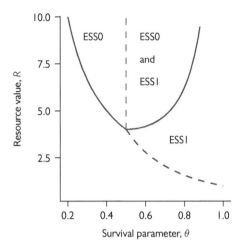

Fig. 10.6 Effect of environmental parameters on the ESSs for the reproductive scheduling model. Above the solid line the strategy ESS0 of maximizing reproduction at age 0 is both demographically viable and evolutionarily stable. Above and to the right of the dashed line the strategy ESS1 of delaying reproduction until age 1 and then maximizing reproduction is both demographically viable and evolutionarily stable.

conditions are $n_1 = n_0\theta$ and $n_0 = n_1 r_1\theta$, so that $r_1 = \theta^{-2}$. By eq (10.39) $n_1 = R\theta^2 - 1$, so that $r_0 = 0.5\theta^{-2}$. By eq (10.40), the resident strategy is the unique best response to itself if $r_0 < r_1\theta$. Thus if $\theta > 0.5$ then the strategy of delaying reproduction until age 1 and then reproducing maximally is evolutionarily stable. We refer to this strategy as ESS1. In this case we require that $R\theta^2 > 1$ to ensure population viability.

Figure 10.6 plots the dependence of the ESSs (for viable populations) on environmental parameters. As can be seen there are regions of parameter space when there are two alternative ESSs. These can both occur because of the effects of competition. When there are no age 1 individuals age 0 individuals get sufficient resources to make maximal reproduction optimal, resulting in no age 1 individuals; when reproduction is delayed to age 1 individuals are outcompeted at age 0 and it is best for them to delay reproduction until age 1. McNamara (1994) obtained multiple ESSs for similar reasons.

10.7 Dispersal

An environment that is composed of local patches or habitats linked by dispersal is referred to as a metapopulation. We previously encountered a metapopulation in Section 4.3, where two habitats with different characteristics were linked by a dispersal. In that model we took the dispersal probability as fixed and considered the evolution of the trait that determined reproductive success on each habitat. In this section we illustrate how the location of an individual acts as a state variable in a metapopulation context by describing how projection matrices are constructed for the two-habitat example. We then briefly describe a complication that can occur when the dispersal rate is low.

Recall that the two-habitat model of Section 4.3 assumes an asexual species with non-overlapping generations and an annual cycle of reproduction. An individual with trait x in habitat i ($i = 1, 2$) leaves $Ng_i(x)$ offspring where N is large. Each newborn individual either remains in its birth habitat (with probability $1 - d$) or disperses to the other habitat (with probability d). After the dispersal phase, all individuals currently in a habitat compete for the K territories in the habitat. Those individuals left without territories die, while the K that possess territories grow to maturity, so that the breeding population in a habitat has size K.

Let the resident strategy be x. Then the number of individuals that compete for territories in habitat 1 is $KNc_1(x)$, where $c_1(x) = (1 - d)g_1(x) + dg_2(x)$. Similarly $KNc_2(x)$ compete in habitat 2, where $c_2(x) = dg_1(x) + (1 - d)g_2(x)$. Let x' be a rare mutant strategy. A mutant in habitat 1 produces $Ng_1(x')$ offspring. Of these, a proportion $1 - d$ compete for the K territories in habitat 1. Thus on average $a_{11}(x', x) = \frac{KN(1-d)g_1(x')}{KNc_1(x)} = \frac{(1-d)g_1(x')}{c_1(x)}$ secure territories. The remaining offspring disperse to habitat 2 and secure $a_{21}(x', x) = \frac{KNdg_1(x')}{KNc_2(x)} = \frac{dg_1(x')}{c_2(x)}$ territories on average. Similar formulae hold for the offspring of mutants in habitat 2. The projection matrix for the mutant strategy is thus

$$\mathbf{A}(x', x) = \begin{bmatrix} \frac{(1-d)g_1(x')}{c_1(x)} & \frac{dg_2(x')}{c_1(x)} \\ \frac{dg_1(x')}{c_2(x)} & \frac{(1-d)g_2(x')}{c_2(x)} \end{bmatrix}. \tag{10.42}$$

The methods in Box 10.1 can then be used to find the invasion fitness of the mutant strategy, $\lambda(x', x)$. Exercise 10.6 illustrates this calculation.

The definition of the invasion fitness of a mutant strategy is as an appropriate measure of the per-capita rate of increase in the number of mutants when the mutation is rare in the environment. The assumption of rarity is to ensure that one can assume that the background environment does not change during this growth phase. Consider a metapopulation in which there are many local habitats linked by dispersal, but the carrying capacity of each local habitat is low. Then for low dispersal probability, a mutation that arises in a local habitat can quickly become common in this particular habitat while it is still rare in the environment as a whole. As a mutation becomes locally common it can change the local habitat and so affect other mutants. When this is the case, the approach to invasion fitness based on projection matrices that we applied to the two-habitat case, is not appropriate. This can be seen in the metapopulation model of McNamara et al. (2011). For moderate dispersal rates in this model, the analysis that takes invasion fitness to be the eigenvalue of the projection matrix for one individual gives results that are in agreement with evolutionary simulations. In contrast for low dispersal rates, predictions based on this measure of invasion fitness fail. One approach to this problem is to enlarge the idea of state, recording the proportion of habitats that have n mutants for each $n = 0, 1, 2, 3, \ldots$. McNamara and Dall (2011) employ this rather cumbersome approach to deal with the case of density dependence in their model of natal philopatry. However, even when it is possible to make analytic predictions in such complex problems, we would always recommend performing careful evolutionary simulations to check the analytic predictions and to test their generality. By themselves, analytic calculations may be of limited use. In the model of the co-evolution of prosociality and dispersal of Section 6.6 predictions were entirely based on simulations.

10.8 Evolutionary Analysis in Structured Populations

A potential difficulty for the analysis of game-theory models for structured populations is that invasion fitness $\lambda(\mathbf{x}',\mathbf{x})$ (from Section 10.1) is an eigenvalue of the projection matrix $\mathbf{A}(\mathbf{x}',\mathbf{x})$ and, except for the case of two demographic states (Box 10.1), tends to be hard to compute analytically. It may then also be difficult to compute derivatives of invasion fitness, which are needed to determine the strength and direction of selection, as well as to characterize the local properties of a singular point, where the selection gradient is zero (see Box 6.1). There are nevertheless a few approaches that can be helpful for the analysis of the general case mentioned in Section 10.1, where the components of a strategy $\mathbf{x} = (x_1, x_2, \ldots, x_L)$ need not correspond to independent actions in each demographic state s_j.

First, assuming that reproductive values $\mathbf{v}(\mathbf{x})$ are normalized such that $\sum_i v_i \rho_i = 1$, where $\rho(\mathbf{x}, \mathbf{x})$ is the stable state distribution, there is a general expression for the selection gradient, given by

$$\frac{\partial \lambda}{\partial x_k'}(\mathbf{x}, \mathbf{x}) = \mathbf{v}(\mathbf{x}) \frac{\partial \mathbf{A}}{\partial x_k'}(\mathbf{x}, \mathbf{x}) \rho(\mathbf{x}, \mathbf{x}), \tag{10.43}$$

where the derivative of the projection matrix has elements $\partial a_{ij}/\partial x_k'$ evaluated at $\mathbf{x}' = \mathbf{x}$. This is a standard result on eigenvalue sensitivity (Caswell, 2001, 2012) and is easy to derive (Exercise 10.8). Because the resident left and right eigenvectors $\mathbf{v}$ and ρ can readily be computed numerically, eq (10.43) can be used to numerically determine the evolutionary change in the strategy vector $\mathbf{x}$. There is, however, no similarly simple expression for the second derivatives of invasion fitness.

Second, let us examine the criteria (10.18) and (10.19) that we used to establish that a strategy is a Nash equilibrium. The left-hand side of the implication (10.18) still works as a sufficient condition in our general case, but the implication (10.19) might be less useful for characterizing Nash equilibria. The reason is that if a mutant strategy component x_k' influences the action in several demographic states s_j, a mutant could in principle do better than the resident in some demographic state but, because of poorer performance in other demographic states, still have lower invasion fitness. This limits the usefulness of reproductive value somewhat, but as eq (10.43) makes clear, the concept can still be helpful for evolutionary analysis in structured populations.

Third, there is another fitness proxy that is sometimes helpful (Metz and Leimar, 2011). It is related to a method for finding eigenvalues, as in eq (10.3), namely to study the characteristic polynomial, defined as

$$P(\lambda; \mathbf{x}', \mathbf{x}) = \det(\lambda \mathbf{I} - \mathbf{A}(\mathbf{x}', \mathbf{x})),$$

where $\mathbf{I}$ is the identity matrix and det denotes the determinant of a matrix. Eigenvalues are solutions to $P(\lambda; \mathbf{x}', \mathbf{x}) = 0$. For instance, from Section 10.1 we know that $P(1; \mathbf{x}, \mathbf{x}) = 0$. The proxy is given by

$$W(\mathbf{x}', \mathbf{x}) = -P(1; \mathbf{x}', \mathbf{x}). \tag{10.44}$$

This fitness proxy was first used by Taylor and Bulmer (1980) for local stability analysis for a sex-ratio model. Its general properties have been explored by Metz and Leimar (2011), showing that it can often be used for global stability analysis, as well as for local conditions such as those in Box 6.1. Still, there is the issue that determinants give rise to complex expressions when there are three or more demographic states, so this proxy also has limitations.

This leads to the conclusion that individual-based evolutionary simulations are important tools for the analysis of structured populations (as well as for the analysis of many game-theory models). Such simulations are sometimes the only manageable approach, and can in any case be a valuable complement to analytical and numerical approaches, because they more readily allow different kinds of genetics and behavioural mechanisms.

10.9 Exercises

Ex. 10.1. Let $(3,3)$ be the resident strategy in the model of Section 10.2. Use the results in Box 10.1 to show that $g_{3,3} = 1.2646$. Show that at demographic stability 27.43% of this resident population is High quality. Show that the ratio of reproductive values is $v_L/v_H = 0.3780$.

Ex. 10.2. Use eq (10.3) to show that eq (10.14) holds.

Ex. 10.3. Use eq (10.1) to verify that $d_j(t) = \mathbf{1}\mathbf{A}(\mathbf{x}, \mathbf{x})^t \mathbf{e}_j$, where $\mathbf{e}_j$ is the column vector with 1 in position j and zeros elsewhere, $\mathbf{1}$ is the row vector of 1s and t is a positive integer. Hence use eqs (10.5), (10.4), and (10.15) to show that the average reproductive value of resident population members (averaged across the steady-state distribution) is 1; i.e.

$$\sum_j \rho_j(\mathbf{x}, \mathbf{x}) v_j(\mathbf{x}) = 1 \tag{10.45}$$

Ex. 10.4. Let $\mathbf{x}$ be the resident strategy and $\mathbf{x}'$ a mutant strategy, and assume that $\mathbf{v}(\mathbf{x})\mathbf{A}(\mathbf{x}', \mathbf{x}) \leq \mathbf{v}(\mathbf{x})$. Use eq (10.3) to show that $\lambda(\mathbf{x}', \mathbf{x}) \leq 1$.

Ex. 10.5. Consider the version of the Fisher process of Section 10.5 in which males have two types and type is genetically determined by the allele at a haploid autosomal locus. On reproduction one type allele mutates to the other allele with probability ϵ.

Assume that the mate-choice strategy of a female can depend on the male type allele she carries. Assume a 1:1 sex ratio of offspring. Consider a rare $\mathbf{x}'$ mutant in a $\mathbf{x}$ population. Define $2a_{ij}(\mathbf{x})$ as the mean number of mutant offspring (half of each sex) that carry the type i allele, left by a type j male. Define $2\alpha_{ij}(\mathbf{x}',\mathbf{x})$ as the mean number of mutant offspring (half of each sex) that carry the type i allele, left by a type j female. Let $\bar{v}_i$ be the sum of the reproductive values of a type i son and a type i daughter in the $\mathbf{x}$ resident population. Set $W_j(\mathbf{x}',\mathbf{x}) = \bar{v}_1\alpha_{1,j}(\mathbf{x}',\mathbf{x}) + \bar{v}_2\alpha_{2,j}(\mathbf{x}',\mathbf{x})$.

(i) Adapt the analysis of Box 10.2 to show that if $W_1(\mathbf{x}',\mathbf{x}) \le W_1(\mathbf{x},\mathbf{x})$ and $W_2(\mathbf{x}',\mathbf{x}) \le W_2(\mathbf{x},\mathbf{x})$ then $\lambda(\mathbf{x}',\mathbf{x}) \le \lambda(\mathbf{x},\mathbf{x})$.

(ii) Show that $W_j(\mathbf{x}',\mathbf{x}) = m_{1j}(x'_j,\mathbf{x})r_{1j} + m_{2j}(x'_j,\mathbf{x})r_{2j}$ for suitably defined payoffs r_{ij}.

(iii) Show that eq (10.36) holds when $N_1 = N_2$.

Ex. 10.6. In the model of two habitats linked by dispersal of Sections 4.3 we have $g_1(0.25) = 0.9375$, $g_2(0.25) = 0.4375$ and $g_1(0.75) = 0.4375$, $g_2(0.75) = 0.9375$ (Fig. 4.6a). Consider a population in which the resident strategy is $x = 0.25$. Let $x = 0.75$ be a rare mutant strategy. Use eq (10.42) to write down the projection matrix for this mutant strategy when $d = 0.3$. Hence show that $\lambda(x',x) = 1.2137$. Note that this means that x' can invade x, and by symmetry, x can invade x'. This can also be seen from Fig. 4.7a.

Ex. 10.7. In a population individuals are characterized by their age at the start of a year. A mutant strategy specifies that all resources are put into growth until age n is reached, at which time the mutant reproduces and dies. The probability that a mutant survives from age 0 to age n is 0.85^n. If it does survive it leaves n age 0 recruits at the start of the next year. Let $\lambda(n)$ denote the invasion fitness of the mutant; i.e. the per-capita annual growth rate in mutant numbers. Let $w_0(n)$ denote the mean LRS of the mutant. Plot $\lambda(n)$ and $w_0(n)$ as functions of n, and hence find the values $\hat{n}$ and $\tilde{n}$ maximizing these two measures. Note that $\hat{n} < \tilde{n}$. Deduce that mean LRS cannot be a strong fitness proxy in the sense given by condition (2.7).

Ex. 10.8. Verify eq (10.43). It is helpful to use the following. Let $\lambda(\mathbf{x}',\mathbf{x})$ and $\rho(\mathbf{x}',\mathbf{x})$ be the eigenvalue and right eigenvector from eq (10.3). Start by taking the derivative of this equation with respect to x'_k and evaluate at $\mathbf{x}' = \mathbf{x}$. Then multiply this with the left eigenvector $\mathbf{v}(\mathbf{x})$ to the projection matrix $\mathbf{A}(\mathbf{x},\mathbf{x})$. From eq (10.16) the left eigenvector has eigenvalue $\lambda(\mathbf{x},\mathbf{x}) = 1$. Normalize the left eigenvector such that $\sum_i v_i\rho_i = 1$ holds at $\mathbf{x}' = \mathbf{x}$, and the result follows.

11

Future Perspectives

Game theory has typically used simple schematic models to expose the logic behind the action of frequency dependence. One reason for the success of the theory is precisely because the models have been simple. We contend, however, that the limitations of these simple models have not always been appreciated. For instance, adding richness to the models can alter their predictions (McNamara, 2013). Much of this book has explored the consequences of moving beyond the simple models of Chapter 3.

One issue is the way traits are modelled. In the simplest models, such as the Hawk–Dove game, there are just two actions and the trait is the probability of choosing one of these actions. As we and others have argued, rather than a trait being a probability, decisions are more likely to be based on some underlying state variable and the evolved trait is a decision threshold. In Section 3.11 we saw that this perspective changed predictions: in contrast to the case of a trait being a probability, threshold decisions predicted that at evolutionary stability those individuals taking one action gained a greater payoff than those taking the other action.

In Chapter 6 we argued that some traits are fundamentally multidimensional, and even when they are not, a focal trait can have a strong interaction with other traits. When this is the case, analysing the co-evolution of traits highlights feedback and the possibility of disruptive selection, and can lead to insights that would not be obtained by analysing a single trait in isolation.

There is extensive trait variation in most natural populations, but many of the standard game-theoretic models in biology ignore it, instead assuming that all individuals are the same. Variation plays a crucial role in many situations analysed in this book. In Chapter 7 we outlined how equilibria, for example signalling equilibria, can be stabilized by variation. Variation is also central to the existence of reputation effects and the workings of biological markets. In these contexts, the existence of variation selects for other traits such as social sensitivity or choosiness. Since these other traits interact strongly with the focal trait, this provides another reason to consider co-evolution rather than the evolution of an isolated trait.

In most of the standard games there is simultaneous choice, i.e. individuals choose their action without knowing the action of a partner and cannot later change their mind once their partner's action becomes known. This is not a realistic representation of most animal interactions. Instead, individuals show some flexibility in what they do, responding to aspects of their partner as the interaction progresses. Such interactions

Game Theory in Biology: Concepts and Frontiers. John M. McNamara and Olof Leimar,
Oxford University Press (2020). © John M. McNamara and Olof Leimar (2020).
DOI: 10.1093/oso/9780198815778.003.00011

can be regarded as a process in which the proponents learn about aspects of one another and are prepared to base their own behaviour on what they have learnt. Here variation is again crucial. If there were no variation, so that all possible partners always acted in the same way, there would be nothing to learn. Once there are real-time interactions there is a change in perspective; rather than regarding a strategy as specifying a single action, a strategy specifies the rule to respond to information as the interaction progresses. As we illustrate in Chapter 8, this changed perspective changes predictions about the outcome of the interaction.

In Chapter 9 we highlighted that many standard games are considered in isolation rather than being embedded in a wider model that specifies the ecological and life-history context. In particular, the payoffs are specified in advance rather than arising in a consistent manner from these outside considerations. As we have demonstrated, adopting a more holistic view can restrict the range of phenomena that are predicted.

Many of the above enrichments to game theory take place in a small world. But real organisms are adapted to large worlds, and have evolved strategies of limited complexity that perform well on average but may perform badly in specific circumstances, especially if circumstance of this type are rare. Almost all current models in biological game theory fail to take account of this. As with all enrichments, they are only worth pursuing if they tell us something new and lead to new predictions. We have introduced some approaches based on a large-worlds perspective that show that predictions can change radically: for example, we predicted biased decision rules in Section 8.5. Still, much more needs to be done. This is not necessarily an easy task as models have to specify all the details of the model environment, and hence are intrinsically small-worlds models. Taking a large-worlds perspective leads to the consideration of general learning rules and their consequences, and the limitations and consequences of behavioural mechanisms. We have argued that for many important categories of behaviour, including contests and social dominance, game theory needs to make use of large-worlds models. We further discuss these issue in Sections 11.2 and 11.3.

We can view possible future directions for biological game theory through the lens of Tinbergen's four questions (Tinbergen, 1963). These questions are phrased in Bateson and Laland (2013) as: What is it for? How did it evolve? How does it work? How did it develop? In the past game theory has largely focused on explaining traits as adaptations to the current circumstances, i.e. on the first question. Much of the book has also been in this tradition.

The second question of 'How did it evolve?' refers to phylogeny. Game theory has traditionally not been applied to evolutionary histories, but it could be a fruitful direction because the theory concerns itself with evolutionary trajectories and whether particular ESSs can be reached. The theory also reveals the possibility of multiple ESSs as well as changes to the stability of strategies as circumstances change, which we might regard as alternatives that are found by different species. We elaborate on these ideas in Section 11.1.

The third question of 'How does it work?' is concerned with the underlying neural and physiological mechanisms that bring about behaviour. Since it is necessary to specify a limited class of rules to implement when modelling evolution in large worlds, this question is central in applying a large-worlds perspective. In modelling learning we have focused on actor–critic reinforcement learning rules, which are rules that have some justification in terms of our knowledge of the action of dopamine in the brain. Section 11.2 discusses what can be gained by using such models. It also illustrates an interesting consequence of limited flexibility by individuals; it is worth noting that limited flexibility is a characteristic feature of behavioural mechanisms in large-worlds models.

We have so far not explicitly addressed the final question, but our focus on learning is related to development in that learning is about how current behaviour is acquired as a result of experience. This topic does not necessarily require a large-worlds perspective. For example, Bayesian methods (Section 3.12) are applicable in small-worlds models. Still, large-worlds perspectives are likely to be important for understanding cognitive development. In Section 11.3 we argue that game theory has the potential to address questions about cognitive development and cognitive sophistication in animals.

11.1 Phylogeny

Phylogeny is concerned with evolutionary history, recording the relationship between species through their lines of decent from common ancestors. Although physical features of the environment, such as resource abundance and the degree of seasonality, and external biotic factors, such as predator abundance, can affect what evolves, there is overwhelming empirical evidence that phylogeny is also important: closely related species tend to be similar. Assuming that each species that evolves is at an ESS, this suggests that for given environmental factors there can be more than one evolutionary outcome; i.e. multiple ESSs. Some of the models that we have presented certainly have more than one ESS. For example, in the model of the co-evolution of prosociality and dispersal of Section 6.6 there exist two very different ESSs: at one individuals were prosocial and had low dispersal, at the other they were not prosocial and dispersed more. In the model of competition between age classes presented in Section 10.6, there can be two ESSs, depending on environmental conditions: at ESS0 all population members reproduce in their first year of life (age 0), while at ESS1 all delay reproduction until a year older. In both of these examples, the strong dependence of the best response strategy on the resident strategy is responsible for the existence of more than one ESS. For other modelling examples of strong feedbacks that lead to multiple ESSs, see, for example, Alonzo (2010) and Lehtonen and Kokko (2012).

Parental care involves the co-evolution of multiple traits that have strong inter-dependencies (see e.g., Royle et al. 2016). So, for example, the extent of male care may influence male mortality and hence the adult sex ratio, which then exerts an influence on all individuals since one sex is more common than the other. How much a male should care also depends on his certainty of paternity and on remating opportunities, both of which depend on the behaviour of other males and females in the population. Thus one might expect that the physical environment does not uniquely determine the care strategy that evolves, and there is empirical evidence to support this. For example, Remeš et al. (2015) use the data on 659 bird species to examine the extent of cooperation between the parents during biparental care. They focus on the effect of three factors on the degree of cooperation: (A) the strength of sexual selection, as indicated by measures such as the level of extra-pair copulations and polygamy; (B) the adult sex ratio; and (C) the physical environment. They found that both factor A and factor B are correlated with the degree of cooperation. However, aspects of the physical environment, such as ambient temperature, are poor predictors of the degree of cooperation. Furthermore, there can be more than one form of care for given environmental conditions; even to the extent that reversed and conventional sex role species may breed side by side. (For more on this topic see Liker et al., 2013.) Remeš et al. (2015) conclude that several evolutionarily stable parental cooperation strategies may be adaptive in a given set of climatic conditions, although these strategies are likely to have co-evolved with the species' mating systems and demographic structure. They see this as a consequence of feedbacks among sexual selection, the social environment, and parental care, which are linked together in eco-evolutionary feedback loops (Alonzo, 2010).

In models, the range of ESSs that are possible in an environment can vary with environmental conditions. For example, in the model of competition between age classes presented in Section 10.6, there are three distinct regions of environmental space in which the population is viable (Fig. 10.6). In one region ESS0 is the unique ESS, in one ESS1 is the unique ESS, and in one both ESS0 and ESS1 are possible. One can consider the effect of a gradual change in the environment in such cases. For example, consider the effect of reducing the resource value, R, while holding the survival parameter fixed at $\theta = 0.75$ in Fig. 10.6. Suppose that initially we have $R = 10$ and that the resident strategy is ESS0. As R decreases this will remain the resident strategy until R falls below the value at which ESS0 ceases to be an ESS (this critical value can be seen to be approximately $R = 4.5$ in the figure). The population will then rapidly evolve to the only ESS, which is ESS1. If R were later to increase, the population would remain at ESS1 even when R returned to its initial value of $R = 10$. We might refer to this phenomenon of path dependence as a form of hysteresis.

Figure 11.1 shows two hypothetical cases where transitions from one ESS to another can occur. In both cases the environment is characterized by a single parameter, and there are two ESS for some values of the parameter and only one ESS for other values. Transitions between one ESS and the other occur when the environmental parameter changes so that the current ESS ceases to be an ESS. In Fig. 11.1a there would be an alternation between the two ESSs as the environmental parameter fluctuated over time. The length of time spent in each ESS would depend on how the environment fluctuated. For example, if low values of this parameter were common and higher values rare the lineage would spend most of its time following ESS1. In Fig. 11.1b transitions are one way; once the population follows ESS1 it can never transition to ESS2.

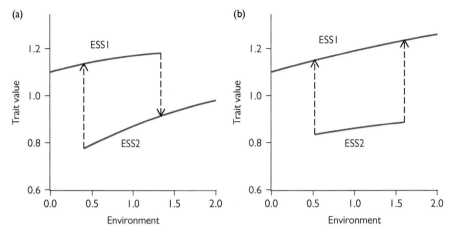

Fig. 11.1 Schematic representation of the transitions from one ESS to another under environmental change. Solid lines show the ESS trait values. Arrows show transitions between these ESSs as the environment changes.

The above suggest a modelling venture that would be used to predict the likelihood of the various changes that occur during evolution, and hence predict likely phylogenetic trees. If we could build a realistic model that specified the degrees of freedom in the physical environment, the relevant biological traits, and their influence on one another, then we could in principle identify what ESSs are possible for each physical environment. This model would allow us to predict how the ESS trait of a species changes as the environment changes. In particular, we would see to what extent the ESS reached in a particular environment depends on the (i) the starting environment and ESS, and (ii) the sequence of environmental steps that lead to the current environment. For a given ancestral environment and ESS (species) we could examine different isolated subpopulations. As these are subject to different environmental changes they would move apart, forming a range of new species. Thus even if the environmental conditions later become the same for all of these subpopulations, they will follow different ESSs (hysteresis).

As an example, the approach might have applications to understanding the phylogeny of mammalian social organization. Lukas and Clutton-Brock (2013) classified 2545 mammal species as solitary (68%), socially monogamous (9%), or group living (23%). The data suggest that for the common ancestor females were solitary with each male occupying a range that overlapped several females. Lukas and Clutton-Brock (2013) suggest that there have been 61 transitions for this solitary lifestyle to social monogamy in which males and females are paired. They also suggest that there has been at most one transition from group living to social monogamy, although the reverse may have occurred. Thus certain transitions are more likely to occur than others. The resource abundance may be a key environmental factor in what form of social organization evolves. In particular, when this abundance implies that females are at low density, males may be unable to defend more than one female, so that social monogamy evolves. Of course any application would require a lot of detail to be specified, and might be strongly dependent on the traits that are allowed to co-evolve.

11.2 Behavioural Mechanisms in Large Worlds

In this book we have developed and exemplified a large-worlds perspective that could be a useful direction for game theory in biology to pursue (see Chapter 5 and Sections 1.5, 8.5–8.8). Our concept of large worlds is similar to but in some ways different from that used in work on rational human decision-making and learning under ignorance (Savage, 1972; Binmore, 2009; Jones and Love, 2011; Huttegger, 2017). Thus, we are not aiming to provide a philosophical basis for how decision-makers might act in situations of limited knowledge of the nature of their world, but rather to develop the currently well-established tradition in behavioural ecology of integrating function and mechanism (McNamara and Houston, 2009). As pointed out by McNamara and Houston (2009), even if we are unable to predict from first principles which particular kinds of mechanisms should evolve, we can still analyse the evolution of parameters or component traits of an empirically established kind of mechanism. This is in fact a variant of the standard approach in evolutionary biology (Section 8.8). Instead of aiming to predict the time course of a system *a priori*, as might for instance be done in physics, in biology one typically reasons against the background of an already observed or inferred history of evolutionary changes in organisms and traits.

So what can be gained by focusing on relatively simple behavioural mechanisms in complex environments, rather than on small-worlds models with complex states and strategies involving Bayesian updating of probability distributions representing an individual's information about its environment? We see three major advantages. The first is that mechanisms often act as constraints on strategies. Sometimes evolutionary equilibria in a constrained strategy space have qualitatively distinct properties compared with unconstrained equilibria, thus giving rise to potentially novel predictions. We give an example of this dealing with mechanisms and flexibility (see below), illustrating that limited flexibility can make it possible to communicate characteristics such as trustworthiness.

The second and often overlooked advantage is that behavioural mechanisms can allow modellers to analyse important phenomena, such as learning about the characteristics of the members of a social group, that in effect are too difficult to study using traditional small-worlds models. As an illustration, game theory in biology started with attempting to answer the question of why animals often show limited aggression in contests (Maynard Smith and Price, 1973). If, as seems to be true, assessment of characteristics relating to fighting ability is a major aspect of the explanation, and if we wish to extend the analysis to interactions in social groups with dominance hierarchies, developing models with empirically inspired mechanisms for how individuals process information and select actions might be a workable option.

If we use this approach, the space of strategies to consider is a set of behavioural mechanisms spanned by variation in parameters or traits that can be tuned by evolution (Section 8.8). To the extent that these traits have a correspondence to the characteristics of real animals, there is the third and crucial advantage of a potentially closer link between model and observation. In effect, incorporating mechanisms can be a tool for game theory to achieve greater biological relevance. A consequence can be certain changes to the traditional methods of game theory in biology, including a greater reliance on individual-based evolutionary simulations.

11.2.1 Which Parameters are Tuned by Evolution?

If one incorporates mechanisms into game-theory models, one is faced with the question of which parameters should be treated as tunable by evolution. There is no ready answer, but there are some guidelines. First, it is probably not a good idea to assume that a great number of parameters will be perfectly tuned for every particular situation. Second, there is a long-standing and currently active study of the relative importance of special adaptations and general processes for animal cognition, including behavioural mechanisms. Associative learning could be the most studied aspect of cognition. There is a tradition in animal psychology maintaining that individuals of different species learn in more or less the same way. This was challenged by, among others, early ethologists who promoted the idea that psychological traits are special adaptations that should be expected to differ between species. The currently established view is that reality is somewhere between these positions (e.g. Bitterman, 1975; Shettleworth, 2010), but there is still the question of which cognitive traits are most likely to be adapted to special circumstances.

For social interactions, it is often assumed that general cognitive sophistication is important for assessing and evaluating the characteristics of group members. Nevertheless, the potentially great fitness impact of social interactions makes it likely that there are special adaptations to these circumstances. An example, from Section 8.6, could be the special neural processing of dominance interactions (Kumaran et al., 2016; Qu and Dreher, 2018; Zhou et al., 2018). Taborsky and Oliveira (2012) introduced the concept of social competence, meaning an individual's ability to display adequate behaviour using available information about a social situation, such as information about the interacting individuals. This would include the traits of behavioural mechanisms that are tuned by evolution. They note that both general and specialized cognitive capacities are likely to contribute to social competence. Varela et al. (2020) use the concept to outline when cognition is predicted to be adapted to social in comparison with non-social domains.

An idea we have illustrated in Chapters 5 and 8 is that species and populations can differ in how readily they perceive and take into account different stimuli, which is referred to as stimulus salience in animal psychology and is known to influence learning rates, and how strongly individuals value, both positively and negatively, certain outcomes of interactions, which falls under the heading of primary rewards, including penalties. O'Connell and Hofmann (2011) give an overview of the comparative neuroscience of salience and reward, with a focus on social behaviour. In the end it will be observations and experiments in animal psychology and neuroscience that decide the relative importance of special adaptations and general cognitive capacity for behavioural mechanisms. In Box 11.1 we give a few examples of observations on comparative cognition. So far, most work has been on the ability of different animals to solve various cognitive tasks, and there is much less information about the specific cognitive traits that underlie the differences. This is an important topic for future studies, with implications for game-theory modelling.

Box 11.1 Comparative cognition

Here we briefly describe a few studies on comparative cognition, spanning from species differences, over individual developmental plasticity, and to the role of general cognitive capacity in performance in cognitive tasks.

Species differences. Templeton et al. (1999) showed that the highly social pinion jay (a corvid bird) was relatively better at social (copying a conspecific) than non-social learning of tasks, compared with the less social Clark's nutcracker. Gingins and Bshary (2016) showed that a specialized cleaner fish species (*Labroides dimidiatus*) outperformed other and less specialized species in tasks linked to cleaner–client fish interactions, but performed at a similar level in other tasks. Munger et al. (2010) found that mice have a special olfactory processing of stimuli that can help them identify and learn about foods consumed by social group members, which ought to be safe to eat.

Developmental plasticity. Ashton et al. (2018) showed that Australian magpies that grew up in big groups performed better at a battery of cognitive tasks compared with individuals from small groups. Wismer et al. (2014) and Triki et al. (2018) found that *L. dimidiatus* cleaner fish living in socially simple environments, including those with reduced population density from environmental perturbations, performed more poorly in cognitive tasks linked to cleaner–client fish interactions, compared with cleaners from socially complex and unperturbed environments. Furthermore, Triki et al. (2019) showed that cleaners from high-density locations, with presumably more social interactions, had larger forebrain size, which is associated with cognitive function, compared with cleaners from low-density locations.

The importance of general cognitive capacity. Kotrschal et al. (2013) found that females in guppies artificially selected for larger brain size performed better in a numeric learning task than those selected for smaller brain size. MacLean et al. (2014) compared many different species in how they performed at two problem-solving tasks and found the performance to be positively correlated with brain size, using a phylogenetic comparative method.

11.2.2 Mechanisms and Flexibility

Game theoretic models typically predict behaviour in a single situation; sometimes there is only a single decision, sometimes the situation is experienced repeatedly and requires the same decision each time. In the real world organisms face a range of circumstances that are similar to but distinct from each other. Individuals often show behavioural consistency over these related but distinct circumstances (Sih et al., 2004). There may be cases in which there is selection pressure to be consistent (e.g. Wolf et al., 2007, 2011). However, in many cases simple evolutionary arguments suggest that limited plasticity is maladaptive; individuals could do better if they were more flexible (Wolf et al., 2011).

Strategies are implemented by physiological and psychological mechanisms. Any mechanism has finite complexity, so that the strategy it implements is necessarily of limited flexibility. Greater flexibility presumably requires more neural machinery, which is costly to maintain (Bullmore and Sporns, 2012). The number of subtly distinct circumstances encountered in the real world is vast. Thus the cost of increasing flexibility is one reason that we cannot expect a strategy that responds flexibly and appropriately to every different circumstance to evolve. Instead, we expect the evolution of strategies implemented by physiological and psychological mechanisms of limited complexity that perform well on average but may not be exactly optimal in any circumstance (McNamara and Houston, 2009). For example, Section 8.5 provides an illustration of a learning mechanism of limited complexity that is of bounded rationality and so is not optimal but can perform well if a suitable cognitive bias is introduced.

The implications of limited flexibility for game theory are underexplored. In Box 11.2 we describe a simple model (developed by McNamara and Barta, in preparation)

Box 11.2 A model of trust and the flexibility of behaviour

Each population member pairs up to play a sequence of trust games, each with a different partner, as in Section 7.7. There are two circumstances, O and U, in which pair members make their trust decisions. The interaction is observed by other population members in O but not in U (observations are cost-free). Individuals are aware of their circumstance. Since the previous behaviour of the partner in circumstance U is not known to Player 1, the decision to trust or reject the partner is based solely on the partner's reputation in circumstance O.

There are four genetically determined parameters. p_o and p_u specify the probabilities of cooperating if trusted in the two circumstances, when in the role of Player 2. t_o and t_u determine behaviour when in the role of Player 1. The partner is trusted if $\tilde{p}_o \geq t_o$ when observed and trusted if $\tilde{p}_o \geq t_u$ when unobserved, where $\tilde{p}_o$ is the probability the partner is trustworthy when observed. The fitness of a strategy is proportional to the average payoff per round discounted by the factor $e^{-c(p_o - p_u)^2}$, where $c > 0$ determines the cost of flexibility.

Continued

Box 11.2 *Continued*

When observed, a Player 1 should accept a Player 2 if and only if their probability of being trustworthy satisfies $\tilde{p}_o \geq \frac{s}{r}$, eq (7.5). Thus in order to be trusted in future, the Player 2 should have a probability of being trustworthy such that $\tilde{p}_o \geq \frac{s}{r}$.

Figure 11.2 illustrates the effect of the cost parameter c on evolved behaviour. For small c, Player 2s are only sufficiently trustworthy to be trusted when observed. They are rather untrustworthy when unobserved, and are rarely trusted. For large c, Player 2s are again only sufficiently trustworthy to be trusted when observed. Since Player 2s pay a high cost for being flexible, most have approximately the same probability of being trustworthy when not observed, so Player 2s tend to be trusted when not observed and are indeed trustworthy. When the cost to Player 2 is intermediate, a Player 2 with $\tilde{p}_o$ just greater than $\frac{s}{r}$ is liable to have $\tilde{p}_u < \frac{s}{r}$ and should not be trusted when unobserved. However, if $\tilde{p}_o$ is much greater than $\frac{s}{r}$, then we might expect $\tilde{p}_u > \frac{s}{r}$ since otherwise the Player 2 would incur high flexibility costs. Thus a Player 2 with high $\tilde{p}_o$ should be trusted when unobserved. Consequently, for intermediate costs of flexibility, Player 2s are very trustworthy when observed in order to convince Player 1s that they are also trustworthy when unobserved.

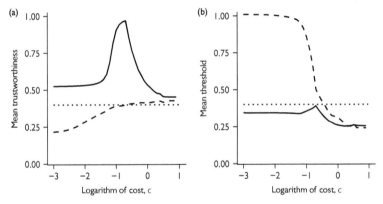

Fig. 11.2 Effect of the cost of flexibility on trust behaviour. Plots show evolved mean values of the four genetically determined parameters as a function of $\log_{10}(c)$. (a) p_o (solid line) and p_u (dashed line). (b) t_o (solid line) and t_u (dashed line). Each trust interaction is observed with probability 0.3. Rewards are $s = 0.24, r = 0.6$: the plots also show the value of the critical ratio $\frac{s}{r} = 0.4$ (dotted lines). Evolutionary simulation of an infinite population for 50,000 generations with values of each parameter constrained to lie on a discrete grid of 40 evenly separated points.

that attempts to capture the consequences of the costs of being flexible. As in the model of Section 7.7, individuals decide whether to trust their partner, and decide whether to be trustworthy themselves if they are trusted by their partner. Their choice of action can depend on whether others are observing them or not, although being flexible across circumstances is costly. The key finding of the analysis of this model is that predictions when there are intermediate levels of flexibility differ qualitatively to predictions when individuals are completely inflexible or completely flexible. Specifically, when the cost of flexibility is intermediate individuals are very trustworthy when observed in order to convince future partners that they are also trustworthy when unobserved (Fig. 11.2).

The above model is highly schematic and does not include a realistic mechanism that determines behaviour. In nature, emotions such as fear or affection may play an important role in limiting the flexibility of individuals. This is because emotions are to some extent generalized responses, meaning they lump together similar circumstances. Future work should consider whether it is biologically realistic and useful to explicitly incorporate these and other aspects of mental states in game-theoretical models.

11.3 Ontogeny and the Acquisition of Behaviour

Developmental plasticity arises from divergence of developmental trajectories during ontogeny and is an established field of investigations in game theory and evolutionary ecology. We have covered several examples, such as alternative mating types and environmental sex determination (Section 3.11), and cases of signalling (Section 3.6) where the signal is determined during development. Here we are instead concerned with the possible influence of the social environment during ontogeny on cognitive development, with possible relevance for game theory.

Studies on the cleaning mutualism between the coral reef fish *L. dimidiatus* (referred to in Box 11.1) and other fish illustrate how this might work. Cleaners occupy small territories or cleaning stations on coral reefs where they remove and feed on ectoparasites from so-called client fish. Clients differ in many respects, including in whether they are 'residents' with small home ranges, giving them limited choice between cleaning stations, or 'visitors' with large home ranges, encompassing several cleaning stations. A cognitive challenge for cleaners is to be able to discriminate between residents and visitors. Being able to do so can be advantageous for a cleaner, because visitor clients tend to leave for another station if not serviced right away, whereas resident clients are willing to wait.

By taking cleaners into the lab and exposing them to cognitive tasks showing some similarity to natural cleaning interactions, it was discovered that cleaners from different field locations differed in how readily they would learn these tasks (Wismer et al., 2014; Triki et al., 2018). In one of the tasks, the so-called biological market task, individual cleaners in the lab were presented with a choice between two 'artificial clients', in the form of Plexiglas plates with food on them. One type of plate would immediately be withdrawn if the cleaner selected the other plate first, whereas the other type would remain until the cleaner had eaten the food on it. The studies showed that cleaners from locations where they would continually be exposed to choices in the natural biological market, and where those choices would influence their feeding rate, more readily learnt to first eat from the 'impatient' type of plate. As cleaners are open-water spawners with pelagic eggs and larvae, spending time in the ocean before settling on a reef, it is not likely that the differences can be explained as local adaptations. Instead it appears that growing up in a suitable social environment causes cleaners to develop the needed cognitive sophistication.

Quiñones et al. (2020) analysed a game-theory model of how cleaners might learn to discriminate between resident and visitor clients, using actor–critic reinforcement learning. The model is simplified in that it does not explicitly represent the different traits of client fish that cleaners might use for discrimination. The analysis still makes the qualitative point that, in order to succeed in the discrimination task, cleaners need to learn to associate the current situation at the cleaning station, in terms of which clients are present, with rewards from the consumption of ectoparasites in the near future. Crucially, it is not enough for cleaners to learn that clients differ in how much reward they represent and to choose the immediately most rewarding client. Instead, they need to take into account the current configuration of clients, and this could be difficult to learn. The model thus suggests that only in certain social environments will a cleaner have sufficient incentive and experiences to learn to discriminate between and efficiently respond to visitors and residents.

The example indicates that game theory could contribute to the study of cognitive development. There might well be many instances where animals need to be exposed to the right kind of social environment in order to become competent at handling the complexities of social life. For game theory to assist in throwing light on such questions, it needs to incorporate the requisite mechanisms from cognitive psychology and neuroscience.

Appendix A

Summary of Notation

x	a resident strategy
x'	a mutant strategy
$\lambda(x',x)$	invasion fitness of the mutant strategy x' in a resident x population
$W(x',x)$	payoff of the mutant strategy x' in a resident x population, where the payoff function W is a fitness proxy
$\hat{b}(x)$	best response to the resident strategy x
$\frac{\partial W}{\partial x'}$	partial derivative of the payoff $W(x',x)$ with respect to its first argument; i.e. the rate of change of $W(x',x)$ as x' changes while x is held fixed
$\frac{\partial W}{\partial x'}(a,b)$	partial derivative of $W(x',x)$ with respect to its first argument evaluated at $x' = a$ and $x = b$
$D(x)$	strength of selection: $\frac{\partial W}{\partial x'}(x,x)$
s_i	state
u_i	action performed by individual (or role) i, or in state s_i
a_{ij}	element of projection matrix giving demographic transitions from s_j to s_i
R_{it}	actual reward for individual i from round t of actor–critic learning
w_{it}	estimated reward by individual i at the start of round t of actor–critic learning
δ_{it}	temporal-difference error for individual i from round t of actor–critic learning: $R_{it} - w_{it}$
θ_{it}	preference by individual i at the start of round t for the first of its two actions
p_{it}	probability for individual i of selecting the first of its two actions in round t: $\frac{1}{1+\exp(-\theta_{it})}$
α_w, α_θ	learning rates in actor–critic learning
ζ_{it}	eligibility of the action used by individual i in round t of actor–critic learning: the derivative with respect to the action preference θ_{it} of the logarithm of the probability of selecting that action

Appendix B

Solutions to Exercises

Solutions for Chapter 3

Ex. 3.1. For resident strategy x the payoff to the mutant strategy x' is

$$W(x',x) = 2(x'+x) - (x'+x)^2 - x'^2.$$

[Strictly speaking, this is only valid for $x'+x \leq 1$, but the best response will never be greater than $1-x$ as there is no extra benefit for this but there is an extra cost.] Differentiating $W(x',x)$ partially with respect to x' (holding x fixed) gives

$$\frac{\partial W}{\partial x'} = 2\left[1 - (x'+x) - x'\right].$$

Let $\hat{b}(x) = \frac{1-x}{2}$. Then the partial derivative is positive for $x' < \hat{b}(x)$ and negative for $x' > \hat{b}(x)$, so that $W(x',x)$ has a unique maximum at $x' = \hat{b}(x)$, and this is the unique best response to the resident strategy.

To find a Nash equilibrium x^* we set $\hat{b}(x^*) = x^*$. This gives $\frac{1-x^*}{2} = x^*$, so that $x^* = \frac{1}{3}$.

Ex. 3.2. Let $x \geq 0$ be the resident strategy. Differentiating $W(x',x)$ partially with respect to x' (holding x fixed) gives

$$\frac{\partial W}{\partial x'}(x',x) = B'(x'+x) - C'(x').$$

Let $\hat{b}(x)$ be the value of x' at which this derivative is zero, so that

$$B'(\hat{b}(x)+x) = C'(\hat{b}(x)). \tag{S1}$$

This is our candidate best response, but we must check some details. We first note that, by our assumptions about B and C, the second derivative

$$\frac{\partial^2 W}{\partial x'^2}(x',x) = B''(x'+x) - C''(x')$$

is strictly negative for all $x' \geq 0$. Thus $W(x',x)$ has a unique maximum for $x' \geq 0$. Furthermore, evaluating the partial derivative at $x' = 0$ we get $\frac{\partial W}{\partial x'}(0,x) = B'(x) - C'(0) > 0$ since $C'(0) = 0$. Thus the maximum is an internal maximum and the partial derivative must equal zero at this value. This shows that $\hat{b}(x)$, given by eq (S1), is the unique best response.

Differentiating both sides of eq (S1) gives

$$B''(\hat{b}(x)+x)\left[\hat{b}'(x) + 1\right] = C''(\hat{b}(x))\hat{b}'(x).$$

Rearranging this equation

$$\hat{b}'(x) = \frac{B''(\hat{b}(x) + x)}{C''(\hat{b}(x)) - B''(\hat{b}(x) + x)}.$$

By our assumptions about B and C, the numerator of this expression is negative and the denominator is positive and of greater absolute value than the numerator. Thus $1 < \hat{b}'(x) < 0$.

Ex. 3.3. Let the resident strategy be x and consider a mutant strategy x'. We are given that $W(x',x) = (1 - x'^4 x^{4(n-1)}/n)x'$. Differentiating $W(x',x)$ partially with respect to x' (holding x fixed) gives

$$\frac{\partial W}{\partial x'}(x',x) = 1 - \left(\frac{5}{n}\right)x'^4 x^{4(n-1)}.$$

Consider first the case where $n \geq 5$. Then $\frac{\partial W}{\partial x'}(x',x) > 0$ for all $x' < 1$, so that $W(x',x)$ is a strictly increasing function of x'. Thus the best response is to use the maximum value of x'; i.e. $\hat{b}(x) = 1$. Since this is true for all resident strategies the unique Nash equilibrium is $x^*(n) = 1$.

Now let $n \leq 4$. Let $\tilde{b}(x)$ satisfy $\frac{\partial W}{\partial x'}(\tilde{b}(x),x) = 0$, so that

$$\tilde{b}(x) = \left(\frac{n}{5}\right)^{\frac{1}{4}}\frac{1}{x^{n-1}}.$$

Then it can be seen that the best response is $\hat{b}(x) = \min(\tilde{b}(x), 1)$. Since $\hat{b}(1) < 1$ any Nash equilibrium must satisfy $x^* < 1$, so that

$$\left(\frac{n}{5}\right)^{\frac{1}{4}}\frac{1}{x^{*(n-1)}} = x^*,$$

which gives

$$x^* = \left(\frac{n}{5}\right)^{\frac{1}{4n}}.$$

This formula gives $x^*(1) = 0.669$, $x^*(2) = 0.892$, $x^*(3) = 0.958$, $x^*(4) = 0.986$.

Ex. 3.4. The payoff to the female when she contributes effort x and the male contributes y is $W_f(x,y) = 2(x+y) - (x+y)^2 - x^2$ (for $x + y < 1$). Partially differentiating with respect to x

$$\frac{\partial W_f}{\partial x}(x,y) = 2 - 2(x+y) - 2x.$$

Setting this derivative equal to zero at $x = \hat{b}_f(y)$ gives the best response function of the female:

$$\hat{b}_f(y) = \frac{1-y}{2}. \tag{S2}$$

The payoff to the male when he contributes effort y and the female contributes x is $W_m(x,y) = 2(x+y) - (x+y)^2 - 2y^2$ (for $x + y < 1$). Partially differentiating with respect to y

$$\frac{\partial W_m}{\partial y}(x,y) = 2 - 2(x+y) - 4y.$$

Setting this derivative equal to zero at $y = \hat{b}_m(x)$ gives the best response function of the male:

$$\hat{b}_m(x) = \frac{1-x}{3}. \tag{S3}$$

The pair (x^*, y^*) of Nash equilibrium strategies satisfies $x^* = \hat{b}_f(y^*)$ and $y^* = \hat{b}_m(x^*)$. By eqs (S2) and (S3)

$$\frac{1 - y^*}{2} = x^* \quad \text{and} \quad \frac{1 - x^*}{3} = y^*.$$

Solving these simultaneous equations gives $x^* = \frac{2}{5}$ and $y^* = \frac{1}{5}$.

At the Nash equilibrium the female pays cost $x^{*2} = 0.16$ and the male pays cost $2y^{*2} = 0.08$.

Ex. 3.5. Equation (3.8) is

$$w(x, q; \bar{x}) = \left(1 - b + b\frac{x}{\bar{x}}\right)v(x, q).$$

Here $v(x, q) = u(q)\left(1 - c(x - q)^2\right)$, where $u(q)$ is positive and increasing with q. The precise form of the multiplicative factor $u(q)$ does not influence the optimum signal. We get that

$$\frac{\partial w}{\partial x} = u(q)\left(\frac{b}{\bar{x}}\left(1 - c(x - q)^2\right) - \left(1 - b + b\frac{x}{\bar{x}}\right)2c(x - q)\right). \tag{S4}$$

We can note that this derivative is quadratic in x with a negative coefficient of x^2 and that it is positive for $x = q$. This means that the maximum of w will occur for the larger of the two roots of the quadratic equation $\partial w/\partial x = 0$. Collecting terms this quadratic corresponds to

$$-3bcx^2 + \left(4bcq - 2(1 - b)c\bar{x}\right)x + b(1 - cq^2) + 2(1 - b)cq\bar{x} = 0,$$

with solution

$$x = \frac{2bcq - (1 - b)c\bar{x} + \sqrt{\left(2bcq - (1 - b)c\bar{x}\right)^2 + 3b^2c(1 - cq^2) + 6b(1 - b)c^2q\bar{x}}}{3bc}.$$

We can simplify this expression to

$$x = \frac{2bcq - (1 - b)c\bar{x} + \sqrt{3b^2c + b^2c^2q^2 + 2b(1 - b)c^2q\bar{x} + (1 - b)^2c^2\bar{x}^2}}{3bc}$$

$$= \frac{2}{3}q - \frac{(1 - b)\bar{x}}{3b} + \sqrt{\frac{1}{3c} + \frac{q^2}{9} + \frac{2(1 - b)q\bar{x}}{9b} + \frac{(1 - b)^2\bar{x}^2}{9b^2}}.$$

Ex. 3.6. From eq (S4) in the previous derivation we get that

$$\frac{\partial}{\partial q}\frac{\partial w}{\partial x} = \frac{u(q)}{\bar{x}}c\left(4b + 2(1 - b)\bar{x} - 2bq\right) = \frac{u(q)}{\bar{x}}c\left(2b + 2b(1 - q) + 2(1 - b)\bar{x}\right),$$

which is positive.

Ex. 3.7. Let the resident investment strategy be x. Then a resident female contributes $f(x) = \frac{(1-x)R}{r_f}$ daughters and $m(x) = \frac{xR}{r_m}$ sons to the breeding population in the next generation. The ratio of females to males in the breeding population is thus

$$s(x) = \frac{f(x)}{m(x)} = \frac{r_m}{r_f}\left(\frac{1 - x}{x}\right).$$

The payoff to a mutant strategy x' is

$$W(x',x) = f(x') + m(x')s(x)$$

$$= \frac{R}{xr_f}(x + x' - 2xx')$$

$$= \frac{R}{r_f} + \frac{R}{xr_f}(1 - 2x)x'.$$

From the last line of this equation it can be seen that if $x < \frac{1}{2}$ then the payoff is maximized when $x' = 1$. If $x > \frac{1}{2}$ it is maximized when $x' = 0$. If $x = \frac{1}{2}$ all values of x' are best responses. Thus the unique Nash equilibrium is $x^* = \frac{1}{2}$.

Ex. 3.8. Consider an individual that waits for time t before giving up. Let the opponent give up after time s. If $s < t$ the payoff to the focal individual is $v - cs$. If $s > t$ the payoff to the focal individual is $-ct$. Suppose that the waiting time in the population has probability density function $f(.)$. Then the average payoff to the focal individual is

$$R(t) = \int_0^t (v - cs)f(s)ds + \int_t^\infty (-ct)f(s)ds.$$

Let the resident strategy be to persist for an exponential time with parameter x. Then $f(s) = xe^{-xs}$. By integration we obtain

$$R(t) = (v - \frac{c}{x})(1 - e^{-xt}).$$

Consider a mutant that persists for an exponential time with parameter x' in this resident population. The payoff to this mutant is

$$W(x',x) = \int_0^\infty R(t)x'e^{-x't}dt$$

$$= \frac{xv - c}{x + x'}.$$

From this formula it can be seen that:

- If $xv > c$ then the best response is $x' = 0$; i.e. to remain indefinitely.
- If $xv = c$ then $W(x',x) = 0$ for all x', so that all x' are best responses.
- if $xv < c$ then the larger x' the better, so that the best response is to give up immediately.

Thus $x^* = \frac{c}{v}$ is the unique Nash equilibrium. At this equilibrium all population members have a payoff of zero.

Ex. 3.9. From eq (3.13)

$$W_D(q';x) = \int_0^x \frac{V}{2}dq = \frac{Vx}{2}.$$

From eq (3.14)

$$W_H(q';x) = \int_0^x Vdq + \int_x^1 \left[\alpha(q',q)V - (1 - \alpha(q',q))C\right]dq$$

$$= Vx + \frac{1}{2}\int_x^1 \left[(1 + q' - q)V - (1 + q - q')C\right]dq \quad \text{from eq (3.12)}$$

$$= Vx + \frac{1}{2}\left[(1 + q')V - (1 - q')C\right](1 - x) - \frac{1}{4}(V + C)(1 - x^2).$$

Thus

$$4W_H(q';x) = V - 3C + 2(V+C)x + (V+C)x^2 + 2(V+C)(1-x)q'.$$

From eq (3.17) $W_D(\hat{b}(x);x) = W_H(\hat{b}(x);x)$, so that on rearrangement

$$\hat{b}(x) = \frac{3C - V - 2Cx - (V+C)x^2}{2(1-x)(V+C)}.$$

We can express this as

$$\hat{b}(x) = \frac{3 - r - 2x - (r+1)x^2}{2(1-x)(r+1)},$$

where $r = \frac{V}{C}$. Setting $\hat{b}(x^*) = x^*$ gives

$$(r+1)x^{*2} - 2(r+2)x^* + (3-r) = 0.$$

This quadratic equation has two roots

$$x^* = \frac{(r+2) \pm \sqrt{2r^2 + 2r + 1}}{(r+1)},$$

We require the root that lies in the interval $[0,1]$, so take the minus sign of the $\pm$.

Ex. 3.10. Let $20 < x < 30$. For resident strategy x, $F(x)$ is an average over the number of recruits left by resident females, so that

$$F(x) = \int_x^{30} f(q)\frac{dq}{10}$$

$$= \int_x^{30} \frac{(q-20)}{10}\frac{dq}{10}$$

$$= \int_\beta^1 z\,dz \quad \text{where } z = (q-20)/10$$

$$= \frac{1}{2}(1-\beta^2).$$

Similarly

$$M(x) = \int_{20}^x m(q)\frac{dq}{10}$$

$$= \int_x^{30} \frac{1}{2}\frac{dq}{10}$$

$$= \int_0^\beta \frac{1}{2}dz \quad \text{where } z = (q-20)/10$$

$$= \frac{1}{2}\beta.$$

By eq (3.20) at a Nash equilibrium threshold x^* we have

$$\frac{0.1(x^* - 20)}{0.5(1 - \beta^{*2})} = \frac{0.5}{0.5\beta^*},$$

where $\beta^* = (x^* - 20)/10$. Thus

$$\frac{\beta^*}{0.5(1-\beta^{*2})} = \frac{1}{\beta^*},$$

so that $3\beta^{*2} = 1$ and hence $\beta^* = 1/\sqrt{3}$. This gives $x^* = 20 + 10\beta^* = 20 + \frac{10}{\sqrt{3}}$.

Solutions for Chapter 4

Ex. 4.1. From Table 3.3

$$\begin{aligned} W(p',p) &= (1-p')(1-p) \times 1 + (1-p')p \times (-1) + p'(1-p) \times (-1) + p'p \times 1 \\ &= 1 - 2p' - 2p + 4p'p \\ &= (1-2p')(1-2p). \end{aligned}$$

Let the resident strategy be $p^* = 0.5$. Then $W(p',p^*) = 0$ for all p', so that p^* is a best response to itself and hence a Nash equilibrium.

Let p' be a mutant strategy with $p' \neq p^* = 0.5$. Then $W(p',p^*) = 0 = W(p^*,p^*)$ so that condition (ES1) fails but (ES2)(i) holds. We also have

$$\begin{aligned} W(p^*,p') - W(p',p') &= (1-2p^*)(1-2p') - (1-2p')(1-2p') \\ &= (2p' - 2p^*)(1-2p') \\ &= -(1-2p')^2 \quad \text{since } p^* = 0.5 \\ &< 0. \end{aligned}$$

Thus condition (ES2)(ii) fails, so that p^* is not an ESS.

Ex. 4.2. To simplify notation let $\alpha = w_0 + w(\epsilon(t))$ and $\alpha^* = w_0 + w^*(\epsilon(t))$. Then $n(t+1) = K\alpha n(t)$ and $n^*(t+1) = K\alpha^* n(t)$. Thus

$$\epsilon(t+1) = \frac{n(t+1)}{n(t+1) + n^*(t+1)} = \frac{\alpha n(t)}{\alpha n(t) + \alpha^* n^*(t)}. \tag{S5}$$

Since $\epsilon(t) = n(t)/(n(t) + n^*(t))$ we have $n^*(t) = n(t)(1 - \epsilon(t))/\epsilon(t)$. By eq (S5)

$$\epsilon(t+1) = \frac{\alpha\epsilon(t)}{\alpha\epsilon(t) + \alpha^*(1 - \epsilon(t))}.$$

Thus

$$\begin{aligned} \epsilon(t+1) - \epsilon(t) &= \epsilon(t)\left[\frac{\alpha}{\alpha\epsilon(t) + \alpha^*(1 - \epsilon(t))} - 1\right] \\ &= \frac{\epsilon(t)(1 - \epsilon(t))(\alpha - \alpha^*)}{\alpha\epsilon(t) + \alpha^*(1 - \epsilon(t))}, \end{aligned}$$

and the required result follows immediately.

Ex. 4.3. Let $W_i(p)$ denote the payoff for action u_i when the resident strategy is to play action u_2 with probability p. Then $W_i(p) = (1-p)a_{i1} + pa_{i2}$. Set $\Delta(p) = W_2(p) - W_1(p)$. We note that for any strategies p and p' we have

$$W(p',p) = (1-p')W_1(p) + p'W_2(p) = W_1(p) + \Delta(p)p'. \tag{S6}$$

We also note that a direct calculation gives $\Delta(p) = (a_{21} - a_{11}) + \alpha p$, where $\alpha = (a_{11} + a_{22}) - (a_{12} + a_{21})$, so that $\Delta(p)$ is a linear function of p with non-zero slope.

Let p^* be a Nash equilibrium satisfying $0 < p^* < 1$. Then $W(p',p^*) = W_1(p^*) + \Delta(p^*)p'$ by eq (S6). Thus if $\Delta(p^*) \neq 0$ this function of p' would have a unique maximum at either 0 or 1, so that p^* would not be a best response to itself. We conclude that $\Delta(p^*) = 0$. Since $\Delta(p)$ is a linear function of p with a zero at $p = p^*$ and slope α we have $\Delta(p) = \alpha(p - p^*)$. We can therefore rewrite eq (S6) as

$$W(p',p) = W_1(p) + \alpha(p - p^*)p'. \tag{S7}$$

(i) We first consider best responses to the two pure strategies. By eq (S7) we have $W(p',0) = W_1(0) - \alpha p^* p'$. Thus the pure strategy $p' = 1$ is the unique best response to $p = 0$ if and only if $\alpha < 0$. Similarly, the pure strategy $p' = 0$ is the unique best response to $p = 1$ if and only if $\alpha < 0$.

(ii) We now examine the conditions for p^* to be an ESS. Let $p' \neq p^*$ be a rare mutant strategy. By eq (S7) we have $W(p',p^*) = W_1(p^*) = W(p^*,p^*)$, so that condition (ES1) fails but condition (ES2)(i) holds. By eq (S7)

$$W(p',p') - W(p^*,p') = \alpha(p' - p^*)^2.$$

Thus condition (ES2)(ii) holds if and only if $\alpha < 0$.

(iii) By eq (S7), $D(p) \equiv \frac{\partial W}{\partial p'}(p,p) = \alpha(p - p^*)$. Thus $D'(p) = \alpha$. By inequality (4.8), p^* is convergence stable if and only if $\alpha < 0$.

Ex. 4.4. Let x^* be a resident strategy and x' a rare mutant strategy. Assume that condition (ES2)(i) holds; i.e. $W(x',x^*) = W(x^*,x^*)$. In condition (ES2′)(ii) we use the temporary notation that $W(x,x_\epsilon)$ denotes the payoff to an x strategist when a proportion $1 - \epsilon$ of the population is x^* strategists and a proportion ϵ is x' strategists. For a two-player game the probabilities that an individual partners each of these strategists are $1 - \epsilon$ and ϵ, respectively. Thus

$$W(x,x_\epsilon) = (1 - \epsilon)W(x,x^*) + \epsilon W(x,x').$$

Condition (ES2′)(ii) states that $W(x',x_\epsilon) < W(x^*,x_\epsilon)$ for sufficient small, but positive, ϵ. By the above we have

$$W(x',x_\epsilon) < W(x^*,x_\epsilon)$$

$$\Longleftrightarrow (1 - \epsilon)W(x',x^*) + \epsilon W(x',x') < (1 - \epsilon)W(x^*,x^*) + \epsilon W(x^*,x')$$

$$\Longleftrightarrow \epsilon W(x',x') < \epsilon W(x^*,x') \text{ since } W(x',x^*) = W(x^*,x^*)$$

$$\Longleftrightarrow W(x',x') < W(x^*,x') \text{ since } \epsilon > 0.$$

This says that conditions (ES2′)(ii) and (ES2)(ii) are equivalent.

Ex. 4.5. Suppose that the resident strategy in the sex-allocation game is to produce sons with probability p. Then the ratio of females to males in the breeding population is $s(p) = (1 - p)/p$. Consider the mutant strategy of producing sons with probability p'. The payoff to a female (number of matings obtained by her offspring) following this strategy is

$$W(p',p) = (1-p')N + p's(p)N$$
$$= N + \frac{N}{p}(1-2p)p'.$$

For the resident strategy $p^* = 0.5$ we have $W(p',p^*) = N$ for all p'. Thus for a rare mutant strategy $p' \neq p^*$ we have $W(p',p^*) = W(p^*,p^*)$, so that condition (ES1) fails but condition (ES2)(i) holds.

The sex ratio in a population in which a proportion $1 - \epsilon$ of females use strategy p^* and a proportion ϵ use p' is the same as in a population in which all females use strategy $p(\epsilon) = (1-\epsilon)p^* + \epsilon p'$. Thus, in the notation of condition (ES2')(ii),

$$W(p,p_\epsilon) = W(p,p(\epsilon)) = N + \frac{N}{p(\epsilon)}(1-2p(\epsilon))p \quad \text{for all } p.$$

Thus

$$W(p',p_\epsilon) - W(p^*,p_\epsilon) = \frac{N}{p(\epsilon)}(1-2p(\epsilon))(p'-p^*). \tag{S8}$$

Since $p^* = 0.5$ we have $p' - p^* = \frac{1}{2}(2p' - 1)$. Since $p(\epsilon) = (1-\epsilon)p^* + \epsilon p'$ we also have $1 - 2p(\epsilon) = \epsilon(1 - 2p')$. Equation (S8) then gives

$$W(p',p_\epsilon) - W(p^*,p_\epsilon) = -\frac{N}{p(\epsilon)}\frac{\epsilon}{2}(1-2p')^2$$
$$< 0 \quad \text{for } \epsilon > 0.$$

Thus condition (ES2')(ii) holds, so that p^* is an ESS.

Ex. 4.6. In the notation of Exercise 3.10, we have $W(x',x) = N[F(x') + s(x)M(x')]$, where $s(x) = F(x)/M(x)$. From Exercise 3.10 we have $F(x) = (1 - \beta^2(x))/2$ and $M(x) = \beta(x)/2$. Thus

$$2W(x',x) = 1 - \beta^2(x') + \frac{(1-\beta^2(x))}{\beta(x)}\beta(x').$$

Differentiating partially with respect to x' and using the fact that the derivative of $\beta(x')$ is $\frac{1}{10}$ then gives

$$\frac{\partial W}{\partial x'}(x',x) = \frac{N}{20\beta(x)}\left[-2\beta(x)\beta(x') + 1 - \beta^2(x)\right].$$

Let $x^* = 20 + 10/\sqrt{3}$, so that $\beta(x^*) = \frac{1}{\sqrt{3}}$. Then $\frac{\partial W}{\partial x'}(x^*,x^*) = 0$. It can also be verified that $\frac{\partial^2 W}{\partial x'^2}(x',x^*) < 0$ for all x'. Thus $W(x',x^*)$ has a unique maximum at $x' = x^*$, and is hence an ESS. We have

$$D(x) = \frac{\partial W}{\partial x'}(x,x) = \frac{N}{20\beta(x)}\left[1 - 3\beta^2(x)\right].$$

$\beta(x)$ is an increasing function of x, with $1 - 3\beta^2(x^*) = 0$. Thus $D(x) > 0$ for $x < x^*$ and $D(x) < 0$ for $x > x^*$, so that x^* is convergence stable.

Ex. 4.7. Differentiating partially with respect to x' gives

$$\frac{\partial W}{\partial x'}(x',x) = x - \left[1 + \frac{4}{25}x'^2\right]. \tag{S9}$$

Let $0 \leq x < 1$. Then $\frac{\partial W}{\partial x'}(x',x) < 0$ for all x'. Thus there is a unique best response at $\hat{b}(x) = 0$. Now suppose that $x \geq 1$. Setting $\frac{\partial W}{\partial x'}(x',x) = 0$ at $x' = \hat{b}(x)$ gives $\hat{b}(x) = \frac{5}{2}\sqrt{x-1}$. Since the second derivative of W is negative this turning point is a unique maximum. Overall, for each $x \geq 0$ there is a unique best response at

$$\hat{b}(x) = \frac{5}{2}\sqrt{\max(0, x-1)}.$$

It can be seen that $x_1^* = 0$ is a best response to itself and therefore a Nash equilibrium. It can also be seen that any other Nash equilibria must be in the range $x \geq 1$. Setting $\hat{b}(x) = x$ for $x \geq 1$ gives

$$4x^2 - 25x + 25 = 0.$$

The two roots of this quadratic equation are $x_2^* = \frac{5}{4}$ and $x_3^* = 5$. All three Nash equilibria are ESSs since best responses are unique.

By eq (S9) we have

$$D(x) \equiv \frac{\partial W}{\partial x'}(x,x) = x - \left[1 + \frac{4}{25}x^2\right]$$

$$= -\frac{4}{25}(x - \frac{5}{4})(x - 5).$$

Thus $D(x)$ is negative for $x_1^* \leq x < x_2^*$, positive for $x_2^* < x < x_3^*$ and negative for $x > x_3^*$. This shows that x_1^* and x_3^* are convergence stable, whereas x_2^* is an evolutionary repeller.

Ex. 4.8. The definition of fitness in eq (4.15) is

$$w_i(t) = \frac{1}{n_i(t)}\frac{dn_i(t)}{dt},$$

so if we introduce $n_i(t) = v_i(t)n(t)$ we can write this as

$$w_i(t) = \frac{1}{v_i(t)}\frac{dv_i(t)}{dt} + \frac{1}{n(t)}\frac{dn(t)}{dt}.$$

Thus

$$\frac{dv_i(t)}{dt} = v_i(t)\left(w_i(t) - \frac{1}{n(t)}\frac{dn(t)}{dt}\right).$$

This is the replicator equation because, using that $\bar{w}(t) = \sum_i v_i(t)w_i(t)$, we get from the next-to-last equation that

$$\bar{w}(t) = \sum_i \frac{dv_i(t)}{dt} + \frac{1}{n(t)}\frac{dn(t)}{dt} = \frac{1}{n(t)}\frac{dn(t)}{dt}.$$

This also shows that $\bar{w}(t) = 0$ if $dn(t)/dt = 0$. For the Price equation (4.17), $\bar{x}(t) = \sum_i x_i v_i(t)$, so that

$$\frac{d\bar{x}(t)}{dt} = \sum_i v_i(t) x_i \left(w_i(t) - \sum_j v_j(t) w_j(t) \right)$$

$$= \sum_i v_i(t) w_i(t) x_i - \sum_j v_j(t) w_j(t) \bar{x}(t) - \bar{w}(t)\bar{x}(t) + \bar{w}(t)\bar{x}(t)$$

$$= \mathrm{Cov}(w., x.).$$

Finally, note that the replicator and Price equations presuppose that we already have $n_i(t)$ given, for instance as a solution to some differential equation $dn_i(t)/dt = f_i(n., t)$. Thus, these equations do not model the actual population dynamics of the types, but they can still tell us something about the dynamics.

Ex. 4.9. A strategy is a vector $\mathbf{p} = (p_R, p_S, p_P)$ that specifies the probabilities of the three actions, R, S, P. Let $\mathbf{p}$ be the resident strategy. Let $H_R(\mathbf{p}) = p_S a - p_P b$ be the payoff from playing R in this population. $H_S(\mathbf{p})$ and $H_P(\mathbf{p})$ are similarly defined. Then for any rare mutant strategy $\mathbf{p}'$ we have $W(\mathbf{p}', \mathbf{p}) = p_R' H_R(\mathbf{p}) + p_S' H_S(\mathbf{p}) + p_P' H_S(\mathbf{p})$. Let $\mathbf{p}^* = (\frac{1}{3}, \frac{1}{3}, \frac{1}{3})$, so that $H_R(\mathbf{p}^*) = H_S(\mathbf{p}^*) = H_P(\mathbf{p}^*) = \frac{1}{3}(a - b)$. Then $W(\mathbf{p}', \mathbf{p}^*) = (p_R' + p_S' + p_P') \frac{1}{3}(a - b) = \frac{1}{3}(a - b)$. Thus $W(\mathbf{p}', \mathbf{p}^*) = \frac{1}{3}(a - b) = W(\mathbf{p}^*, \mathbf{p}^*)$, so that condition (ES1) fails, but (ES2)(i) holds. To check condition (ES2)(ii) we examine the sign of

$$W(\mathbf{p}', \mathbf{p}') - W(\mathbf{p}^*, \mathbf{p}') = (a - b)\left[p_R' p_S' + p_S' p_P' + p_P' p_R' - \frac{1}{3} \right].$$

Suppose that $a > b$. We note that since $p_R' + p_S' + p_P' = 1$ then

$$p_R' p_S' + p_S' p_P' + p_P' p_R' \leq \frac{1}{3}, \tag{S10}$$

with equality if and only if $p_R' = p_S' = p_P' = \frac{1}{3}$. Thus $W(\mathbf{p}', \mathbf{p}') < W(\mathbf{p}^*, \mathbf{p}')$ for $\mathbf{p}' \neq \mathbf{p}^*$, so that condition (ES2)(ii) holds. We deduce that $\mathbf{p}^*$ is an ESS.

Suppose that $a \leq b$. Let $\mathbf{p}' = (1, 0, 0)$. Then $W(\mathbf{p}', \mathbf{p}') = 0$ whereas $W(\mathbf{p}^*, \mathbf{p}') = \frac{1}{3}(a - b) \leq 0$. Thus condition (ES2)(ii) fails for this mutant, so that $\mathbf{p}^*$ is not an ESS. Since an ESS is necessarily a Nash equilibrium, and this is the only Nash equilibrium, there is no ESS.

We have $w_i = H_i((v_R, v_S, v_P))$; so that for example $w_R = a v_S - b v_P$. Thus $\bar{w} = v_R H_R + v_S H_S + v_P H_P = (a - b)M$. By eq (4.16) we thus have $\dot{v}_R(t) = v_R(t)[a v_S(t) - b v_P(t) - (a - b)M(t)]$. Differentiating y gives

$$\dot{y} = \dot{v}_R v_S v_P + v_R \dot{v}_S v_P + v_R v_S \dot{v}_P$$

$$= v_R v_S v_P [(a v_S - b v_P) + (a v_P - b v_R) + (a v_R - b v_S) - 3M]$$

$$= (a - b)(1 - 3M)y.$$

Suppose that $a > b$. By inequality (S10) we have $\dot{y}(t) \geq 0$ for all t. Since y is bounded above, $y(t)$ tends monotonically to a limit as $t \to \infty$. Assuming that $y(0) > 0$ this limit is positive. Thus $\dot{y}(t) \to 0$ as $t \to \infty$, which implies that $M(t) \to \frac{1}{3}$ as $t \to \infty$. This implies that $(v_R(t), v_S(t), v_P(t)) \to (\frac{1}{3}, \frac{1}{3}, \frac{1}{3})$ as $t \to \infty$.

Suppose that $a = b$. Then $\dot{y} = 0$ so that $y(t)$ is constant. In this case, assuming that $y(0) > 0$ and the initial proportion are not all equal to $\frac{1}{3}$, there is a cycle in which each pure strategy R, P, S, R, P, S, and so on, predominates, in that order.

Finally, suppose that $a < b$. Then, providing initial proportions are not all equal to $\frac{1}{3}$, we have $y(t) \to 0$ as $t \to \infty$. The cycle R, P, S, R, P, S, ..., thus gets more pronounced over time. (See Hofbauer and Sigmund (2003).)

Ex. 4.10. Since each population member is equally likely to be chosen we have

$$E(X_i) = P(X_i = 1) = \alpha \text{ for } i = 1,2$$
$$E(X_i^2) = P(X_i^2 = 1) = \alpha \text{ for } i = 1,2.$$

Hence

$$\text{Var}(X_i) = E(X_i^2) - (E(X_i))^2 = \alpha(1-\alpha) \quad \text{for } i = 1,2.$$

We also have

$$E(X_1 X_2) = P(X_1 = 1, X_1 = 2)$$
$$= P(X_1 = 1)P(X_2 = 1|X_1 = 1)$$
$$= \alpha p_C.$$

Thus

$$\text{Cov}(X_1, X_2) = E(X_1 X_2) - E(X_1)E(X_2) = \alpha(p_C - \alpha).$$

We now use conditional probabilities to note that

$$P(X_2 = 1) = P(X_2 = 1|X_1 = 0)P(X_1 = 0) + P(X_2 = 1|X_1 = 1)P(X_1 = 1)$$
$$= p_D(1-\alpha) + p_C\alpha.$$

Thus $\alpha = p_D(1-\alpha) + p_C\alpha$, so that $p_C - \alpha = (p_C - p_D)(1-\alpha)$. We can then write $\text{Cov}(X_1, X_2) = \alpha(p_C - p_D)(1-\alpha)$. Putting these preliminary results together we have

$$\rho(X_1, X_2) = \frac{\text{Cov}(X_1, X_2)}{\sqrt{\text{Var}(X_1)\text{Var}(X_2)}}$$
$$= p_C - p_D.$$

Solutions for Chapter 5

Ex. 5.1. There is a single state and K actions. The preference by individual i for action k at time t is θ_{ikt}, with the probability that the individual chooses action k as

$$p_{ikt} = \frac{\exp(\theta_{ikt})}{\sum_l \exp(\theta_{ilt})}.$$

We have the eligibilities

$$\zeta_{iklt} = \frac{\partial \log p_{ikt}}{\partial \theta_{ilt}} = \delta[k,l] - p_{ilt},$$

where $\delta[k,l]$ is the Kronecker delta, which is equal to one if $k = l$ and zero otherwise. With w_{it} and R_{it} the estimated and actual reward and $\delta_{it} = R_{it} - w_{it}$ the TD error, we have the updates $w_{i,t+1} = w_{it} + \alpha_w \delta_{it}$ and, if action k is chosen,

$$\theta_{il,t+1} = \theta_{ilt} + \alpha_\theta \delta_{it} \zeta_{iklt},$$

for $l = 1, \ldots, K$. Consider now the case of two actions ($K = 2$) and define $\theta_{it} = \theta_{i1t} - \theta_{i2t}$ as a combined preference for the first action and the eligibility $\zeta_{it} = \zeta_{i11t} - \zeta_{i12t}$ if the first action is chosen and $\zeta_{it} = \zeta_{i21t} - \zeta_{i22t}$ if the second action is chosen. The update for this combined preference is then

$$\theta_{i,t+1} = \theta_{it} + 2\alpha_\theta \delta_{it} \zeta_{it},$$

which is the same as eq (5.5) except for a factor of two.

Ex. 5.2. Instead of an individual choosing its action from the normal distribution in eq (5.9), we use a log-normal distribution. A reason for doing this might be to constrain actions to positive values by representing them on a logarithmic scale in the learning model, i.e. using $z = \log(u)$. If one does this, a log-normal distribution for u becomes a normal distribution for z. So let us assume a log-normal distributions for the actions:

$$P(u|\theta_{it}) = \frac{1}{u\sqrt{2\pi\sigma^2}} \exp\left(-\frac{(\log(u) - \theta_{it})^2}{2\sigma^2}\right)$$

$$= \frac{1}{\sqrt{2\pi\sigma^2}} \exp\left(-z - \frac{(z - \theta_{it})^2}{2\sigma^2}\right).$$

We thus express the policy as parameterized by a mean value θ_{it} of the density of z, and a standard deviation σ. The eligibility is then the derivative of the log of the probability density with respect to this action preference:

$$\zeta_{it} = \frac{\partial \log P(u|\theta_{it})}{\partial \theta_{it}} = \frac{\log(u) - \theta_{it}}{\sigma^2} = \frac{z - \theta_{it}}{\sigma^2}.$$

Ex. 5.3. We have a one-shot game where player 1 knows its own quality q_1 but only knows that the partner's quality q_2 has a uniform distribution on the interval $[0, 1]$. The payoff to player 1 is

$$W_1(a_1, q_1) = E[B(a_1 + a_2)] - C(a_1, q_1),$$

with B and C from eqs (5.8, 5.14) and the expectation is with respect to a_2, which is a stochastic variable with a distribution that depends on the strategy of player 2. We have the derivative

$$\frac{\partial W_1}{\partial a_1} = b_1 - b_2(a_1 + \bar{a}_2) - c_1(1 - q_1) - c_2 a_1,$$

where $\bar{a}_2$ is the mean investment by player 2. For an optimal investment by player 1 the derivative should be zero, leading to

$$a_1 = \frac{b_1 - c_1 - b_2 \bar{a}_2 + c_1 q_1}{b_2 + c_2}.$$

The optimal investment is linear in q_1 and we can write is as $a_1 = k_{10} + k_{11} q_1$. Let us assume that player 2 also has this kind of strategy, i.e. $a_2 = k_{20} + k_{21} q_2$. For optimality for both players we must have $k_{11} = k_{21} = k_1 = c_1/(b_2 + c_2)$. The mean investment by player 2 is then $\bar{a}_2 = k_{20} + k_1 \bar{q}_2$, with $\bar{q}_2 = 0.5$. The condition that $\partial W_1/\partial a_1 = 0$ then determines k_{10} as a function of k_{20}. Setting $k_{10} = k_{20} = k_0$ we can solve for k_0:

$$k_0 = \frac{(b_1 - c_1)(b_2 + c_2) - 0.5 b_2 c_1}{(2b_2 + c_2)(b_2 + c_2)}.$$

The dark-grey lines in Fig. 5.6a, b show this ESS strategy, $a = k_0 + k_1 q$, with payoff parameters $b_1 = 2$, $b_2 = 0.5$, $c_1 = c_2 = 1.5$.

Ex. 5.4. The one-shot game where individuals are unaware of their own quality q_i and there is no assessment is similar to the Hawk–Dove game. The expected payoff from the resource is V, and the expected cost from an AA interaction is

$$E\big(C\exp(-q_i + q_j)\big) = C\exp(\sigma_q^2),$$

which follows by noting that q_i and q_j have independent normal distributions with standard deviation σ_q. Let p_A be the probability for the resident strategy to use A. The payoff for a mutant to use A is then

$$(1 - p_A)V - p_A C\exp(\sigma_q^2).$$

A and S give equal payoff for the mutant when this is equal to

$$(1 - p_A)\frac{1}{2}V,$$

which is the payoff for S. The one-shot ESS is thus

$$p_A^* = \frac{V}{V + 2C\exp(\sigma_q^2)},$$

as given in eq (5.17).

Ex. 5.5. We now assume that an individual knows its own q_i and we examine threshold strategies where a resident individual uses A when $q_i > \hat{q}$. We note that $p_A(\hat{q}) = 1 - \Phi(\hat{q}/\sigma_q)$, where Φ is the standard normal CDF. The expected benefit for a mutant using A is $V(1 - p_A(\hat{q}))$, and the expected cost is

$$E\big(C\exp(-q_i + q_j)|q_j > \hat{q}\big)p_A(\hat{q}) = C\exp\!\Big(-q_i + \frac{1}{2}\sigma_q^2\Big)\big[1 - \Phi\big(\hat{q}/\sigma_q - \sigma_q\big)\big]. \tag{S11}$$

For an ESS, at the threshold ($q_i = \hat{q}$) the benefit minus cost of using A should be equal the expected payoff of using S, which is $(1 - p_A(\hat{q}))V/2$, so we have

$$V\Phi(\hat{q}/\sigma_q) = 2C\exp\!\Big(-\hat{q} + \frac{1}{2}\sigma_q^2\Big)\big[1 - \Phi\big(\hat{q}/\sigma_q - \sigma_q\big)\big], \tag{S12}$$

which is the requested equation. Note that the left-hand side increases from zero with $\hat{q}$ and the right-hand side decreases towards zero, so there is a unique solution $\hat{q} = \hat{q}^*$, which can readily be found numerically. For very small σ_q, this ESS becomes the same as the one from eq (5.17). It is worth noting that for this model, eq (S12) gives a somewhat lower probability of using A than eq (5.17). This is similar to what we found for the Hawk–Dove game in Section 3.11.

Ex. 5.6. With $p_A(\hat{q}) = 1 - \Phi(\hat{q}/\sigma_q)$ for the resident strategy, the expected benefit for a mutant individual using A is v (it is v regardless of whether the opponent uses A or S). The expected cost is given by eq (S11) in the solution to Exercise 5.5. For an ESS, the benefit minus cost of using A should be equal the expected payoff of using S, which is zero, at $q_i = \hat{q}$, so we have the equation

$$v = C\exp\!\Big(-\hat{q} + \frac{1}{2}\sigma_q^2\Big)\big[1 - \Phi\big(\hat{q}/\sigma_q - \sigma_q\big)\big], \tag{S13}$$

which is the requested equation. There is a unique solution $\hat{q} = \hat{q}^*$, which can readily be found numerically.

Ex. 5.7. The thresholds from eqs (S12, S13) in Exercises 5.5 and 5.6 are equal when

$$v = \frac{V}{2}(1 - p_A(\hat{q}^*)).$$

A way to think about this is that the perceived reward v from Exercise 5.6 could evolve to produce the ESS from Exercise 5.5.

Solutions for Chapter 6

Ex. 6.1. The eigenvalue equation is $(\lambda - b_{11})(\lambda - b_{22}) - b_{12}b_{21} = 0$, which simplifies to give

$$\lambda^2 - T\lambda + \Delta = 0.$$

Let λ_1 and λ_2 be the two eigenvalues. Then since $(\lambda - \lambda_1)(\lambda - \lambda_2) = 0$ we have

$$T = \lambda_1 + \lambda_2 \quad \text{and} \quad \Delta = \lambda_1 \lambda_2. \tag{S14}$$

The eigenvalues are either both real or not real and complex conjugates of each other.

Consider first the case of two real eigenvalues. Then if both eigenvalues are negative we have $T < 0$ and $\Delta > 0$ by eq (S14). Conversely, $\Delta > 0$ implies that the eigenvalues are not zero and have the same sign, so that this sign is negative if and only if $T < 0$.

Now suppose that the eigenvalues are not real but are complex conjugates of each other. This implies that $\Delta = \lambda_1 \lambda_2 > 0$. Both eigenvalues have the same real part, which equals $\frac{T}{2}$. Thus the eigenvalues have negative real parts if and only if $T < 0$.

Ex. 6.2. Partially differentiating invasion fitness gives

$$\frac{\partial \lambda}{\partial x_1'}(x_1', x_2) = x_1 + 2x_2 - 2x_1',$$

so that $D_1(x_1, x_2) = \frac{\partial \lambda}{\partial x_1'}(x_1, x_2) = 2x_2 - x_1$. Similarly $D_2(x_1, x_2) = x_2 - x_1$. The Taylor series expansion of the D_is about $(0,0)$ is thus exact, so that in the notation of Box 6.1 the Jacobian matrix is

$$\mathbf{A} = \begin{bmatrix} -1 & 2 \\ -1 & 1 \end{bmatrix}.$$

Since

$$\mathbf{C} = \begin{bmatrix} r_1 & 0 \\ 0 & r_2 \end{bmatrix},$$

we have

$$\mathbf{B} = \begin{bmatrix} -r_1 & 2r_1 \\ -r_2 & r_2 \end{bmatrix}.$$

In the notation of Exercise 6.1 we thus have $T = r_2 - r_1$ and $\Delta = r_1 r_2$. Since $\Delta > 0$, by the results of Exercise 6.1 we see that $(0,0)$ is convergence stable if $r_2 < r_1$ and is not convergence stable if $r_2 > r_1$.

Ex. 6.3. We have

$$\frac{\partial W_1}{\partial x_1'}(x_1';x_1,x_2) = RM(x_1,x_2)\left[\frac{-1}{x_1'^2}f(x_1'+x_2)+\frac{1}{x_1'}f'(x_1'+x_2)\right].$$

Thus

$$D_1(x_1,x_2) = \frac{\partial W_1}{\partial x_1'}(x_1;x_1,x_2) = \frac{RM(x_1,x_2)}{x_1^2}[x_1f'(x_1+x_2)-f(x_1+x_2)].$$

At an isogamous ESS (x^*,x^*), we must have $D_1(x^*,x^*) = 0$, which gives $x^*f'(2x^*) = f(2x^*)$. We must also have $D_2(x^*,x^*) = 0$, which gives the same equation. This shows that eq (6.14) must hold.

For $f(z) = e^{-1/z^2}$ we have $f'(z) = \frac{2}{z^3}e^{-1/z^2}$. Set $z^* = 2x^*$, then eq (6.14) becomes $z^*f'(z^*) = 2f(z^*)$, which has solution $z^* = 1$. Thus $x^* = \frac{1}{2}$. We also have $f''(z) = \left[\frac{4}{z^6}-\frac{6}{z^4}\right]e^{-1/z^2}$. Thus $f''(z^*) = -2e^{-1} < 0$.

Now let $f(z) = z/(1+z)$. Then $f'(z) = 1/(1+z)^2$, so that $z^*f'(z^*) = 2f(z^*)$ implies $2(1+z^*) = 1$, which has no positive solution.

Ex. 6.4. As in the solution to Exercise 6.3 we have

$$D_1(x_1,x_2) = \frac{\partial W_1}{\partial x_1'}(x_1;x_1,x_2) = \frac{RM(x_1,x_2)}{x_1^2}[x_1f'(x_1+x_2)-f(x_1+x_2)].$$

Thus

$$\frac{\partial D_1}{\partial x_1}(x_1,x_2) = \frac{\partial\left(\frac{RM}{x_1^2}\right)}{\partial x_1}(x_1,x_2)[x_1f'(x_1+x_2)-f(x_1+x_2)]$$

$$+ \frac{RM(x_1,x_2)}{x_1^2}\left[x_1f''(x_1,x_2)\right].$$

Thus by eq (6.14)

$$a_{11} \equiv \frac{\partial D_1}{\partial x_1}(x^*,x^*) = H\frac{f''(2x^*)}{x^*},$$

where $H = RM(x^*,x^*)$. Similar calculations give $b_{12} = H[f''(2x^*)/x^* - f'(2x^*)/x^{*2}]$, $b_{21} = H[f''(2x^*)/x^* - f'(2x^*)/x^{*2}]$ and $b_{22} = Hf''(2x^*)/x^*$. Thus

$$T = 2Hf''(2x^*)/x^*,$$

$$\Delta = -H\left(f'(2x^*)/x^{*2}\right)\left(f'(2x^*)/x^{*2} - 2f''(2x^*)/x^*\right).$$

Thus $T < 0$ implies $f''(2x^*) < 0$, which implies $\Delta < 0$ since $f'(2x^*) > 0$.

Ex. 6.5. (i) We are given that $\frac{\partial W_1}{\partial x_1'}(\hat{b}_1(x_2),x_2) = 0$. Differentiating with respect to x_2 gives

$$\hat{b}_1'(x_2)\frac{\partial^2 W_1}{\partial x_1'^2}(\hat{b}_1(x_2),x_2) + \frac{\partial^2 W_1}{\partial x_1'\partial x_2}(\hat{b}_1(x_2),x_2) = 0.$$

Evaluating this at $x_2 = x_2^*$ and using the fact that $\hat{b}_1(x_2^*) = x_1^*$ we obtain

$$\hat{b}_1'(x_2^*) \frac{\partial^2 W_1}{\partial x_1'^2}(x_1^*, x_2^*) + \frac{\partial^2 W_1}{\partial x_1' \partial x_2}(x_1^*, x_2^*) = 0.$$

Similarly

$$\hat{b}_2'(x_1^*) \frac{\partial^2 W_2}{\partial x_2'^2}(x_1^*, x_2^*) + \frac{\partial^2 W_2}{\partial x_2' \partial x_1}(x_1^*, x_2^*) = 0.$$

(ii) From these equations we deduce that

$$(1 - \hat{b}_1'(x_2^*)\hat{b}_2'(x_1^*)) \frac{\partial^2 W_1}{\partial x_1^2} \frac{\partial^2 W_2}{\partial x_2^2} = \frac{\partial^2 W_1}{\partial x_1^2} \frac{\partial^2 W_2}{\partial x_2^2} - \frac{\partial^2 W_1}{\partial x_1' \partial x_2} \frac{\partial^2 W_2}{\partial x_2' \partial x_1},$$

where the partial derivative are evaluated at (x_1^*, x_2^*). By eq (6.8), $\frac{\partial^2 W_1}{\partial x_1^2} \frac{\partial^2 W_2}{\partial x_2^2}$ is the product of two negative terms, and so is positive. Thus

$$\hat{b}_1'(x_2^*)\hat{b}_2'(x_1^*) < 1 \iff \frac{\partial^2 W_1}{\partial x_1^2} \frac{\partial^2 W_2}{\partial x_2^2} - \frac{\partial^2 W_1}{\partial x_1' \partial x_2} \frac{\partial^2 W_2}{\partial x_2' \partial x_1} > 0. \tag{S15}$$

We have $D_i(x_1, x_2) = \frac{\partial W_i}{\partial x_i}(x_1, x_2)$, $i = 1, 2$. Thus the Jacobian matrix is

$$\mathbf{A} = \begin{bmatrix} \frac{\partial^2 W_1}{\partial x_1^2} & \frac{\partial^2 W_1}{\partial x_1' \partial x_2} \\ \frac{\partial^2 W_2}{\partial x_2' \partial x_1} & \frac{\partial^2 W_2}{\partial x_2^2} \end{bmatrix}.$$

Thus the right-hand side of eq (S15) is just Δ. Hence

$$\hat{b}_1'(x_2^*)\hat{b}_2'(x_1^*) < 1 \iff \Delta > 0.$$

(iii) If the covariance matrix is diagonal, then since the diagonal entries are necessarily positive it is easy to show that the determinate of $\mathbf{B}$ is positive if and only if the determinate of $\mathbf{A}$ is positive. Furthermore the trace of $\mathbf{B}$ is negative since $\frac{\partial^2 W_1}{\partial x_1^2}$ and $\frac{\partial^2 W_2}{\partial x_2^2}$ are negative.

Ex. 6.6. For given x the function

$$K(x, \theta) + C(\theta) = \beta(1 - \theta)x + \frac{1}{2}\alpha x^2 + \frac{1}{2}\theta^2$$

has a minimum at $\hat{\theta}(x) = \beta x$. Thus

$$\tilde{C}(x) = \beta(1 - \hat{\theta}(x))x + \frac{1}{2}\alpha x^2 + \frac{1}{2}\hat{\theta}(x)^2$$

$$= \beta x + \frac{1}{2}(\alpha - \beta^2)x^2.$$

Since $\tilde{C}''(x) = \alpha - \beta^2$, $\tilde{C}$ is concave if $\beta^2 > \alpha$.

For the specific values of β and α the effective cost function is $\tilde{C}(x) = 0.8x - 0.25x^2$. Then for fixed x_m the payoff $B(x_f + x_m) - \tilde{C}(x_f)$ is maximized when $x_f = \hat{b}_f(x_m)$, where $\hat{b}_f(x_m) = \max(0, 0.8 - \frac{4}{3}x_m)$. Similarly, $\hat{b}_m(x_f) = \max(0, 0.8 - \frac{4}{3}x_f)$. Setting $x_f^* = \hat{b}_f(x_m^*)$ and $x_m^* = \hat{b}_m(x_f^*)$, we find there are three equilibrium pairs of efforts: $(0, 0.8)$, $(\frac{2.4}{7}, \frac{2.4}{7})$, and $(0.8, 0)$.

Solutions for Chapter 7

Ex. 7.1. Suppose that neither the mutant nor its partner have defected on the first $n-1$ rounds. If the mutant cooperates on round n its payoff is

$$H_C(n) = F(n)\,[b-c+V(n+1)] + [1-F(n)]\,(-c)$$
$$= bF(n) + F(n)V(n+1) - c.$$

If the mutant defects on round n its payoff is $H_D = bF(n)$. Since the mutant makes the best choice of action we have $V(n) = \max(H_C(n), H_D(n))$. Thus

$$V(n) = bF(n) + \max(F(n)V(n+1) - c, 0). \tag{S16}$$

Suppose that all $N+1$ strategies are equally likely. Then if $n-1$ rounds have elapsed without the partner defecting, then partner's strategy is equally likely to be each of $n-1, n, n+1, \ldots, N$. Thus $F(n) = \frac{N+1-n}{N+2-n}$. On the Nth round the mutant will defect, so that $V(N) = \frac{b}{2}$. Thus by eq (S16), $V(N-1) = \frac{2}{3}b + \max(\frac{2}{3}\frac{b}{2} - c, 0)$. Since $c > \frac{b}{3}$ the maximum occurs by defecting and $V(N-1) = \frac{2b}{3}$. Applying eq (S16) again we have $V(N-2) = \frac{3}{4}b + \max(\frac{3}{4}\frac{2b}{3} - c, 0)$. Since $c > \frac{b}{2}$ it is best to defect and $V(N-2) = \frac{3b}{4}$. For any n we have $V(n) \geq bF(n)$. Thus for $n \leq N-3$ we have

$$F(n)V(n+1) \geq bF(n)F(n+1)$$
$$= b\left(\frac{N+1-n}{N+2-n}\right)\left(\frac{N-n}{N+1-n}\right)$$
$$= b\frac{N-n}{N+2-n}$$
$$\geq \frac{3}{5}b > c.$$

Thus it is best to cooperate for $n \leq N-3$.

Ex. 7.2. First suppose that the female strategy is to lay two eggs if and only if signal s_2 is received. Let the male be low ability. If he signals s_1 the female receives signal s_2 with probability ϵ. The male's payoff is thus $(1-\epsilon) + 2\epsilon = 1 + \epsilon$. If he signals s_2 his payoff is $\epsilon + 2(1-\epsilon) - c_L = 2 - \epsilon - c_L$. Thus his best action is to signal s_1 if $c_L > 1 - 2\epsilon$, which holds by assumption.

Let the male be high ability. If he signals s_1 his payoff is $(1-\epsilon) + 2\epsilon = 1 + \epsilon$. If he signals s_2 his payoff is $\epsilon + 2(1-\epsilon) - c_H = 2 - \epsilon - c_H$. Thus his best action is to signal s_2 if $c_H < 1 - 2\epsilon$, which holds by assumption. The male strategy of signalling s_2 if and only if high ability is thus the unique best response to the female egg-laying strategy.

Now consider the female's best egg-laying strategy. Suppose that the probability the male is high ability is α. If the female lays one egg her payoff is 1. If she lays two eggs her payoff is $\alpha 2 + (1-\alpha)(2-K)$. After some manipulation we deduce that laying two eggs is strictly best if and only if

$$K - 1 < \frac{\alpha}{1-\alpha}.$$

Suppose that the male strategy is to signal s_2 if and only if of high ability. Let α_i be the probability the male is high ability given that the female receives signal s_i. Then her strategy of laying i eggs on receipt of signal s_i is the unique best response strategy to the male strategy if and only if

$$\frac{\alpha_1}{1-\alpha_1} < K - 1 < \frac{\alpha_2}{1-\alpha_2}. \tag{S17}$$

By Bayes theorem the probability that the male is high ability given signal s_1 is

$$\alpha_1 = \frac{\rho\epsilon}{\rho\epsilon + (1-\rho)(1-\epsilon)},$$

so that

$$\frac{\alpha_1}{1-\alpha_1} = \frac{\rho\epsilon}{(1-\rho)(1-\epsilon)}.$$

Similarly

$$\frac{\alpha_2}{1-\alpha_2} = \frac{\rho(1-\epsilon)}{(1-\rho)\epsilon}.$$

By eq (S17) the female strategy of laying i eggs on receipt of signal s_i is the unique best response strategy to the male signalling strategy.

Ex. 7.3. Suppose that the population is at a Nash equilibrium. Let the random variable Y be the effort of a randomly selected population member. Consider an individual that has quality q. The payoff to this individual if it chooses effort x is

$$w(x;q) = E[B(x, Y)] - C_q(x)$$
$$= 2(x + E(Y)) - (x^2 + 2xE(Y) + E(Y^2)) - (1-q)x - x^2.$$

This is maximized at $x = \hat{x}(q)$ where

$$\hat{x}(q) = \frac{1}{4}(1 + q - 2E(Y)).$$

Since the population is at a Nash equilibrium, all residents use this strategy. Thus $E(Y) = E(\hat{x}(Q))$, where Q is the quality of a randomly selected population members. Thus $E(Y) = \frac{1}{4}(1 + E(Q) - 2E(Y))$, so that $E(Y) = \frac{1}{4}$ since $E(Q) = \frac{1}{2}$. It follows that $\hat{x}(q) = \frac{1}{8} + \frac{1}{4}q$.

Ex. 7.4. Suppose that a proportion $\alpha(n)$ have reputation Good after round n. Then in round $n+1$ a proportion $\alpha(n)$ of DISC individuals will have Good partners and cooperate. Since all ALLC individuals cooperate the proportion cooperating on round $n+1$ is $\rho_C + \rho_{DISC}\alpha(n)$. So this is the proportion that has reputation Good after this round; i.e. $\alpha(n+1) = \rho_C + \rho_{DISC}\alpha(n)$. Assuming $\alpha(n)$ tends to α as n tends to infinity we have $\alpha = \rho_C + \rho_{DISC}\alpha$, so that

$$\alpha = \frac{\rho_C}{1 - \rho_{DISC}} = \frac{\rho_C}{\rho_C + \rho_D}.$$

Assuming that the proportion of Good individuals has reached this limit, the payoff to a cooperator in a round is

$$W_C = [\rho_C + \rho_{DISC}]b - c.$$

The probability a DISC individual is Good is the probability their last partner was Good; i.e. α. Thus the payoff to a DISC strategist is

$$W_{DISC} = [\rho_C + \alpha\rho_{DISC}]b - \alpha c.$$

Thus

$$W_C > W_{DISC} \iff \rho_{DISC}b > c.$$

Ex. 7.5. (i) If the P2 individual defected on the last round (in which it was trusted), then it will be rejected by all samplers on this round, obtaining a payoff of s. If it cooperated on the last round it will be trusted by all samplers and will obtain payoff $(1 - p) \times 1 + p \times r$. Since the probability it cooperated on the last round is p we have

$$V_{\text{sampler}}(p) = (1 - p)s + p\left[(1 - p)1 + pr\right]$$
$$= s + (1 - s)p - (1 - r)p^2.$$

The local payoff to a P2 that cooperates with probability p in the mixed P1 population is

$$V(p) = \beta_{\text{UA}}\left[(1 - p)1 + pr\right]$$
$$+ \beta_S V_{\text{sampler}}(p)$$
$$+ (1 - (\beta_{\text{UA}} + \beta_S))s.$$

Differentiating this expression and setting the derivative equal to 0 at $p = \hat{p}$ gives

$$2\hat{p} = \left(\frac{1 - s}{1 - r}\right) - \left(\frac{\beta_{\text{UA}}}{\beta_S}\right).$$

(ii) In case (iii) in Fig. 7.7 we have $s = 0.2$ and $r = 0.5$ and $c = 0.005$. From the numerical values of β_{UA} and β_S, the above gives $\hat{p} = 0.4703$. This value is close to the mean value $E[p] = 0.4676$. From eq 7.6 we have $W_{\text{sampler}} = 0.2340$. The payoff to UA is $W_{\text{UA}} = E(p)r = 0.2336$. Overall, P2s are doing approximately the best (with variation due to mutation) given the mixture of P1s, and the P1s in the mixture are doing as well as each other given the distribution of P2s.

Ex. 7.6. Suppose the current server has value v. Let w be the expected future net payoff under an optimal strategy given that the current server is rejected. Then the payoff given that the optimal choice is made with the current server is $\max(v, w)$. Averaging over v, the maximum expected payoff on encountering a server is thus $\int_0^\infty \max(w, v)f(v)dv$. w equals this expected payoff minus the cost of searching for the next server, so that

$$w = -c + \int_0^\infty \max(w, v)f(v)dv. \tag{S18}$$

For the given f we have

$$\int_0^\infty \max(w, v)f(v)dv = \frac{1}{2\delta}\int_{1-\delta}^{1+\delta} \max(w, v)dv.$$

Since $\max(v, w) \geq v$ the right-hand integral is at least 1. Thus since $c < \delta$, eq (S18) shows that $w \geq 1 - \delta$. We can then write

$$\int_{1-\delta}^{1+\delta} \max(w, v)dv = \int_{1-\delta}^w wdv + \int_w^{1+\delta} vdv$$
$$= \frac{1}{2}\left[w^2 + 2(\delta - 1)w + (1 + \delta)^2\right].$$

Equation (S18) then becomes

$$w^2 - 2(1 + \delta)w + (1 + \delta)^2 - 4\delta c = 0.$$

The root of this equation in the range $[1 - \delta, 1 + \delta]$ is $w = 1 + \delta - 2\sqrt{\delta c}$.

Ex. 7.7. (i) We have $\frac{dz}{dt} = -z(t)$. Thus if the initial nectar level at time 0 is x, then $z(t) = xe^{-t}$. The pollinator leaves when the intake rate $-\frac{dz}{dt}$ falls to γ, so that $z(T) = \gamma$. Thus $T = \log(x/\gamma)$. Since the amount of nectar obtained is $x - z(T) = x - \gamma$ we have $g(x, \gamma) = \frac{x - \gamma}{\log(x/\gamma) + \tau}$. The optimal foraging strategy is to leave when the intake rate falls to $\gamma^*(x)$ where $g(x, \gamma^*(x)) = \gamma^*(x)$. Thus

$$\gamma^*(x) = \frac{x - \gamma^*(x)}{\log(x/\gamma^*(x)) + \tau}.$$

This rate equals the amount left on leaving.

(ii) Let x' be a mutant flower strategy. Since foragers leave when there is $\gamma^*(x)$ left on a flower, the time spent on a mutant flower is $T' = \log \frac{x'}{\gamma^*(x)}$. The payoff to the mutant is

$$1 - e^{-3T'} - (x' - \gamma^*(x)) = 1 - \left(\frac{\gamma^*(x)}{x'}\right)^3 - (x' - \gamma^*(x)).$$

Differentiating this with respect to x' and setting the derivative equal to zero at $x' = \hat{b}(x)$ gives $\hat{b}(x) = 3^{\frac{1}{4}}(\gamma^*(x))^{\frac{3}{4}}$.

Ex. 7.8. [Note on notation in the proof below: U and U' are random variables, whereas u and u' are numbers.]

Suppose that a mutant's contribution is u'. Then the mean payoff to this individual is $v(u') = E(H(u', U)|U' = u')$. Thus since $E(U|U' = u') = x + \rho(u' - x)$, eq (7.8) gives

$$v(u') = u' + E(U|U' = u') + \frac{R}{2}(1 - u'^2)$$

$$= (1 + \rho)u' + (1 - \rho)x + \frac{R}{2}(1 - u'^2).$$

Note that $E(U') = x'$ and $E(U'^2) = \text{Var}(U') + (E(U'))^2 = \sigma^2 + x'^2$. Thus

$$W(x', x) = E(v(U')) = (1 + \rho)x' + (1 - \rho)x + \frac{R}{2}(1 - (\sigma^2 + x'^2)).$$

Ex. 7.9. An individual that rejects the outside option a proportion ρ' of the time will reject when the reward r satisfies $r < \rho'R$ and take the outside option otherwise. Thus the mean reward to the individual from its own actions is

$$\rho' + \int_{\rho'R}^{R} \frac{r}{R} dr = \rho' + \frac{1}{2}R(1 - \rho'^2).$$

Taking the contribution of the partner into account we have $W(\rho', \rho) = \rho' + \rho + \frac{R}{2}(1 - \rho'^2)$.

Solutions for Chapter 8

Ex. 8.1. Let $\hat{b}_1(x_2)$ be the best response by Player 1 to Player 2 strategy x_2. Similarly, let $\hat{b}_2(x_1)$ be the best response by Player 2 to Player 1 strategy x_1. Let (x_1^*, x_2^*) be a Nash equilibrium, so that $x_1^* = \hat{b}_1(x_2^*)$ and $x_2^* = \hat{b}_2(x_1^*)$. Let $V_1(x_1) = W_1(x_1, \hat{b}_2(x_1))$ be the payoff to Player 1 if this player chooses strategy x_1 and Player 2 responds optimally to this strategy. Let x_1^s be Player 1's strategy at the Stackelberg equilibrium. Then

$$V_1(x_1^s) = \max_{x_1} V_1(x_1) \geq V_1(x_1^*) = W_1(x_1^*, \hat{b}_2(x_1^*)) = W_1(x_1^*, x_2^*) = W_1^*.$$

Ex. 8.2. If we put $u_{1t} = \tilde{u}_1$ and $u_{2,t-1} = u_{2t} = \tilde{u}_2$ in eq (8.1) we get

$$\tilde{u}_1 = m_1 + k_1 q_1 - l_1 \tilde{u}_2$$
$$\tilde{u}_2 = m_2 + k_2 q_2 - l_2 \tilde{u}_1.$$

The solution to this system is as in eq (8.4), which thus is an equilibrium for the iteration in eq (8.1). Examining a deviation from the equilibrium one finds directly that it is multiplied by a factor $l_1 l_2$ per investment round, so the equilibrium is stable when $|l_1 l_2| < 1$.

Ex. 8.3. Using eq (8.4) we can readily compute the following derivatives:

$$\frac{\partial \tilde{u}_1(q_1, q_2)}{\partial m_1} = \frac{1}{1 - l_1 l_2}$$
$$\frac{\partial \tilde{u}_2(q_2, q_1)}{\partial m_1} = -\frac{l_2}{1 - l_1 l_2}$$

and

$$\frac{\partial \tilde{u}_1(q_1, q_2)}{\partial k_1} = \frac{q_1}{1 - l_1 l_2}$$
$$\frac{\partial \tilde{u}_2(q_2, q_1)}{\partial k_1} = -\frac{l_2 q_1}{1 - l_1 l_2}.$$

With a bit more effort we also find these derivatives:

$$\frac{\partial \tilde{u}_1(q_1, q_2)}{\partial l_1} = \frac{-m_2 + l_2 m_1 - k_2 q_2 + l_2 k_1 q_1}{(1 - l_1 l_2)^2} = -\frac{\tilde{u}_2(q_2, q_1)}{1 - l_1 l_2}$$
$$\frac{\partial \tilde{u}_2(q_2, q_1)}{\partial l_1} = l_2 \frac{m_2 - l_2 m_1 + k_2 q_2 - l_2 k_1 q_1}{(1 - l_1 l_2)^2} = l_2 \frac{\tilde{u}_2(q_2, q_1)}{1 - l_1 l_2}.$$

Inserting these results in the left-hand side of eq (8.5) the equality follows.

Ex. 8.4. The expression in brackets in eq (8.5) should be zero for all q_1 and q_2, so that

$$(1 - l_2) B'(\tilde{u}_1, \tilde{u}_2) - \frac{\partial C(\tilde{u}_1, q_1)}{\partial \tilde{u}_1} =$$
$$(1 - l_2)\Big(b_1 - b_2(\tilde{u}_1 + \tilde{u}_2)\Big) - c_1(1 - q_1) - c_2 \tilde{u}_1 = 0.$$

From eq (8.4), this expression depends linearly on q_1 and q_2. Collecting terms and setting the coefficient of q_2 equal to zero we get

$$\frac{1}{1 - l_1 l_2}\Big[(1 - l_2)\big(-b_2(-l_1 k_2 + k_2)\big) - c_2(-l_1 k_2)\Big] = 0.$$

Because k_2 appears as a common factor, this becomes

$$-(1 - l_2)(1 - l_1) b_2 + l_1 c_2 = 0.$$

We can now solve l_1 as a function of l_2, yielding the left-hand part of eq (8.6). Next, to get k_1 we can set the coefficient of q_1 equal to zero:

$$\frac{1}{1 - l_1 l_2}\left[(1 - l_2)\left(-b_2(k_1 - l_2 k_1)\right) - c_2 k_1\right] + c_1 = 0,$$

so that

$$k_1 = c_1 \frac{1 - l_1 l_2}{b_2(1 - l_2)^2 + c_2} = \frac{c_1}{b_2} \frac{1}{1 - l_2 + c_2/b_2},$$

where we used the left-hand part of eq (8.6) to replace $1 - l_1 l_2$. This is now the same as the middle part of eq (8.6). Finally, we can use a trick by setting $q_1 = -m_1/k_1$ and $q_2 = -m_2/k_2$, in which case $\tilde{u}_1 = \tilde{u}_2 = 0$, so the condition that the expression in brackets should be zero becomes

$$(1 - l_2)b_1 - c_1(1 + m_1/k_1) = 0,$$

from which we get that

$$m_1 = k_1(1 - l_2)\frac{b_1}{c_1} - k_1.$$

If we now use the middle part of eq (8.6) to replace the first k_1 on the right-hand side, we end up with the right-hand part of eq (8.6).

Ex. 8.5. For simplicity, assuming that $b_2 = c_2$, we have the local best response $l_1 = \hat{l}_1(l_2) = (1 - l_2)/(2 - l_2)$ to l_2, and by symmetry the local best response $l_2 = \hat{l}_2(l_1) = (1 - l_1)/(2 - l_1)$ to l_1. The iterated local best response is then $\hat{l}_1(\hat{l}_2(l_1)) = 1/(3 - l_1)$. Plotting this together with the 45° line shows that there is only one intersection, and thus only one Nash equilibrium.

Ex. 8.6. Suppose that the strategies $x_f = (m_f, k_f, l_f)$ and $x_m = (m_m, k_m, l_m)$ are best responses to one another. As in Box 8.2 the negotiated efforts are

$$\tilde{u}_f = \frac{1}{1 - l_f l_m}\left(m_f - l_f m_m + k_f q_f - l_f k_m q_m\right),$$

$$\tilde{u}_m = \frac{1}{1 - l_m l_f}\left(m_m - l_m m_f + k_m q_m - l_m k_f q_f\right).$$

As in the box, the condition that the female's rule is the best given that of the male can be written as

$$\frac{\partial W_f}{\partial u_f}(\tilde{u}_f, \tilde{u}_m; q_f) = l_m \frac{\partial W_f}{\partial u_m}(\tilde{u}_f, \tilde{u}_m; q_f).$$

For the given functions this yields

$$2\left[1 - (\tilde{u}_f + \tilde{u}_m)\right] - (1 - q_f) - 2\alpha_f \tilde{u}_f = 2l_m\left[1 - (\tilde{u}_f + \tilde{u}_m)\right].$$

Substituting for $\tilde{u}_f$ and $\tilde{u}_m$ we arrive at an expression of the form $A + B_f q_f + C_m q_m = 0$, where A, B_f, B_m are constants. Since this expression holds for all q_f and q_m, all the constants are zero. Setting $B_m = 0$ gives $l_f = \frac{1 - l_m}{1 - l_m + \alpha_f}$. Similarly, $l_m = \frac{1 - l_f}{1 - l_f + \alpha_m}$.

Ex. 8.7. Using eqs (5.8, 5.14), the conditions $B'(u_i + u_j) = \partial C(u_i, q_i)/\partial u_i$ for $i, j = 1, 2$ and $i, j = 2, 1$ become

$$b_1 - b_2(u_1 + u_2) = c_1(1 - q_1) - c_2 u_1$$
$$b_1 - b_2(u_2 + u_1) = c_1(1 - q_2) - c_2 u_2.$$

The solution is

$$u_i^*(q_i, q_j) = \frac{b_1 c_2 - c_1 c_2 + c_1(b_2 + c_2)q_i - c_1 b_2 q_j}{c_2(2b_2 + c_2)}. \tag{S19}$$

For the cases we consider, all payoff parameters are positive, so $u_i^*(q_i, q_j)$ increases with q_i and decreases with q_j.

Ex. 8.8. Just as for the derivation of the Nash equilibrium in eq (S19) in Exercise 8.7, with perceived qualities p_i and p_j, the slow-learning limit outcome of learning is

$$u_i^*(p_i, p_j) = \frac{b_1 c_2 - c_1 c_2 + c_1(b_2 + c_2)p_i - c_1 b_2 p_j}{c_2(2b_2 + c_2)}.$$

We can note that $u_i^*(p_j, p_i)$ decreases with p_i. From the Nash equilibrium conditions this learning outcome satisfies $B'(u_i + u_j) = \partial C(u_i, p_i)/\partial u_i$ for $i, j = 1, 2$ and $i, j = 2, 1$. Taking the partial derivative with respect to p_i of the fitness payoff $W_i(u_i^*(p_i, p_j), u_j^*(p_j, p_i), q_i) = B(u_i^*(p_i, p_j)) + u_j^*(p_j, p_i)) - C(u_i^*(p_i, p_j), q_i)$ and using the Nash equilibrium condition we get, as stated, that

$$\left[\frac{\partial C(u_i^*, p_i)}{\partial u_i} - \frac{\partial C(u_i^*, q_i)}{\partial u_i} \right] \frac{\partial u_i^*}{\partial p_i} + B'(u_i^* + u_j^*) \frac{\partial u_j^*}{\partial p_i}. \tag{S20}$$

For $p_i = q_i$ the term in square brackets of the expression for the derivative is zero and the other term is negative (B' will be positive at a Nash equilibrium), so a mutant can increase fitness by shifting from zero bias towards negative bias.

Solutions for Chapter 9

Ex. 9.1. Suppose that the resident strategy is (z, z). Then the payoff for a single mutant (male or female) that cares is $W_C(z) = (1 - z) \times 1.25 + z \times 1.0$ and the payoff for desertion is $W_D(z) = (1 - z) \times 1.0 + z \times 0 + \frac{z}{(0.15 + 2z)} \times 1.25$. Let $\Delta(z) = W_D(z) - W_C(z)$. Then after some manipulation we find that

$$\Delta(z) = -f(z) \left[6z^2 - 2.55z + 0.15 \right],$$

where $f(z)$ is a positive function of z. The quadratic function of z in the square brackets has roots at $z_1^* = 0.0705$ and $z_2^* = 0.3545$. $\Delta(z) < 0$ for $0 \le z < z_1^*$. Thus when the resident strategy is $(0, 0)$, care by both parents is best, so this is a Nash equilibrium. Similarly, (z_1^*, z_1^*) and (z_2^*, z_2^*) are Nash equilibria, since for either of these resident strategies the resident strategy is a best response to itself. Note that $(1, 1)$ is not a Nash equilibrium since $\Delta(1) < 0$, so that care is better than desertion.

Ex. 9.2. (i) Suppose the resident population plays Hawk with probability p. A p' mutant in this population wins his contest over a female with probability

$$B(p',p) = (1-\theta) + \theta\left[(1-p')(1-p)\times\frac{1}{2} + p'(1-p)\times 1 + p'p\times\frac{1}{2}\right]$$

$$= 1 - \frac{\theta}{2}(1+p-p').$$

He dies in an encounter with a contested female with probability $\frac{1}{2}p'pz$. Thus he survives until the meeting with the next female with probability

$$S(p',p) = r(1 - \theta\frac{1}{2}p'pz).$$

Let $W(p',p)$ be the mean number of matings obtained by the p' male, given he survives to meet his first female. Then $W(p',p) = B(p',p) + S(p',p)W(p',p)$, so that $W(p',p) = B(p',p)/(1 - S(p',p))$, which gives

$$W(p',p) = \frac{2 - \theta(1+p-p')}{2 - r(2 - \theta p'pz)}. \tag{S21}$$

(ii) Differentiating eq (S21) with respect to p' gives

$$[2 - r(2 - \theta p'pz)]^2 \frac{\partial W}{\partial p'}(p',p) = \theta\Delta(p), \tag{S22}$$

where $\Delta(p) = \theta zrp^2 - zr(2-\theta)p + 2(1-r)$. Thus $D(p) = \frac{\partial W}{\partial p'}(p,p)$ has the same sign as $\Delta(p)$. When $\theta = 0.8$, $r = \frac{8}{9}$ and $z = \frac{4}{7}$ we have

$$\Delta(p) = \frac{2}{5\times 63}\left[64p^2 - 96p + 35\right]$$

$$= \frac{2}{5\times 63}(8p - 5)(8p - 7)$$

$$= \frac{128}{5\times 63}(p - \frac{5}{8})(p - \frac{7}{8}).$$

Thus $\Delta(p) = 0$ has roots at $p_1^* = \frac{5}{8}$ and $p_2^* = \frac{7}{8}$. Furthermore $\Delta(p) > 0$ for $p < p_1^*$, $\Delta(p) < 0$ for $p_1^* < p < p_2^*$, and $\Delta(p) > 0$ for $p > p_2^*$. Thus p_1^* is convergence stable while p_2^* is an evolutionary repeller.

(iii) Let $p' \neq p^*$. Since $\Delta(p^*) = 0$ we have $\frac{\partial W}{\partial p'}(p',p^*) = 0$ for all p' by eq (S22). Thus $W(p',p^*) = W(p^*,p^*)$, so that condition (ES1) fails, but condition (ES2)(i) holds. This is a game against the field, and we proceed in the same way as for the sex-allocation game (Solution to Exercise 4.5). The probability an opponent plays Hawk in a population in which a proportion $1-\epsilon$ of males uses strategy p^* and a proportion ϵ uses p' is the same as in a population in which all males use strategy $p(\epsilon) = (1-\epsilon)p^* + \epsilon p'$. Thus, in the notation of condition (ES2′)(ii), $W(p,p_\epsilon) = W(p,p(\epsilon))$ for all p. Consider first the Nash equilibrium at $p^* = \frac{5}{8}$. Let $p' > p^*$. Then for small ϵ we have $\frac{5}{8} < p(\epsilon) < \frac{7}{8}$ so that $\Delta(p(\epsilon)) < 0$. By eq (S22) we have $\frac{\partial W}{\partial p'}(p,p(\epsilon)) < 0$ for all p, so that the function $W(p,p(\epsilon))$ is a strictly decreasing function of p. Thus $W(p',p(\epsilon)) < W(p^*,p(\epsilon))$ since $p' > p^*$. Similarly, $W(p',p(\epsilon)) < W(p^*,p(\epsilon))$ for $p' < p^*$. Thus condition (ES2′)(ii) holds, so that $p^* = \frac{5}{8}$ is an ESS. An analogous argument shows that condition (ES2′)(ii) fails when $p^* = \frac{7}{8}$, so this Nash equilibrium is not an ESS.

Ex. 9.3. Suppose the resident population plays Hawk with probability p. A p' mutant in this population wins his contest over a female with probability

$$B(p',p) = (1-p')(1-p) \times \frac{1}{2} + p'(1-p) \times 1 + p'p \times \frac{1}{2}$$

$$= \frac{1}{2}(1+p'-p).$$

He survives until the meeting with the next female with probability

$$S(p',p) = r(1 - \frac{1}{2}p'p).$$

Let $W(p',p)$ be the mean number of matings obtained by the p' male, given he survives to meet his first female. Then $W(p',p) = B(p',p) + S(p',p)W(p',p)$, so that

$$W(p',p) = \frac{1+p'-p}{2-r(2-p'p)}. \tag{S23}$$

(i) By eq (S23) we have

$$W(p,p) = \frac{1}{2-r(2-p^2)}. \tag{S24}$$

A newborn resident survives to meet his first female with probability $\frac{1}{2}$. Thus his expected number of matings is $\frac{1}{2}W(p,p)$. Since each mating results in one son, the population is of stable size if $\frac{1}{2}W(p,p) = 1$. From eq (S24) this condition gives

$$r = \frac{3}{2(2-p^2)}. \tag{S25}$$

In order for the population to be viable we must have $r \le 1$. From eq (S25) we have $p \le \frac{1}{\sqrt{2}}$.

(ii) Differentiating eq (S23) with respect to p' gives

$$[2-r(2-p'p)]^2 \frac{\partial W}{\partial p'}(p',p) = 2 - r(2+p-p^2).$$

Substituting for r from eq (S25) then gives

$$\frac{\partial W}{\partial p'}(p',p) \propto 2 - 3p - p^2.$$

The equation $p^2 + 3p - 2 = 0$ has two roots, p_1 and p_2, where $p_1 < 0$ and $p_2 = \frac{\sqrt{17}-3}{2}$. For $0 \le p < p_2$ we have $\frac{\partial W}{\partial p'}(p',p) > 0$ for all p'. Thus the unique best response to p is $\hat{b}(p) = 1$. Similarly, if $p_2 < p \le 1$ then $\hat{b}(p) = 0$. When $p = p_2$ we have $\frac{\partial W}{\partial p'}(p',p) = 0$ for all p'. Thus p_2 is a best response to itself and is the unique Nash equilibrium.

Ex. 9.4. (i) On day 2 the individual must gain an item. The probability it does so is $H_2(u) = (1-u)up_2$ if it forages with intensity u. Differentiating with respect to u and setting the derivative equal to 0 gives $u_2^* = 0.5$. The reproductive value at the start of this day is thus $V_2 = H(u_2^*) \times 1 = \frac{p_2}{4}$. On day 1 the individual either gains an item, and hence has a reproductive value of 1 at the end of the day, or lives and gains no item, and so has reproductive value V_2

at the end of the day (assuming optimal behaviour on day 2), or is dead. Thus the payoff from foraging intensity u on this day is

$$H_1(u) = (1-u)up_1 \times 1 + (1-u)(1-up_1) \times V_2 + u \times 0.$$

Setting $p_1 = 0.5$, differentiating with respect to u and setting the derivative equal to 0 at $u = u_1^*$ gives $u_1^* = \frac{1}{2}\frac{(1-3V_2)}{(1-V_2)}$. On setting $V_2 = p_2/4$ we obtain

$$u_1^* = \frac{1}{2}\frac{(4-3p_2)}{(4-p_2)}. \tag{S26}$$

(ii) The only population members foraging on day 2 are those that survived day 1 and failed to get an item. For each individual this occurs with probability $(1-u_1^*)(1-\frac{u_1^*}{2})$, so that

$$p_2 = (1-u_1^*)(1-\frac{u_1^*}{2}). \tag{S27}$$

We are given that

$$p_2 = 1 - \frac{p_2}{2} \tag{S28}$$

Thus given a value for u_1^* we can find p_2 from eq (S27), then find p_2 from eq (S28), and hence find a new value for u_1^* from eq (S26). Denote the new value as $F(u_1^*)$. We seek a solution of the equation $F(u_1^*) = u_1^*$. This can be found by simply plotting the function $F(u_1^*) - u_1^*$ and observing its zeros. This procedure reveals a single zero at $u_1^* = 0.2893$.

Solutions for Chapter 10

Ex. 10.1. In the notation of Box 10.1 we have $a_{11} = 1.0$, $a_{12} = 0.1$, $a_{21} = 0.7$, $a_{22} = 1.0$. Thus $B = 2$ and $C = 0.93$. By eq (10.11), $\lambda = 1.2646$. By eq (10.12), $\rho = 0.2743$. By eq (10.13), $v_H/v_L = 0.3780$.

Ex. 10.2. Summing over i in eq (10.3) gives

$$\lambda(x',x)\sum_i \rho_i(x',x) = \sum_i \left[\sum_j a_{ij}(x',x)\rho_j(x',x)\right]. \tag{S29}$$

By eq (10.4) $\sum_i \rho_i(x',x) = 1$. We can also change the order of summation on the right-hand side of eq (S29) to give

$$\sum_i \left[\sum_j a_{ij}(x',x)\rho_j(x',x)\right] = \sum_j \left[\sum_i a_{ij}(x',x)\right]\rho_j(x',x)$$

$$= \sum_j a_j(x',x)\rho_j(x',x),$$

where $a_j(x',x) = \sum_i a_{ij}(x',x)$. Thus $\lambda(x',x) = \sum_j a_j(x',x)\rho_j(x',x)$.

Ex. 10.3. Taking the mutant strategy x' equal to the resident strategy x, eq (10.1) can be written in vector notation as $n(t) = A(x,x)n(t-1)$. Since we have $n(t-1) = A(x,x)n(t-2)$, etc., we deduce that

$$n(t) = A(x,x)^t n(0).$$

Suppose there is just one individual in state s_j in generation 0. Then $\mathbf{n}(0) = \mathbf{e}_j$. In this case the vector $\mathbf{n}(t) = \mathbf{A}(\mathbf{x}',\mathbf{x})^t \mathbf{e}_j$ gives the mean number of descendants of this one founder individual in each state in t years' time. The mean total number of descendants, $d_j(t)$, is the sum of the elements of $\mathbf{n}(t)$; i.e. $d_j(t) = \mathbf{1}\mathbf{n}(t)$. Thus $d_j(t) = \mathbf{1}\mathbf{A}(\mathbf{x},\mathbf{x})^t \mathbf{e}_j$.

We now note that $\sum_j \mathbf{e}_j \rho_j(\mathbf{x},\mathbf{x}) = \boldsymbol{\rho}(\mathbf{x},\mathbf{x})$. Thus

$$\sum_j d_j(t)\rho_j(\mathbf{x},\mathbf{x}) = \mathbf{1}\mathbf{A}(\mathbf{x},\mathbf{x})^t \sum_j \mathbf{e}_j \rho_j(\mathbf{x},\mathbf{x})$$

$$= \mathbf{1}\mathbf{A}(\mathbf{x},\mathbf{x})^t \boldsymbol{\rho}(\mathbf{x},\mathbf{x})$$

$$= \mathbf{1}\boldsymbol{\rho}(\mathbf{x},\mathbf{x}) \text{ by eq (10.5)}$$

$$= 1 \text{ by eq (10.4)}.$$

By eq (10.15) we have

$$\sum_j v_j(\mathbf{x})\rho_j(\mathbf{x},\mathbf{x}) = 1.$$

Ex. 10.4. Assume that $\mathbf{v}(\mathbf{x})\mathbf{A}(\mathbf{x}',\mathbf{x}) \leq \mathbf{v}(\mathbf{x})$. Multiplying both sides of this equation on the right by the column vector $\boldsymbol{\rho}(\mathbf{x}',\mathbf{x})$ gives

$$\mathbf{v}(\mathbf{x})\mathbf{A}(\mathbf{x}',\mathbf{x})\boldsymbol{\rho}(\mathbf{x}',\mathbf{x}) \leq \mathbf{v}(\mathbf{x})\boldsymbol{\rho}(\mathbf{x}',\mathbf{x}). \tag{S30}$$

Using eq (10.3) we have

$$\mathbf{v}(\mathbf{x})\mathbf{A}(\mathbf{x}',\mathbf{x})\boldsymbol{\rho}(\mathbf{x}',\mathbf{x}) = \mathbf{v}(\mathbf{x})\lambda(\mathbf{x}',\mathbf{x})\boldsymbol{\rho}(\mathbf{x}',\mathbf{x}) \tag{S31}$$

$$= \lambda(\mathbf{x}',\mathbf{x})\mathbf{v}(\mathbf{x})\boldsymbol{\rho}(\mathbf{x}',\mathbf{x}). \tag{S32}$$

By eqs (S30) and (S32) we have $\lambda(\mathbf{x}',\mathbf{x}) \leq 1$ since $\mathbf{v}(\mathbf{x})\boldsymbol{\rho}(\mathbf{x}',\mathbf{x}) > 0$.

Ex. 10.5. There are four states for an individual: type 1 male, type 2 male, type 1 female, type 2 female. Taking account of the 1:1 offspring sex ratio, the projection matrix for the mutant cohort can then be written as

$$\mathbf{A}(\mathbf{x}',\mathbf{x}) = \begin{bmatrix} a_{11}(\mathbf{x}) & a_{12}(\mathbf{x}) & \alpha_{11}(\mathbf{x}',\mathbf{x}) & \alpha_{12}(\mathbf{x}',\mathbf{x}) \\ a_{21}(\mathbf{x}) & a_{22}(\mathbf{x}) & \alpha_{21}(\mathbf{x}',\mathbf{x}) & \alpha_{22}(\mathbf{x}',\mathbf{x}) \\ a_{11}(\mathbf{x}) & a_{12}(\mathbf{x}) & \alpha_{11}(\mathbf{x}',\mathbf{x}) & \alpha_{12}(\mathbf{x}',\mathbf{x}) \\ a_{21}(\mathbf{x}) & a_{22}(\mathbf{x}) & \alpha_{21}(\mathbf{x}',\mathbf{x}) & \alpha_{22}(\mathbf{x}',\mathbf{x}) \end{bmatrix}.$$

(i) Let $\mathbf{v}(\mathbf{x}) = (v_1^m, v_2^m, v_1^f, v_2^f)$ denote the row vector of reproductive values in the resident population. Let $\boldsymbol{\rho}(\mathbf{x}',\mathbf{x}) = (\rho_1', \rho_2', \rho_1', \rho_2')^T$ denote the column vector of stable proportions in the mutant cohort. [Here we have used the 1:1 sex ratio of offspring.] As in Box 10.2 we have

$$\mathbf{v}\big[\mathbf{A}(\mathbf{x},\mathbf{x}) - \mathbf{A}(\mathbf{x}',\mathbf{x})\big]\boldsymbol{\rho}' = \big[\lambda(\mathbf{x},\mathbf{x}) - \lambda(\mathbf{x}',\mathbf{x})\big]\mathbf{v}\boldsymbol{\rho}'. \tag{S33}$$

All the entries in the first and second columns of the matrix $\mathbf{A}(\mathbf{x},\mathbf{x}) - \mathbf{A}(\mathbf{x}',\mathbf{x})$ are zero, so that

$$\mathbf{v}\big[\mathbf{A}(\mathbf{x},\mathbf{x}) - \mathbf{A}(\mathbf{x}',\mathbf{x})\big]\boldsymbol{\rho}' = \big[W_1(\mathbf{x},\mathbf{x}) - W_1(\mathbf{x}',\mathbf{x})\big]\rho_1'$$
$$+ \big[W_2(\mathbf{x},\mathbf{x}) - W_2(\mathbf{x}',\mathbf{x})\big]\rho_2'.$$

From this equation and eq (S33) we deduce that if $W_1(\mathbf{x}',\mathbf{x}) \leq W_1(\mathbf{x},\mathbf{x})$ and $W_2(\mathbf{x}',\mathbf{x}) \leq W_2(\mathbf{x},\mathbf{x})$ then $\lambda(\mathbf{x}',\mathbf{x}) \leq \lambda(\mathbf{x},\mathbf{x})$.

(ii) If a type 1 mutant female mates with a type 1 resident male, one half of the N_1 offspring are mutant and one half of these are female. Of these a proportion $1 - \epsilon$ is type 1 and a proportion ϵ is type 2. If the type 1 female mates with a type 2 male, one half of the N_2 offspring are mutant and one half of these are female. Of these one half are type 1 and one half are type 2. We thus have

$$\alpha_{11}(\mathbf{x}',\mathbf{x}) = \frac{N_1}{4}m_{11}(x_1',\mathbf{x})(1-\epsilon) + \frac{N_2}{8}m_{21}(x_1',\mathbf{x}),$$

$$\alpha_{21}(\mathbf{x}',\mathbf{x}) = \frac{N_1}{4}m_{11}(x_1',\mathbf{x})\epsilon + \frac{N_2}{8}m_{21}(x_1',\mathbf{x}).$$

Hence

$$W_1(\mathbf{x}',\mathbf{x}) = \bar{v}_1\alpha_{1,1}(\mathbf{x}',\mathbf{x}) + \bar{v}_2\alpha_{2,1}(\mathbf{x}',\mathbf{x})$$

$$= m_{11}(x_1',\mathbf{x})\frac{N_1}{4}[(1-\epsilon)\bar{v}_1 + \epsilon\bar{v}_2] + m_{21}(x_1',\mathbf{x})\frac{N_2}{8}(\bar{v}_1 + \bar{v}_2).$$

It follows that we can write $W_1(\mathbf{x}',\mathbf{x}) = m_{11}(x_1',\mathbf{x})r_{11} + m_{21}(x_1',\mathbf{x})r_{21}$, where

$$r_{11} = \frac{N_1}{4}[(1-\epsilon)\bar{v}_1 + \epsilon\bar{v}_2]$$

$$r_{21} = \frac{N_2}{8}(\bar{v}_1 + \bar{v}_2).$$

Similarly,

$$r_{22} = \frac{N_2}{4}[\epsilon\bar{v}_1 + (1-\epsilon)\bar{v}_2],$$

$$r_{21} = \frac{N_1}{8}(\bar{v}_1 + \bar{v}_2).$$

(iii) Now let $N_1 = N_2 = N$. Then by the above,

$$r_{21} - r_{11} = \frac{N(1-2\epsilon)}{8}(\bar{v}_2 - \bar{v}_1),$$

$$r_{22} - r_{12} = \frac{N(1-2\epsilon)}{8}(\bar{v}_2 - \bar{v}_1),$$

which proves eq (10.36).

Ex. 10.6. Using the formulae of Section 10.7 we have $c_1(x) = 0.7875$ and $c_2(x) = 0.5875$. Equation (10.42) then gives

$$\mathbf{A}(x',x) = \begin{bmatrix} 0.3889 & 0.3571 \\ 0.2234 & 1.1170 \end{bmatrix}.$$

In the terminology of Box 10.1 we then have $B = 1.5059$ and $C = 0.3546$, giving $\lambda(x',x) = 1.2137$.

Ex. 10.7. We have $\lambda(n) = 0.85n^{\frac{1}{n}}$ and $w_0(n) = 0.85^n n$. We plot these below:

n	$\lambda(n)$	$w_0(n)$
1	0.8500	0.8500
2	1.2021	1.4450
3	1.2259	1.8424
4	1.2021	2.0880
5	1.1728	2.2185
6	1.1458	2.2629
7	1.1224	2.2440
8	1.1023	2.1799

Thus $\lambda(n)$ is maximized at $n = 3$ and $w_0(n)$ is maximized at $n = 6$. Since $\lambda(3) > \lambda(6)$ and $w_0(3) < w_0(6)$ we infer that w_0 is not a strong fitness proxy.

Ex. 10.8. The derivative of eq (10.3) with respect to x'_k, evaluated at $\mathbf{x}' = \mathbf{x}$ is

$$\frac{\partial\lambda(\mathbf{x},\mathbf{x})}{\partial x'_k}\rho_i(\mathbf{x},\mathbf{x}) + \frac{\partial\rho_i(\mathbf{x},\mathbf{x})}{\partial x'_k} = \sum_j \frac{\partial a_{ij}(\mathbf{x},\mathbf{x})}{\partial x'_k}\rho_j(\mathbf{x},\mathbf{x}) + \sum_j a_{ij}(\mathbf{x},\mathbf{x})\frac{\partial\rho_j(\mathbf{x},\mathbf{x})}{\partial x'_k},$$

where we used that $\lambda(\mathbf{x},\mathbf{x}) = 1$. Multiply this equation with $v_i(\mathbf{x})$ and sum over i, and use that $v_i(\mathbf{x})$ is a left eigenvector. The terms with sums of $v_i(\mathbf{x})\partial\rho_i(\mathbf{x},\mathbf{x})/\partial x'_k$ then cancel, and using our normalization we obtain eq (10.43).

References

Abrams, P. A. 2000. The evolution of predator-prey interactions: theory and evidence. *Annual Review of Ecology and Systematics*, 31(1):79–105.

Abrams, P. A., Matsuda, H., and Harada, Y. 1993. Evolutionarily unstable fitness maxima and stable fitness minima of continuous traits. *Evolutionary Ecology*, 7(5):465–487.

Alexander, R. D. 1987. *The biology of moral systems*. Aldine de Gruyter, New York.

Alizon, S., de Roode, J. C., and Michalakis, Y. 2013. Multiple infections and the evolution of virulence. *Ecology Letters*, 16(4):556–567.

Allen, B. J. and Levinton, J. S. 2007. Costs of bearing a sexually selected ornamental weapon in a fiddler crab. *Functional Ecology*, 21(1):154–161.

Alonzo, S. H. 2002. State-dependent habitat selection games between predators and prey: the importance of behavioural interactions and expected lifetime reproductive success. *Evolutionary Ecology Research*, 4:759–778.

Alonzo, S. H. 2010. Social and coevolutionary feedbacks between mating and parental investment. *Trends in Ecology & Evolution*, 25(2):99–108.

Alonzo, S. H., Switzer, P. V., and Mangel, M. 2003. Ecological games in space and time: the distribution and abundance of antarctic krill and penguins. *Ecology*, 84:1598–1607.

Andersson, M. 1982. Female choice selects for extreme tail length in a widowbird. *Nature*, 299:818–820.

Aplin, L. M., Farine, D. R., Morand-Ferron, J., Cockburn, A., Thornton, A., and Sheldon, B. C. 2015. Experimentally induced innovations lead to persistent culture via conformity in wild birds. *Nature*, 518(7540):538–541.

Aplin, L. M. and Morand-Ferron, J. 2017a. Data from: Stable producer-scrounger dynamics in wild birds: sociability and learning speed covary with scrounging behaviour. *Dryad Digital Repository*, http://dx.doi.org/10.5061/dryad.n7c42

Aplin, L. M. and Morand-Ferron, J. 2017b. Stable producer-scrounger dynamics in wild birds: sociability and learning speed covary with scrounging behaviour. *Proceedings of the Royal Society B*, 284(1852):20162872.

Argasinski, K. and Broom, M. 2013. Ecological theatre and the evolutionary game: how environmental and demographic factors determine payoffs in evolutionary games. *Journal of Mathematical Biology*, 67:935–962.

Argasinski, K. and Broom, M. 2018. Interaction rates, vital rates, background fitness and replicator dynamics: how to embed evolutionary game structure into realistic population dynamics. *Theory in Biosciences*, 137:33–50.

Arnott, G. and Elwood, R. W. 2009. Assessment of fighting ability in animal contests. *Animal Behaviour*, 77(5):991–1004.

Ashton, B. J., Ridley, A. R., Edwards, E. K., and Thornton, A. 2018. Cognitive performance is linked to group size and affects fitness in Australian magpies. *Nature*, 554(7692):364–367.

Austad, S. N. 1983. A game theoretical interpretation of male combat in the bowl and doily spider (*Frontinella pyramitela*). *Animal Behaviour*, 31(1):59–73.

Axelrod, R. and Hamilton, W. D. 1981. The evolution of cooperation. *Science*, 211(4489): 1390–1396.

Baerends, G. P. and Baerends-Van Roon, J. M. 1950. An introduction to the study of the ethology of the cichlid fishes. *Behaviour, Supplement*, 1:1–243.

Balogh, A. C., Gamberale-Stille, G., Tullberg, B. S., and Leimar, O. 2010. Feature theory and the two-step hypothesis of Müllerian mimicry evolution. *Evolution*, 64(3):810–822.

Balogh, A. C. and Leimar, O. 2005. Müllerian mimicry: an examination of Fisher's theory of gradual evolutionary change. *Proceedings of the Royal Society B*, 272(1578):2269–2275.

Barnard, C. J. and Sibly, R. M. 1981. Producers and scroungers: a general model and its application to captive flocks of house sparrows. *Animal Behaviour*, 29:543–550.

Barta, Z., Houston, A. I., McNamara, J. M., and Székely, T. 2002. Sexual conflict about parental care: the role of reserves. *American Naturalist*, 159:687–705.

Barta, Z., Székely, T., Liker, A., and Harrison, F. 2014. Social role specialisation promotes cooperation between parents. *American Naturalist*, 183:747–761.

Bateson, P. and Laland, K. N. 2013. Tinbergen's four questions: an appreciation and an update. *Trends in Ecology & Evolution*, 28:712–718.

Bell, A. M., Hankison, S. J., and Laskowski, K. L. 2009. The repeatability of behaviour: a meta-analysis. *Animal Behaviour*, 77(4):771–783.

Bellhouse, D. R. and Fillion, N. 2015. Le Her and other problems in probability discussed by Bernoulli, Montmort and Waldegrave. *Statistical Science*, 30(1):26–39.

Berger, U. and Grüne, A. 2016. On the stability of cooperation under indirect reciprocity with first-order information. *Games and Economic Behavior*, 98:19–33.

Bergmuller, R. and Taborsky, M. 2010. Animal personality due to social niche specialisation. *Trends in Ecology & Evolution*, 25:504–511.

Bergstrom, C. T., Számadó, S., and Lachmann, M. 2002. Seperating equilibria in continuous signalling games. *Philosophical Transactions of the Royal Society B*, 357:1595–1606.

Berridge, K. C. and Robinson, T. E. 1998. What is the role of dopamine in reward: hedonic impact, reward learning, or incentive salience? *Brain Research Reviews*, 28(3):309–369.

Biernaskie, J. M., Grafen, A., and Perry, J. C. 2014. The evolution of index signals to avoid the cost of dishonesty. *Proceedings of the Royal Society B*, 281:20140876.

Binmore, K. 2009. *Rational decisions*. Princeton University Press, Princeton, NJ.

Binmore, K. 2010. Bargaining in biology? *Journal of Evolutionary Biology*, 23:1351–1363.

Bishop, C. M. 2006. *Pattern recognition and machine learning*. Springer Science, Cambridge, UK.

Bitterman, M. 1975. The comparative analysis of learning: are the laws of learning the same in all animals? *Science*, 188(4189):699–709.

Bogacz, R., Brown, E., Moehlis, J., Holmes, P., and Cohen, J. D. 2006. The physics of optimal decision making: a formal analysis of models of performance in two-alternative forced-choice tasks. *Psychological Review*, 113(4):700–765.

Bonte, D., Van Dyck, H., Bullock, J. M., Coulon, A., Delgado, M., Gibbs, M., ... Travis, J. M. 2012. Costs of dispersal. *Biological Reviews*, 87(2):290–312.

Bradbury, J. W. and Vehrencamp, S. L. 2011. *Principles of animal communication, second edition*. Sinauer Associates, Inc., Oxford, UK.

Broom, M. and Rychtàr, J. 2013. *Game-theoretic models in biology*. CRC Press, Taylor & Francis Group, Boca Raton, FL.

Brown, J. S. and Pavlovic, N. B. 1992. Evolution in heterogeneous environments: effects of migration on habitat specialization. *Evolutionary Ecology*, 6:360–382.

Bründl, A. C., Sorato, E., Sallé, L., Thiney, A. C., Kaulbarsch, S., Chaine, A. S., and Russell, A. F. 2019. Experimentally induced increases in fecundity lead to greater nestling care in blue tits. *Proceedings of the Royal Society B*, 286(1905):20191013.

Bshary, R. and Noë, R. 2003. Biological markets: the ubiquitous influence of partner choice on the dynamics of cleaner fish – client reef fish interactions. In Hammerstein, P., editor, *Genetic and cultural evolution of cooperation*, pages 167–184. MIT Press, Cambridge, MA.

Budaev, S., Jørgensen, C., Mangel, M., Eliassen, S., and Giske, J. 2019. Decision-making from the animal perspective: bridging ecology and subjective cognition. *Frontiers in Ecology and Evolution*, 7:164.

Bullmore, E. and Sporns, O. 2012. The economy of brain network organization. *Nature Reviews Neuroscience*, 13:336–349.

Bulmer, M. G. and Parker, G. A. 2002. The evolution of anisogamy: a game-theoretic approach. *Proceedings of the Royal Society B*, 269:2381–2388.

Bush, R. R. and Mosteller, F. 1955. *Stochastic models for learning*. John Wiley & Sons Inc., New York.

Cadet, C., Ferriére, R., Metz, J. A. J., and van Baalen, M. 2003. The evolution of dispersal under demographic stochasticity. *American Naturalist*, 162:427–441.

Camerer, C. and Ho, T.-H. 1999. Experience-weighted attraction learning in normal form games. *Econometrica*, 67(4):827–874.

Caswell, H. 2001. *Matrix population models, second edition*. Sinauer Associates, Inc., Sunderland, MA.

Caswell, H. 2012. Matrix models and sensitivity analysis of populations classified by age and stage: a vec-permutation matrix approach. *Theoretical Ecology*, 5(3):403–417.

Charnov, E. L. 1982. *The theory of sex allocation*. Princeton University Press, Princeton, NJ.

Charnov, E. L. 1993. *Life history invariants*. Oxford University Press, Oxford, UK.

Charnov, E. L. and Bull, J. J. 1977. When is sex environmentally determined? *Nature*, 266(5605):828–830.

Clements, K. C. and Stephens, D. W. 1995. Testing models of non-kin cooperation: mutualism and the Prisoner's Dilemma. *Animal Behaviour*, 50(2):527–535.

Clutton-Brock, T. 2009. Cooperation between non-kin in animal societies. *Nature*, 462(7269):51–57.

Clutton-Brock, T. H., Albon, S. D., Gibson, R. M., and Guinness, F. E. 1979. The logical stag: adaptive aspects of fighting in red deer (*Cervus elaphus*). *Animal Behaviour*, 27:211–225.

Collins, A. G. E. and Frank, M. J. 2014. Opponent actor learning (OpAL): modeling interactive effects of striatal dopamine on reinforcement learning and choice incentive. *Psychological Review*, 121(3):337–366.

Comins, H. N., Hamilton, W. D., and May, R. M. 1980. Evolutionarily stable dispersal strategies. *Journal of Theoretical Biology*, 82:205–230.

Connor, R. C. 1995. The benefits of mutualism: a conceptual framework. *Biological Reviews*, 70(3):427–457.

Conover, D. O. and Van Voorhees, D. A. 1990. Evolution of a balanced sex ratio by frequency-dependent selection in a fish. *Science*, 250:1556–1558.

Conover, D. O., Voorhees, D. A. V., and Ehtisham, A. 1992. Sex ratio selection and the evolution of environmental sex determination in laboratory populations of *Menidia menidia*. *Evolution*, 46:1722–1730.

Cote, J., Clobert, J., Brodin, T., Fogarty, S., and Sih, A. 2010. Personality-dependent dispersal: characterization, ontogeny and consequences for spatially structured populations. *Philosophical Transactions of the Royal Society B*, 365:4065–4076.

Courville, A. C., Daw, N. D., and Touretzky, D. S. 2006. Bayesian theories of conditioning in a changing world. *Trends in Cognitive Sciences*, 10(7):294–300.

Daan, S., Deerenberg, S., and Dijkstra, C. 1996. Increased daily work precipitates natural death in the kestrel. *Journal of Animal Ecology*, 69:539–544.

Darwin, C. 1871. *The descent of man, and selection in relation to sex*. John Murray, London.

Daw, N. D. 2014. Advanced reinforcement learning. In Glimcher, P. W. and Fehr, E., editors, *Neuroeconomics, decision making and the brain, second edition*, pages 299–320. Academic Press, Amsterdam.

Dawkins, R. 1976. *The selfish gene*. Oxford University Press, Oxford, UK.

Dayan, P., Kakade, S., and Read Montague, P. 2000. Learning and selective attention. *Nature Neuroscience*, 3(11s):1218–1223.

Debarre, F. and Gandon, S. 2010. Evolution of specialization in a spatially continuous environment. *Journal of Evolutionary Biology*, 23:1090–1099.

Dieckmann, U. and Law, R. 1996. The dynamical theory of coevolution: a derivation from stochastic ecological processes. *Journal of Mathematical Biology*, 34:579–612.

Dieckmann, U. and Metz, J. A. J. 2006. Surprising evolutionary predictions from enhanced ecological realism. *Theoretical Population Biology*, 69:263–281.

Doebeli, M. and Dieckmann, U. 2000. Evolutionary branching and sympatric speciation caused by different types of ecological interactions. *American Naturalist*, 156:S77–S101.

Doebeli, M., Dieckmann, U., Metz, J. A. J., and Tautz, D. 2005. What we have also learnt: adaptive speciation is theoretically plausible. *Evolution*, 59:691–695.

Doebeli, M., Hauert, C., and Killingback, T. 2004. The evolutionary origin of cooperators and defectors. *Science*, 306:859–862.

Douhard, M. 2017. Offspring sex ratio in mammals and the Trivers-Willard hypothesis: in pursuit of unambiguous evidence. *BioEssays*, 39(9):1–10.

Dugatkin, L. A. 2001. Bystander effects and the structure of dominance hierarchies. *Behavioral Ecology*, 12(3):348–352.

Düsing, C. 1884. *Die Regulierung des Geschlechtsverhältnisses bei der Vermehrung der Menschen, Tiere und Pflanzen*. Gustav Fischer, Jena, Germany.

Edwards, A. W. F. 2000. Carl Düsing (1884) on the regulation of the sex-ratio. *Theoretical Population Biology*, 58(3):255–257.

Ehrlich, P. R. and Raven, P. H. 1964. Butterflies and plants: a study in coevolution. *Evolution*, 18(4):586–608.

Eliassen, S., Andersen, B. S., Jørgensen, C., and Giske, J. 2016. From sensing to emergent adaptations: modelling the proximate architecture for decision-making. *Ecological Modelling*, 326:90–100.

Emlen, D. J., Warren, I. A., Johns, A., Dworkin, I., and Lavine, L. C. 2012. A mechanism of extreme growth and reliable signaling in sexually selected ornaments and weapons. *Science*, 337(6096):860–864.

Engelmann, D. and Fischbacker, U. 2009. Indirect reciprocity and strategic reputation building in an experimental helping game. *Games and Economic Behavior*, 67:399–407.

Engqvist, L. and Taborsky, M. 2016. The evolution of genetic and conditional alternative reproductive tactics. *Proceedings of the Royal Society B*, 283(1825):20152945.

Enquist, M. 1985. Communication during aggressive interactions with particular reference to variation in choice of behaviour. *Animal Behaviour*, 33(3):1152–1161.

Enquist, M. and Jakobsson, S. 1986. Decision making and assessment in the fighting behaviour of *Nannacara anomala* (Cichlidae, Pisces). *Ethology*, 72(2):143–153.

Enquist, M. and Leimar, O. 1983. Evolution of fighting behaviour: decision rules and assessment of relative strength. *Journal of Theoretical Biology*, 102(3):387–410.

Enquist, M. and Leimar, O. 1987. Evolution of fighting behaviour: the effect of variation in resource value. *Journal of Theoretical Biology*, 127(2):187–205.

Enquist, M. and Leimar, O. 1990. The evolution of fatal fighting. *Animal Behaviour*, 39(1):1–9.

Enquist, M., Leimar, O., Ljungberg, T., Mallner, Y., and Segerdahl, N. 1990. A test of the sequential assessment game: fighting in the cichlid fish *Nannacara anomala*. *Animal Behaviour*, 40(1):1–14.

Erev, I. and Roth, A. E. 1998. Predicting how people play games: reinforcement learning in experimental games with unique, mixed strategy equilibria. *American Economic Review*, 88(4):848–881.

Erev, I. and Roth, A. E. 2014. Maximization, learning, and economic behavior. *Proceedings of the National Academy of Sciences*, 111(Supplement 3):10818–10825.

Eshel, I. 1983. Evolutionary and continuous stability. *Journal of Theoretical Biology*, 103:99–111.

Eshel, I. and Motro, U. 1981. Kin selection and strong evolutionary stability of mutual help. *Theoretic Population Biology*, 19:420–433.

Eshel, I. and Sansone, E. 1995. Owner-intruder conflict, Grafen effect and self-assessment. The bourgeois principle re-examined. *Journal of Theoretical Biology*, 177:341–356.

Fawcett, T. W. and Johnstone, R. A. 2010. Learning your own strength: winner and loser effects should change with age and experience. *Proceedings of the Royal Society B*, 277(1686): 1427–1434.

Ferguson, M. W. and Joanen, T. 1982. Temperature of egg incubation determines sex in *Alligator mississippiensis*. *Nature*, 296:850–853.

Fisher, R. A. 1927. On some objections to mimicry theory: statistical and genetic. *Transactions of the Royal Entomological Society of London*, 75:269–278.

Fisher, R. A. 1930. *The genetical theory of natural selection*. Clarendon Press, Oxford, UK.

Fisher, R. A. 1934. Randomisation, and an old enigma of card play. *The Mathematical Gazette*, 18:294–297.

Fisher, R. A. 1958. Polymorphism and natural selection. *Journal of Ecology*, 46(2):289–293.

Fletcher, J. A. and Doebeli, M. 2009. A simple and general explanation for the evolution of altruism. *Proceedings of the Royal Society B*, 276:13–19.

Frank, R. H. 1988. *Passions within reason*. Norton, London.

Frank, S. A. 1995. George Price's contributions to evolutionary genetics. *Journal of Theoretical Biology*, 175(3):373–388.

Frank, S. A. 1996. Models of parasite virulence. *The Quarterly Review of Biology*, 71(1):37–78.

Fretwell, S. D. and Lucas, H. L. 1970. On territorial behaviour and other factors influencing habitat distribution in birds. I. Theoretical development. *Acta Biotheoretica*, 19:16–36.

Fromhage, L. and Jennions, M. D. 2016. Coevolution of parental investment and sexually selected traits drives sex-role divergence. *Nature Communications*, 7:12517.

Fromhage, L. and Jennions, M. D. 2019. The strategic reference gene: an organismal theory of inclusive fitness. *Proceedings of the Royal Society B*, 286:1904.

Fromhage, L., McNamara, J. M., and Houston, A. I. 2007. Stability and value of male care for offspring: is it worth only half the trouble? *Biology Letters*, 3:234–236.

Fronhofer, E. A., Pasurka, H., Poitrineau, K., Mitesser, O., and Poethke, H.-J. 2011. Risk sensitivity revisited: from individuals to populations. *Animal Behaviour*, 82:875–883.

Fudenberg, D. and Levine, D. K. 1998. *The theory of learning in games*. MIT Press, Cambridge, MA.

Gardner, A., West, S. A., and Wild, G. 2011. The genetical theory of kin selection. *Journal of Evolutionary Biology*, 24(5):1020–1043.

Gavrilets, S. 2005. 'Adaptive speciation'—it is not that easy: a reply to Doebeli et al. *Evolution*, 59:696–699.

Gerhardt, H. C. 1994. The evolution of vocalization in frogs and toads. *Annual Review of Ecology and Systematics*, 25:293–324.

Geritz, S. A. H. 2005. Resident-invader dynamics and the coexistence of similar strategies. *Journal of Mathematical Biology*, 50(1):67–82.

Geritz, S. A. H., Kisdi, E., Meszena, G., and Metz, J. A. J. 1998. Evolutionarily singular strategies and the adaptive growth and branching of the evolutionary tree. *Evolutionary Ecology*, 12: 35–57.

Geritz, S. A. H., Metz, J. A. J., Kisdi, E., and Meszena, G. 1997. Dynamics of adaptation and evolutionary branching. *Physical Review Letters*, 78:2024–2027.

Geritz, S. A. H., van der Meijden, E., and Metz, J. A. J. 1999. Evolutionary dynamics of seed size and seedling competitive ability. *Theoretical Population Biology*, 55:324–343.

Ghirlanda, S. and Enquist, M. 2003. A century of generalization. *Animal Behaviour*, 66(1): 15–36.

Gilley, D. C. 2001. The behaviour of honey bees (*Apis mellifera ligustica*) during queen duels. *Ethology*, 107:601–622.

Gingins, S. and Bshary, R. 2016. The cleaner wrasse outperforms other labrids in ecologically relevant contexts, but not in spatial discrimination. *Animal Behaviour*, 115:145–155.

Giraldeau, L.-A. and Caraco, T. 2000. *Social foraging theory*. Princeton University Press, Princeton, NJ.

Giraldeau, L.-A. and Dubois, F. 2008. Social foraging and the study of exploitative behavior. *Advances in the Study of Behavior*, 38:59–104.

Giraldeau, L.-A., Heeb, P., and Kosfeld, M. 2017. *Investors and exploiters in ecology and economics*. MIT Press, Cambridge, MA.

Godfrey, H. C. J. 1994. *Parasitoids*. Princeton University Press, Princeton, NJ.

Gold, J. I. and Shadlen, M. N. 2007. The neural basis of decision making. *Annual Review of Neuroscience*, 30(1):535–574.

Grafen, A. 1979. The hawk-dove game played between relatives. *Animal Behaviour*, 27(3): 905–907.

Grafen, A. 1984. Natural selection, kin selection and group selection. In Krebs, J. and Davies, N., editors, *Behavioural ecology: an evolutionary approach, second edition*, pages 62–84. Blackwell, Oxford, UK.

Grafen, A. 1987. The logic of divisively asymmetric contests: respect for ownership and the desperado effect. *Animal Behaviour*, 35:462–467.

Grafen, A. 1990. Biological signals as handicaps. *Journal of Theoretical Biology*, 144:517–546.

Grafen, A. 1999. Formal Darwinism, the individual-as-maximizing-agent analogy and bet-hedging. *Proceedings of the Royal Society B*, 266:799–803.

Griggio, M., Matessi, G., and Pilastro, A. 2004. Should I stay or should I go? Female brood desertion and male counterstrategy in rock sparrows. *Behavioral Ecology*, 16:435–441.

Gross, M. R. 1996. Alternative reproductive strategies and tactics: diversity within sexes. *Trends in Ecology & Evolution*, 11:92–98.

Guhl, A. M. 1956. The social order of chickens. *Scientific American*, 194(2):42–47.

Hamilton, W. D. 1964. The genetical evolution of social behaviour. I. *Journal of Theoretical Biology*, 7:1–16.

Hamilton, W. D. 1967. Extraordinary sex ratios. *Science*, 156(3774):477–488.

Hamilton, W. D. and May, R. M. 1977. Dispersal in stable habitats. *Nature*, 269:578–581.

Hammerstein, P. 1981. The role of asymmetries in animal contests. *Animal Behaviour*, 29(1):193–205.

Hammerstein, P. and Noë, R. 2016. Biological trade and markets. *Philosophical Transactions of the Royal Society B*, 371:20150101. 10.1098/rstb.2015.0101

Hammerstein, P. and Riechert, S. E. 1988. Payoffs and strategies in territorial contests: ESS analyses of two ecotypes of the spider *Agelenopsis aperta*. *Evolutionary Ecology*, 2(2): 115–138.

Hardin, G. 1968. The tragedy of the commons. *Science*, 162(3859):1243–1248.

Harley, C. B. 1981. Learning the evolutionarily stable strategy. *Journal of Theoretical Biology*, 89(4):611–633.

Harrison, F., Barta, Z., Cuthill, I., and Székely, T. 2009. How is sexual conflict over parental care resolved? A meta-analysis. *Journal of Evolutionary Biology*, 22(9):1800–1812.

Herrnstein, R. J. 1970. On the law of effect. *Journal of the Experimental Analysis of Behavior*, 13(2):243–266.

Hesse, S., Anaya-Rojas, J. M., Frommen, J. G., and Thünken, T. 2015. Social deprivation affects cooperative predator inspection in a cichlid fish. *Royal Society Open Science*, 2:140451.

Hinde, C. A. 2006. Negotiation over offspring care? A positive response to partner-provisioning rate in great tits. *Behavioral Ecology*, 17(1):6–12.

Hochberg, M. E., Rankin, D. J., and Taborsky, M. 2008. The coevolution of cooperation and dispersal in social groups and its implications for the emergence of multicellularity. *BMC Evolutionary Biology*, 8:238.

Hofbauer, J. and Sigmund, K. 1998. *Evolutionary games and population dynamics*. Cambridge University Press, Cambridge, UK.

Hofbauer, J. and Sigmund, K. 2003. Evolutionary game dynamics. *Bulletin of the American Mathematical Society*, 40:479–519.

Holekamp, K. E. and Smale, L. 1993. Ontogeny of dominance in free-living spotted hyenas: juvenile rank relations with other immature individuals. *Animal Behaviour*, 46:451–466.

Hori, M. 1993. Frequency-dependent natural selection in the handedness of scale-eating cichlid fish. *Science*, 260(5105):216–219.

Houle, D. 1992. Comparing evolvability and variability. *Genetics*, 130(1):195–204.

Houston, A. I. and Davies, N. B. 1985. Evolution of cooperation and life history in dunnocks, *Prunella modularis*. In Sibly, R. and Smith, R. H., editors, *Behavioural ecology*, pages 471–487. Blackwell Scientific Publications, Oxford, UK.

Houston, A. I., Fawcett, T. W., Mallpress, D. E. W., and McNamara, J. M. 2014. Clarifying the relationship between prospect theory and risk-sensitive foraging theory. *Evolution and Human Behavior*, 35:502–507.

Houston, A. I. and McNamara, J. M. 1987. Singing to attract a mate—a stochastic dynamic game. *Journal of Theoretical Biology*, 129:57–68.

Houston, A. I. and McNamara, J. M. 1988. Fighting for food: a dynamic version of the hawk-dove game. *Evolutionary Ecology*, 2:51–64.

Houston, A. I. and McNamara, J. M. 1991. Evolutionarily stable strategies in the repeated hawk-dove game. *Behavioral Ecology*, 2:219–227.

Houston, A. I. and McNamara, J. M. 1999. *Models of adaptive behaviour: an approach based on state*. Cambridge University Press, Cambridge UK.

Houston, A. I. and McNamara, J. M. 2006. John Maynard Smith and the importance of consistency in evolutionary game theory. *Biology and Philosophy*, 20:933–950.

Houston, A. I., Székely, T., and McNamara, J. M. 2005. Conflict between parents over care. *Trends in Ecology & Evolution*, 20:33–38.

Hurd, P. L. 1995. Communication in discrete action-response games. *Journal of Theoretical Biology*, 174:217–222.

Hurd, P. L. and Enquist, M. 2005. A strategic taxonomy of biological communication. *Animal Behaviour*, 70:1155–1170.

Hurst, L. D. and Hamilton, W. D. 1992. Cytoplasmic fusion and the nature of sexes. *Philosophical Transactions of the Royal Society B*, 247:189–194.

Huttegger, S. M. 2017. *The probabilistic foundations of rational learning*. Cambridge University Press, Cambridge, UK.

Indermaur, A., Theis, A., Egger, B., and Salzburger, W. 2018. Mouth dimorphism in scale-eating cichlid fish from Lake Tanganyika advances individual fitness. *Evolution*, 72(9):1962–1969.

Iserbyt, A., Griffioen, M., Eens, M., and Müller, W. 2019. Enduring rules of care within pairs—how blue tit parents resume provisioning behaviour after experimental disturbance. *Scientific Reports*, 9(1):1–9.

Iwasa, Y., Cohen, D., and Leon, J. A. 1985. Tree height and crown shape, as results of competitive games. *Journal of Theoretical Biology*, 112:279–297.

Iwasa, Y., Pomiankowski, A., and Nee, S. 1991. The evolution of costly mate preferences II. The 'handicap' principle. *Evolution*, 45:1431–1442.

Jakobsson, S. 1987. *Male behaviour in conflicts over mates and territories*. PhD thesis, Department of Zoology, Stockholm University.

Jakobsson, S., Brick, O., and Kullberg, C. 1995. Escalated fighting behaviour incurs increased predation risk. *Animal Behaviour*, 49(1):235–239.

Jennions, M. D. and Fromhage, L. 2017. Not all sex ratios are equal: the Fisher condition, parental care and sexual selection. *Philosophical Transactions of the Royal Society B*, 372:20160312.

Johnstone, R. A. 2001. Eavesdropping and animal conflict. *Proceedings of the National Academy of Sciences of the USA*, 98:9177–9180.

Johnstone, R. A. and Grafen, A. 1993. Dishonesty and the handicap principle. *Animal Behaviour*, 46:759–764.

Johnstone, R. A. and Hinde, C. A. 2006. Negotiation over offspring care—how should parents respond to each other's efforts? *Behavioral Ecology*, 17(5):818–827.

Johnstone, R. A., Manica, A., Fayet, A. L., Stoddard, M. C., Rodriguez-Gironés, M. A., and Hinde, C. A. 2014. Reciprocity and conditional cooperation between great tit parents. *Behavioral Ecology*, 25(1):216–222.

Johnstone, R. A. and Norris, K. 1993. Badges of status and the cost of aggression. *Behavioral Ecology and Sociobiology*, 32(2):127–134.

Johnstone, R. A. and Savage, J. L. 2019. Conditional cooperation and turn-taking in parental care. *Frontiers in Ecology and Evolution*, 7:335.

Jones, M. and Love, B. C. 2011. Bayesian fundamentalism or enlightenment? On the explanatory status and theoretical contributions of Bayesian models of cognition. *Behavioral and Brain Sciences*, 34(4):169–188.

Joshi, J., Couzin, I. D., Levin, S. A., and Guttal, V. 2017. Mobility can promote the evolution of cooperation via emergent self-assortment dynamics. *PLoS Computational Biology*, 13:1005732.

Kacelnik, A. 2006. Meanings of rationality. In Nudds, M. and Hurley, S., editors, *Rational animals?*, pages 87–106. Oxford University Press, Oxford, UK.

Kandori, M. 1992. Social norms and community enforcement. *The Review of Economic Studies*, 59:63–80.

Kirmani, A. and Rao, A. R. 2000. No pain, no gain: a critical review of the literature on signaling unobservable product quality. *Journal of Marketing*, 64:66–79.

Kisdi, E. 2004. Conditional dispersal under kin competition: extension of the Hamilton-May model to brood-size dependent dispersal. *Theoretical Population Biology*, 66:369–380.

Kisdi, E. and Geritz, S. A. H. 1999. Adaptive dynamics in allele space: evolution of genetic polymorphism by small mutations in a heterogeneous environment. *Evolution*, 53:993–1008.

Kokko, H. 1999. Competition for early arrival in migratory birds. *Journal of Animal Ecology*, 68(5):940–950.

Kokko, H., Brooks, R., McNamara, J. M., and Houston, A. I. 2002. The sexual selection continuum. *Proceedings of the Royal Society B*, 269:1331–1340.

Kokko, H. and Jennions, M. D. 2008. Competition for early arrival in migratory birds. *Journal of Evolutionary Biology*, 21(4):919–948.

Kokko, H., Johnstone, R. A., and Clutton-Brock, T. H. 2001. The evolution of cooperative breeding through group augmentation. *Proceedings of the Royal Society B*, 268(1463): 187–196.

Kokko, H., Lopez-Sepulchre, A., and Morell, L. J. 2006. From Hawks and Doves to self-consistent games of territorial behaviour. *American Naturalist*, 167:901–912.

Komdeur, J., Daan, S., Tinbergen, J., and Mateman, C. 1997. Extreme adaptive modification in sex ratio of the Seychelles warbler's eggs. *Nature*, 385(6616):522–525.

Kotrschal, A., Rogell, B., Bundsen, A., Svensson, B., Zajitschek, S., Brännström, I., Immler, S., Maklakov, A. A., and Kolm, N. 2013. Artificial selection on relative brain size in the guppy reveals costs and benefits of evolving a larger brain. *Current Biology*, 23(2):168–171.

Krivan, V., Galanthay, T. E., and Cressman, R. 2018. Beyond replicator dynamics: from frequency to density dependent models of evolutionary games. *Journal of Theoretical Biology*, 455:232–248.

Kumaran, D., Banino, A., Blundell, C., Hassabis, D., and Dayan, P. 2016. Computations underlying social hierarchy learning: distinct neural mechanisms for updating and representing self-relevant information. *Neuron*, 92(5):1135–1147.

Küpper, C., Stocks, M., Risse, J. E., dos Remedios, N., Farrell, L. L., McRae, S. B., . . . Burke, T. 2016. A supergene determines highly divergent male reproductive morphs in the ruff. *Nature Genetics*, 48(1):79–83.

Lamichhaney, S., Fan, G., Widemo, F., Gunnarsson, U., Thalmann, D. S., Hoeppner, M. P., . . . Andersson, L. 2016. Structural genomic changes underlie alternative reproductive strategies in the ruff (*Philomachus pugnax*). *Nature Genetics*, 48(1):84–88.

Lande, R. 1981. Models of speciation by sexual selection on polygenic traits. *Proceedings of the National Academy of Science of the USA*, 78:3721–3725.

Langdon, A. J., Sharpe, M. J., Schoenbaum, G., and Niv, Y. 2018. Model-based predictions for dopamine. *Current Opinion in Neurobiology*, 49:1–7.

Lehtonen, J. and Kokko, H. 2011. Two roads to two sexes: unifying gamete competition and gamete limitation in a single model of anisogamy evolution. *Behavioral Ecology and Sociobiology*, 65:445–459.

Lehtonen, J. and Kokko, H. 2012. Positive feedback and alternative stable states in inbreeding, cooperation, sex roles and other evolutionary processes. *Philosophical Transactions of the Royal Society B*, 367:211–221.

Leimar, O. 1996. Life-history analysis of the Trivers and Willard sex-ratio problem. *Behavioral Ecology*, 7:316–325.

Leimar, O. 2001. Evolutionary change and Darwinian demons. *Selection*, 2:65–72.

Leimar, O. 2005. The evolution of phenotypic polymorphism: randomized strategies versus evolutionary branching. *The American Naturalist*, 165:669–681.

Leimar, O. 2009. Multidimensional convergence stability. *Evolutionary Ecology Research*, 11(2):191–208.

Leimar, O., Austad, S., and Enquist, M. 1991. A test of the sequential assessment game: fighting in the bowl and doily spider *Frontinella pyramitela*. *Evolution*, 45(4):862–874.

Leimar, O., Dall, S. R. X., McNamara, J. M., Kuijper, B., and Hammerstein, P. 2019. Ecological genetic conflict: genetic architecture can shift the balance between local adaptation and plasticity. *American Naturalist*, 193:70–80.

Leimar, O. and Enquist, M. 1984. Effects of asymmetries in owner-intruder conflicts. *Journal of Theoretical Biology*, 111(3):475–491.

Leimar, O. and Hammerstein, P. 2001. Evolution of cooperation through indirect reciprocity. *Proceedings of the Royal Society B*, 268:745–753.

Leimar, O. and Hammerstein, P. 2010. Cooperation for direct fitness benefits. *Philosophical Transactions of the Royal Society B*, 365(1553):2619–2626.

Leimar, O., Hammerstein, P., and Van Dooren, T. J. M. 2006. A new perspective on developmental plasticity and the principles of adaptive morph determination. *American Naturalist*, 167:367–376.

Leimar, O. and McNamara, J. M. 2019. Learning leads to bounded rationality and the evolution of cognitive bias in public goods games. *Scientific Reports*, 9(1):16319.

Leimar, O., Tullberg, B. S., and Mallet, J. 2012. Mimicry, saltational evolution, and the crossing of fitness valleys. In Svensson, E. I. and Calsbeek, R., editors, *The adaptive landscape in evolutionary biology*, pages 257–270. Oxford University Press, Oxford, UK.

Leimar, O. and Tuomi, J. 1998. Synergistic selection and graded traits. *Evolutionary Ecology*, 12:59–71.

Lessells, C. M., Snook, R. R., and Hosken, D. J. 2009. The evolutionary origin and maintenance of sperm: selection for a small, motile gamete mating type. In Birkhead, T. R., Hosken, D. J., and Pitnick, S., editors, *Sperm biology: an evolutionary perspective*, pages 43–67. Academic Press, Burlington, MA.

Levins, S. 1970. Community equilibria and stability, and an extension of the competitive exclusion principle. *American Naturalist*, 104:413–423.

Liker, A., Freckleton, R. P., and Székely, T. 2013. The evolution of sex roles in birds is related to adult sex ratio. *Nature Communications*, 4:1587.

Lim, M. M., Wang, Z., Olazábal, D. E., Ren, X., Terwilliger, E. F., and Young, L. J. 2004. Enhanced partner preference in a promiscuous species by manipulating the expression of a single gene. *Nature*, 429:754–757.

Livnat, A., Pacala, S. W., and Levin, S. A. 2005. The evolution of intergenerational discounting in offspring quality. *American Naturalist*, 165:311–321.

Lucas, J. R. and Howard, R. D. 1995. On alternative reproductive tactics in anurans: dynamic games with density and frequency dependence. *American Naturalist*, 146:365–397.

Lucas, J. R., Howard, R. D., and Palmer, J. G. 1996. Callers and satellites: chorus behaviour in anurans as a stochastic dynamic game. *Animal Behaviour*, 51:501–518.

Lukas, D. and Clutton-Brock, T. H. 2013. The evolution of social monogamy in mammals. *Science*, 341:526–530.

MacLean, E. L., Hare, B., Nun, C. L., Addess, E., Amic, F., Anderson, R. C., ... Zhao, Y. 2014. The evolution of self-control. *Proceedings of the National Academy of Sciences*, 111(20):E2140–E2148.

Mallet, J. 1999. Causes and consequences of a lack of coevolution in Müllerian mimicry. *Evolutionary Ecology*, 13(7–8):777–806.

Marshall, J. A. R., Trimmer, P. C., Houston, A. I., and McNamara, J. M. 2013. On evolutionary explanations of cognitive biases. *Trends in Ecology & Evolution*, 28(8):469–473.

Matsuda, H. 1995. Evolutionarily stable strategies for predator switching. *Journal of Theoretical Biology*, 115:351–366.

Matsuda, H. and Abrams, P. A. 1999. Why are equally sized gametes so rare? The instability of isogamy and the cost of anisogamy. *Evolutionary Ecology Research*, 1:769–784.

Maynard Smith, J. 1974. The theory of games and the evolution of animal conflict. *Journal of Theoretical Biology*, 47:209–221.

Maynard Smith, J. 1977. Parental investment: a prospective analysis. *Animal Behaviour*, 25:1–9.

Maynard Smith, J. 1982. *Evolution and the theory of games*. Cambridge University Press, Cambridge, UK.

Maynard Smith, J. 1984. Game theory and the evolution of behaviour. *The Behavioral and Brain Sciences*, 7:95–125.

Maynard Smith, J. 1988. Can a mixed strategy be stable in a finite population? *Journal of Theoretical Biology*, 130(2):247–251.

Maynard Smith, J. 1989. *Evolutionary genetics*. Oxford University Press, Oxford, UK.

Maynard Smith, J. and Harper, D. 2003. *Animal signals*. Oxford University Press, Oxford, UK.

Maynard Smith, J. and Parker, G. A. 1976. The logic of asymmetric animal contests. *Animal Behaviour*, 24(1):159–175.

Maynard Smith, J. and Price, G. R. 1973. The logic of animal conflict. *Nature*, 246:15–18.

McElreath, R. and Boyd, R. 2007. *Mathematical models of social evolution*. The University of Chicago Press, Chicago, IL.

McGraw, L. A. and Young, L. J. 2010. The prairie vole: an emerging model organism for understanding the social brain. *Trends in Neurosciences*, 33:103–109.

McNamara, J. M. 1985. An optimal sequential policy for controlling a Markov renewal process. *Journal of Applied Probability*, 22:324–335.

McNamara, J. M. 1990. The policy which maximises long term survival of an animal faced with the risks of starvation and predation. *Advances in Applied Probability*, 22:295–308.

McNamara, J. M. 1991. Optimal life histories: a generalisation of the Perron-Frobenius theorem. *Theoretical Population Biology*, 40:230–245.

McNamara, J. M. 1993. State-dependent life-history equations. *Acta Biotheoretica*, 41:165–174.

McNamara, J. M. 1994. Multiple stable age distributions in a population at evolutionary stability. *Journal of Theoretical Biology*, 169:349–354.

McNamara, J. M. 1995. Implicit frequency dependence and kin selection in fluctuating environments. *Evolutionary Ecology*, 9:185–203.

McNamara, J. M. 1998. Phenotypic plasticity in fluctuating environments: consequences of the lack of individual optimisation. *Behavioral Ecology*, 9:642–648.

McNamara, J. M. 2013. Towards a richer evolutionary game theory. *Journal of the Royal Society Interface*, 10: 20130544.

McNamara, J. M. and Barta, Z. in preparation. Behavioural flexibility and reputation formation.

McNamara, J. M., Barta, Z., Fromhage, L., and Houston, A. I. 2008. The coevolution of choosiness and cooperation. *Nature*, 451:189–192.

McNamara, J. M., Barta, Z., and Houston, A. I. 2004. Variation in behaviour promotes cooperation in the Prisoner's Dilemma game. *Nature*, 428:745–748.

McNamara, J. M., Binmore, K., and Houston, A. I. 2006a. Cooperation should not be assumed. *Trends in Ecology & Evolution*, 21:476–478.

McNamara, J. M. and Collins, E. J. 1990. The job search problem as an employer-candidate game. *Journal of Applied Probability*, 28:815–827.

McNamara, J. M. and Dall, S. R. X. 2011. The evolution of unconditional strategies via the 'multiplier effect'. *Ecology Letters*, 14:237–243.

McNamara, J. M. and Doodson, P. 2015. Reputation can enhance or suppress cooperation through positive feedback. *Nature Communications*, 6:7134.

McNamara, J. M., Forslund, P., and Lang, A. 1999a. An ESS model for divorce strategies in birds. *Philosophical Transactions of the Royal Society B*, 354:223–236.

McNamara, J. M., Gasson, C. E., and Houston, A. I. 1999b. Incorporating rule for responding into evolutionary games. *Nature*, 401:368–371.

McNamara, J. M. and Houston, A. I. 1986. The common currency for behavioural decisions. *American Naturalist*, 127:358–378.

McNamara, J. M. and Houston, A. I. 1990. State-dependent ideal free distributions. *Evolutionary Ecology*, 4:298–311.

McNamara, J. M. and Houston, A. I. 1992a. Evolutionary stable levels of vigilance as a function of group size. *Animal Behaviour*, 43:641–658.

McNamara, J. M. and Houston, A. I. 1992b. State-dependent life-history theory, and its implications for optimal clutch size. *Evolutionary Ecology*, 6:170–185.

McNamara, J. M. and Houston, A. I. 2005. If animals know their own fighting ability, the evolutionarily stable level of fighting is reduced. *Journal of Theoretical Biology*, 232:1–6.

McNamara, J. M. and Houston, A. I. 2009. Integrating function and mechanism. *Trends in Ecology & Evolution*, 24:670–675.

McNamara, J. M., Houston, A. I., Barta, Z., and Osorno, J.-L. 2003a. Should young ever be better off with one parent than with two? *Behavioral Ecology*, 14:301–310.

McNamara, J. M., Houston, A. I., Dos Santos, M. M., Kokko, H., and Brooks, R. 2003b. Quantifying male attractiveness. *Proceedings of the Royal Society B*, 270(1527):1925–1932.

McNamara, J. M., Houston, A. I., Székely, T., and Webb, J. N. 2002. Do parents make independent decisions about desertion? *Animal Behaviour*, 64:147–149.

McNamara, J. M., Stephens, P. A., Dall, S. R. X., and Houston, A. I. 2009. Evolution of trust and trustworthiness: social awareness favours personality differences. *Proceedings of the Royal Society B*, 276:605–613.

McNamara, J. M., Székely, T., Webb, J. N., and Houston, A. I. 2000. A dynamic game-theoretical model of parental care. *Journal of Theoretical Biology*, 205:605–623.

McNamara, J. M., Trimmer, P., Eriksson, A., Marshall, J., and Houston, A. I. 2011. Environmental variability can select for optimism or pessimism. *Ecology Letters*, 14:58–62.

McNamara, J. M., Webb, J. N., Collins, E. J., Székely, T., and Houston, A. I. 1997. A general technique for computing evolutionarily stable strategies based on errors in decision-making. *Journal of Theoretical Biology*, 189:211–225.

McNamara, J. M., Wilson, E., and Houston, A. I. 2006b. Is it better to give information, receive it or be ignorant in a two-player game? *Behavioral Ecology*, 17:441–451.

McNamara, J. M. and Wolf, M. 2015. Sexual conflict over parental care promotes the evolution of sex differences in care and the ability to care. *Proceedings of the Royal Society B*, 282. 10.1098/rspb.2014.2752

Meszéna, G., Czibula, I., and Geritz, S. A. H. 1997. Adaptive dynamics in a 2-patch environment: a toy model for allopatric and parapatric speciation. *Journal of Biological Systems*, 5:265–284.

Meszéna, G., Gyllenberg, M., Pásztor, L., and Metz, J. A. J. 2006. Competitive exclusion and limiting similarity: a unified theory. *Theoretical Population Biology*, 69:68–87.

Metz, J. A, J., Nisbet, R. M., and Geritz, S. A. H. 1992. How should we define 'fitness' for general ecological scenarios? *Trends in Ecology & Evolution*, 7:198–202.

Metz, J. A. J. and Leimar, O. 2011. A simple fitness proxy for structured populations with continuous traits, with case studies on the evolution of haplo-diploids and genetic dimorphisms. *Journal of Biological Dynamics*, 5(2):163–190.

Metz, J. A. J., Geritz, S. A. H., Meszena, G., Jacobs, F. J. A., and Van Heerwaarden, J. S. 1996. Adaptive dynamics, a geometrical study of the consequences of near faithful reproduction. In Van Strien, S. J. and Verduyn Lunel, S. M., editors, *Stochastic and spatial structures of dynamical systems*, pages 183–231. North Holland, Dordrecht, the Netherlands.

Mock, D. W., Drummond, H., and Stinson, C. H. 1990. Avian siblicide. *American Scientist*, 78(5):438–449.

Molesti, S. and Majolo, B. 2017. Evidence of direct reciprocity, but not of indirect and generalized reciprocity, in the grooming exchanges of wild Barbary macaques (*Macaca sylvanus*). *American Journal of Primatology*, 79. 10.1002/ajp.22679

Möller, M. and Bogacz, R. 2019. Learning the payoffs and costs of actions. *PLoS Computational Biology*, 15(2):1–32.

Morozov, A. and Best, A. 2012. Predation on infected host promotes evolutionary branching of virulence and pathogens' biodiversity. *Journal of Theoretical Biology*, 307:29–36.

Müller, F. 1878. Über die Vortheile der Mimcry bei Schmetterlingen. *Zoologischer Anzeiger*, 1:54–55.

Mullon, C., Keller, L., and Lehmann, L. 2018. Co-evolution of dispersal with social behaviour favours social polymorphism. *Nature Ecology and Evolution*, 2:132–140.

Munger, S. D., Leinders-Zufall, T., McDougall, L. M., Cockerham, R. E., Schmid, A., Wandernoth, P., ... Kelliher, K. R. 2010. An olfactory subsystem that detects carbon disulfide and mediates food-related social learning. *Current Biology*, 20(16):1438–1444.

Murai, M., Backwell, P. R. Y., and Jennions, M. D. 2009. The cost of reliable signaling: experimental evidence for predictable variation among males in a cost-benefit trade-off between sexually selected traits. *Evolution*, 63(9):2363–2371.

Mylius, S. D. and Diekmann, O. 1995. On evolutionarily stable life histories, optimization and the need to be specific about density dependence. *Oikos*, 74:218–224.

Nesse, R. M. 2001. *Evolution and the capacity for commitment*. Russell Sage Foundation, New York.

Noë, R. and Hammerstein, P. 1994. Biological markets: supply and demand determine the effect of partner choice in cooperation, mutualism and mating. *Behavioral Ecology and Sociobiology*, 35:1–11.

Nowak, M. 1990. An evolutionarily stable strategy may be inaccessible. *Journal of Theoretical Biology*, 142:237–241.

Nowak, M. and Sigmund, K. 1998. The dynamics of indirect reciprocity. *Journal of Theoretical Biology*, 194:561–574.

O'Connell, L. A. and Hofmann, H. A. 2011. The vertebrate mesolimbic reward system and social behavior network: a comparative synthesis. *Journal of Comparative Neurology*, 519(18):3599–3639.

Ohtsuki, H. and Iwasa, Y. 2004. How should we define goodness? Reputation dynamics in indirect reciprocity. *Journal of Theoretical Biology*, 231:107–120.

Ohtsuki, H. and Iwasa, Y. 2006. The leading eight: social norms that can maintain cooperation by indirect reciprocity. *Journal of Theoretical Biology*, 239:435–444.

Ohtsuki, H. and Iwasa, Y. 2007. Global analyses of evolutionary dynamics and exhaustive search for social norms that maintain cooperation by reputation. *Journal of Theoretical Biology*, 244:518–531.

Panchanathan, K. and Boyd, R. 2003. A tale of two defectors: the importance of standing for evolution of indirect reciprocity. *Journal of Theoretical Biology*, 224:115–126.

Pangallo, M., Heinrich, T., and Farmer, J. D. 2019. Best reply structure and equilibrium convergence in generic games. *Science Advances*, 5(2):eaat1328.

Parker, G. A. 1974. Assessment strategy and the evolution of fighting behaviour. *Journal of Theoretical Biology*, 47(1):223–243.

Parker, G. A., Baker, R. R., and Smith, V. G. F. 1972. The origin and evolution of gamete dimorphism and the male-female phenomenon. *Journal of Theoretical Biology*, 36:529–553.

Parker, G. A. and Milinski, M. 1997. Cooperation under predation risk: a data-based ESS analysis. *Proceedings of the Royal Society B*, 264:1239–1247.

Parker, G. A., Mock, D. W., and Lamey, T. C. 1989. How selfish should stronger sibs be? *The American Naturalist*, 133(6):846–868.

Pen, I., Uller, T., Feldmeyer, B., Harts, A., While, G. M., and Wapstra, E. 2010. Climate-driven population divergence in sex-determining systems. *Nature*, 468:436–438.

Pen, I. and Weissing, F. J. 2002. Optimal sex allocation: steps towards a mechanistic theory. In Hardy, I. C. W., editor, *Sex ratios: concepts and research methods*, chapter 2, pages 26–45. Cambridge University Press, Cambridge UK.

Persson, O. and Ohrstrom, P. 1989. A new avian mating system: ambisexual polygamy in the penduline tit Remiz pendulinus. *Ornis Scandinavica*, 20:105–111.

Pitcher, T. A., Gree, D. A., and Magurran, A. E. 1986. Dicing with death: predator inspection behaviour in minnow shoals. *Journal of Fish Biology*, 28:439–448.

Pomiankowski, A., Iwasa, Y., and Nee, S. 1991. The evolution of costly mate preferences 1. Fisher and biased mutation. *Evolution*, 45:1422–1430.

Price, K. and Boutin, S. 1993. Territorial bequeathal by red squirrel mothers. *Behavioral Ecology*, 4:144–150.

Pryke, S. and Andersson, S. 2005. Experimental evidence for female choice and energetic costs of male tail elongation in red-collared widowbirds. *Biological Journal of the Linnean Society*, 86:35–43.

Puterman, M. L. 2005. *Markov decision processes: discrete stochastic dynamic programming*. Wiley, NJ.

Qu, C. and Dreher, J. C. 2018. Sociobiology: changing the dominance hierarchy. *Current Biology*, 28(4):R167–R169.

Queller, D. C. 1997. Why do females care more than males? *Proceedings of the Royal Society B*, 264:1555–1557.

Quiñones, A. E., Lotem, A., Leimar, O., and Bshary, R. 2020. Reinforcement learning theory reveals the cognitive requirements for solving the cleaner fish market task. *American Naturalist*, 195(4):664–677.

Quiñones, A. E., van Doorn, G. S., Pen, I., Weissing, F. J., and Taborsky, M. 2016. Negotiation and appeasement can be more effective drivers of sociality than kin selection. *Philosophical Transactions of the Royal Society B*, 371(1687):20150089.

Raffini, F. and Meyer, A. 2019. A comprehensive overview of the developmental basis and adaptive significance of a textbook polymorphism: head asymmetry in the cichlid fish *Perissodus microlepis*. *Hydrobiologia*, 832(1):65–84.

Rands, S. A., Cowlishaw, G., Pettifor, R. A., Rowcliffe, J. M., and Johnstone, R. A. 2003. Spontaneous emergence of leaders and followers in foraging pairs. *Nature*, 423:432–434.

Ratledge, C. and Dover, L. 2000. Iron metabolism in pathogenic bacteria. *Annual Review of Microbiology*, 54:881–941.

Reader, S. M. 2015. Causes of individual differences in exploration and search. *Topics in Cognitive Science*, 7:451–468.

Remeš, V., Freckleton, R. P., Tökölyi, J., Liker, A., and Székely, T. 2015. The evolution of parental cooperation in birds. *Proceedings of the National Academy of Sciences of the USA*, 112.

Ridley, M. 1995. *Animal Behavior, an introduction to behavioral mechanisms, development and ecology, second edition*. Blackwell Science Ltd, Oxford, UK.

Rillich, J., Schildberger, K., and Stevenson, P. A. 2007. Assessment strategy of fighting crickets revealed by manipulating information exchange. *Animal Behaviour*, 74(4):823–836.

Roberts, G. 1998. Competitive altruism: from reciprocity to the handicap principle. *Proceedings of the Royal Society B*, 265:427–431.

Roelfsema, P. R. and Holtmaat, A. 2018. Control of synaptic plasticity in deep cortical networks. *Nature Reviews Neuroscience*, 19(3):166–180.

Roff, D. A. 1996. The evolution of threshold traits in animals. *The Quarterly Review of Biology*, 71(1):3–35.

Roff, D. A. and Fairbairn, D. J. 2007. The evolution and genetics of migration in insects. *BioScience*, 57(2):155–164.

Rohwer, S. 1975. The social significance of avian winter plumage variability. *Evolution*, 29(4):593–610.

Roth, A. E. and Erev, I. 1995. Learning in extensive-form games: experimental data and simple dynamic models in the intermediate term. *Games and Economic Behavior*, 8(1):164–212.

Roughgarden, J. 1978. Coevolution in ecological systems iii. Co-adaptation and equilibrium population size. In Brussard, P. F., editor, *Ecological genetics*, pages 27–48. Springer-Verlag, Berlin.

Roughgarden, J., Oishi, M., and Akçay, E. 2006. Reproductive social behavior: cooperative games to replace sexual selection. *Science*, 311:965–969.

Royle, N. J., Alonzo, S. H., and Moore, A. J. 2016. Co-evolution, conflict and complexity: what have we learned about the evolution of parental care behaviours? *Current Opinion in Behavioral Sciences*, 12:30–36.

Rubenstein, D. R. and Alcock, J. 2018. *Animal behavior, eleventh edition*. Oxford University Press, Oxford, UK.

Rutte, C. and Taborsky, M. 2007. Generalized reciprocity in rats. *PLoS Biology*, 5(7):1421–1425.

Ruxton, G. D., Franks, D. W., Balogh, A. C. V., and Leimar, O. 2008. Evolutionary implications of the form of predator generalization for aposematic signals and mimicry in prey. *Evolution*, 62(11):2913–2921.

Sachs, J. L., Mueller, U. G., Wilcox, T. P., and Bull, J. J. 2004. The evolution of cooperation. *The Quarterly Review of Biology*, 79(2):135–160.

Sánchez-Tójar, A., Nakagawa, S., Sánchez-Fortún, M., Martin, D. A., Ramani, S., Girndt, A., Bókony, V., Kempenaers, B., Liker, A., Westneat, D. F., Burke, T., and Schroeder, J. 2018. Meta-analysis challenges a textbook example of status signalling and demonstrates publication bias. *eLife*, 7:1–26.

Santos, F. P., Santos, F. C., and Pacheco, J. M. 2018. Social norm complexity and past reputations in the evolution of cooperation. *Nature*, 555:242–245.

Sanz, J. J., Kranenbarg, S., and Tinbergen, J. M. 2000. Differential response by males and females to manipulation of partner in the great tit (*Parus major*). *Journal of Animal Ecology*, 69(1): 74–84.

Savage, L. J. 1972. *The Foundation of Statistics*. Dover Publications, Inc., New York.

Schindler, S., Gaillard, J. M., Grüning, A., Neuhaus, P., Traill, L. W., Tuljapurkar, S., and Coulson, T. 2015. Sex-specific demography and generalization of the Trivers-Willard theory. *Nature*, 526(7572):249–252.

Schjelderup-Ebbe, T. 1922. Beiträge zur Sozialpsychologie des Haushuhns. *Zeitschrift für Psychologie*, 88:225–252.

Schwanz, L. E., Cordero, G. A., Charnov, E. L., and Janzen, F. J. 2016. Sex-specific survival to maturity and the evolution of environmental sex determination. *Evolution*, 70(2):329–341.

Schweinfurth, M. K., Aeschbacher, J., Santi, M., and Taborsky, M. 2019. Male Norway rats cooperate according to direct but not generalized reciprocity rules. *Animal Behaviour*, 152:93–101.

Selten, R. 1980. A note on evolutionarily stable strategies in asymmetric animal conflicts. *Journal of Theoretical Biology*, 84:93–101.

Selten, R. 1983. Evolutionary stability in extensive 2-person games. *Mathemtical Social Sciences*, 5:269–363.

Seneta, E. 2006. *Non-negative matrices and Markov chains*. Springer, New York.

Shaw, R. F. and Mohler, J. D. 1953. The selective significance of the sex ratio. *The American Naturalist*, 87(837):337–342.

Sheldon, B. C. and West, S. A. 2004. Maternal dominance, maternal condition, and offspring sex ratio in ungulate mammals. *American Naturalist*, 163(1):40–54.

Sherratt, T. N. and Mesterton-Gibbons, M. 2015. The evolution of respect for property. *Journal of Evolutionary Biology*, 28(6):1185–1202.

Shettleworth, S. J. 2010. *Cognition, evolution, and behavior, second edition*. Oxford University Press, Oxford, UK.

Sih, A., Bell, A., and Johnson, J. C. 2004. Behavioral syndromes: an ecological and evolutionary overview. *Trends in Ecology & Evolution*, 19:372–378.

Silk, J. B., Kaldor, E., and Boyd, R. 2000. Cheap talk about conflicts. *Animal Behavior*, 59: 423–432.

Smith, H. G. and Härdling, R. 2000. Clutch size evolution under sexual conflict enhances the stability of mating systems. *Proceedings of the Royal Society B*, 267:2163–2170.

Smith, R. J. and Moore, F. R. 2005. Arrival timing and seasonal reproductive performance in a long-distance migratory landbird. *Behavioral Ecology and Sociobiology*, 57(3):231–239.

Spence, M. 1973. Job market signaling. *Quarterly Journal of Economics*, 87:355–374.

Staddon, J. 2016. *Adaptive behavior and learning*. Cambridge University Press, Cambridge UK.

Stephens, D. W. and Krebs, J. R. 1986. *Foraging theory*. Princeton University Press, Princeton, NJ.

Stephens, D. W., McLinn, C. M., and Stevens, J. R. 2002. Discounting and reciprocity in an iterated prisoner's dilemma. *Science*, 298(5601):2216–2218.

Stevens, J. R. and Stephens, D. W. 2004. The economic basis of cooperation: tradeoffs between selfishness and generosity. *Behavioral Ecology*, 15(2):255–261.

Sugden, R. 1986. *The economics of rights, co-operation and welfare*. Basil Blackwell, Oxford, UK.

Sutton, R. S. and Barto, A. G. 2018. *Reinforcement learning: an introduction, second edition*. MIT Press, Cambridge, MA.

Számadó, S. 1999. The validity of the handicap principle in discrete action-response games. *Journal of Theoretical Biology*, 198:593–602.

Számadó, S. 2011. The cost of honesty and the fallacy of the handicap principle. *Animal Behaviour*, 81:3–10.

Taborsky, B. and Oliveira, R. F. 2012. Social competence: an evolutionary approach. *Trends in Ecology & Evolution*, 27(12):679–688.

Takada, T. and Kigami, J. 1991. The dynamic attainability of ESS in evolutionary games. *Journal of Mathematical Biology*, 29:513–529.

Takeuchi, Y. and Oda, Y. 2017. Lateralized scale-eating behaviour of cichlid is acquired by learning to use the naturally stronger side. *Scientific Reports*, 7(1):8984.

Taylor, H. M., Gourley, R. S., Lawrence, C. E., and Kaplan, R. S. 1974. Natural selection of life history attributes: an analytic approach. *Theoretical Population Biology*, 5:104–122.

Taylor, P. D. 1990. Allele-frequency change in a class-structured population. *American Naturalist*, 135:95–106.

Taylor, P. D. and Bulmer, M. G. 1980. Local mate competition and the sex ratio. *Journal of Theoretical Biology*, 86(3):409–419.

Taylor, P. D. and Day, T. 2004. Stability in negotiation games and the emergence of cooperation. *Proceedings of the Royal Society B.*, 271:669–674.

Taylor, P. D. and Frank, S. A. 1996. How to make a kin selection model. *Journal of Theoretical Biology*, 180:27–37.

Taylor, P. D., Wild, G., and Gardner, A. 2007. Direct fitness or inclusive fitness: how shall we model kin selection? *Journal of Evolutionary Biology*, 20(1):301–309.

Taylor, P. W. and Elwood, R. W. 2003. The mismeasure of animal contests. *Animal Behaviour*, 65(6):1–8.

Templeton, J. J., Kamil, A. C., and Balda, R. P. 1999. Sociality and social learning in two species of corvids: the pinyon jay (*Gymnorhinus cyanocephalus*) and the Clark's nutcracker (*Nucifraga columbiana*). *Journal of Comparative Psychology*, 113(4):450–455.

Tinbergen, N. 1963. On aims and methods of ethology. *Zeitschrift für Tierpsychologie*, 20:410–433.

Triki, Z., Levorato, E., McNeely, W., Marshall, J., and Bshary, R. 2019. Population densities predict forebrain size variation in the cleaner fish *Labroides dimidiatus*. *Proceedings of the Royal Society B*, 286(1915):20192108.

Triki, Z., Wismer, S., Levorato, E., and Bshary, R. 2018. A decrease in the abundance and strategic sophistication of cleaner fish after environmental perturbations. *Global Change Biology*, 24(1):481–489.

Trimmer, P. C., Paul, E. S., Mendl, M. T., McNamara, J. M., and Houston, A. I. 2013. On the evolution and optimality of mood states. *Behavioral Sciences*, 3(3):501–521.

Trivers, R. L. 1972. Parental investment and sexual selection. In Cambell, B., editor, *Sexual selection and the descent of man*, pages 136–179. Aldine Publishing Company, Chicago, IL.

Trivers, R. L. 1974. Parent-offspring conflict. *American Zoologist*, 14(1):249–264.

Trivers, R. L. and Willard, D. E. 1973. Natural selection of parental ability to vary the sex ratio of offspring. *Science*, 179:90–92.

Ueshima, R. and Asami, T. 2003. Single-gene speciation by left-right reversal. *Nature*, 425:679.

Valera, F., Hoi, H., and Schleicher, B. 1997. Egg burial in penduline tits, *Remiz pendulinus*: its role in mate desertion and female polyandry. *Behavioral Ecology*, 8:20–27.

van Dijk, R., Szentirmai, I., Komdeur, J., and Székely, T. 2007. Sexual conflict over care in penduline tits: the process of clutch desertion. *Ibis*, 149:530–534.

Van Dooren, T. J., Van Goor, H. A., and Van Putten, M. 2010. Handedness and asymmetry in scale-eating cichlids: antisymmetries of different strength. *Evolution*, 64(7):2159–2165.

van Doorn, G. S., Edelaar, P., and Weissing, F. J. 2009. On the origin of species by natural and sexual selection. *Science*, 326:1704–1707.

van Doorn, G. S., Weissing, F. J., and Hengeveld, G. M. 2003. The evolution of social dominance II: Multi-player models. *Behaviour*, 140(10):1333–1358.

van Noort, S., Wang, R., and Compton, S. G. 2013. Fig wasps (Hymenoptera: Chalcidoidea: Agaonidae, Pteromalidae) associated with Asian fig trees (*Ficus*, Moraceae) in southern Africa: Asian followers and African colonists. *African Invertebrates*, 54(2):381–400.

van Tienderen, P. H. and de Jong, H. 1986. Sex ratio under the haystack model: polymorphism may occur. *Journal of Theoretical Biology*, 122:69–81.

Varela, S. A., Teles, M. C., and Oliveira, R. F. 2020. The correlated evolution of social competence and social cognition. *Functional Ecology*, 34:332–343.

Veller, C., Haig, D., and Nowak, M. A. 2016. The Trivers–Willard hypothesis: sex ratio or investment? *Proceedings of the Royal Society B*, 283(1830).

Vickers, D. 1970. Evidence for an accumulator model of psychophysical discrimination. *Ergonomics*, 13(1):37–58.

Warner, D. A. and Shine, R. 2008. The adaptive significance of temperature-dependent sex determination in a reptile. *Nature*, 451(7178):566–568.

Warren, I. A., Gotoh, H., Dworkin, I. M., Emlen, D. J., and Lavine, L. C. 2013. A general mechanism for conditional expression of exaggerated sexually-selected traits. *BioEssays*, 35(10):889–899.

Webb, J. N., Houston, A. I., McNamara, J. M., and Székely, T. 1999. Multiple patterns of parental care. *Animal Behaviour*, 58:983–999.

Weibull, J. W. 1995. *Evolutionary game theory*. MIT Press, Cambridge, MA.

West, S. 2009. *Sex allocation*. Princeton University Press. Princeton, NJ.

West, S. and Buckling, A. 2003. Cooperation, virulence and siderophore production in parasitic bacteria. *Proceedings of the Royal Society B*, 270:37–44.

West, S. A., Griffin, A. S., Gardner, A., and Diggle, S. P. 2006. Social evolution theory for microorganisms. *Nature Reviews Microbiology*, 4(8):597–607.

West, S. A., Murray, M. G., Machado, C. A., Griffin, A. S., and Herre, E. A. 2001. Testing Hamilton's rule with competition between relatives. *Nature*, 409:510–513.

West, S. A. and Sheldon, B. C. 2002. Constraints in the evolution of sex ratio adjustment. *Science*, 295(5560):1685–1688.

Wey, T. W., Spiegel, O., Montiglio, P.-O., and Mabry, K. E. 2015. Natal dispersal in a social landscape: considering individual behavioral phenotypes and social environment in dispersal ecology. *Current Zoology*, 61:543–556.

Wilkinson, G. S., Carter, G. G., Bohn, K. M., and Adams, D. M. 2016. Non-kin cooperation in bats. *Philosophical Transactions of the Royal Society B*, 371:20150095.

Wise, R. A. 2004. Dopamine, learning and motivation. *Nature Reviews Neuroscience*, 5(6): 483–494.

Wismer, S., Pinto, A. I., Vail, A. L., Grutter, A. S., and Bshary, R. 2014. Variation in cleaner wrasse cooperation and cognition: influence of the developmental environment? *Ethology*, 120(6):519–531.

Wolf, M., Van Doorn, G., and Weissing, F. J. 2011. On the coevolution of social responsiveness and behavioural consistency. *Proceedings of the Royal Society B*, 278:440–448.

Wolf, M., van Doorn, G. S., Leimar, O., and Weissing, F. J. 2007. Life-history trade-offs favour the evolution of animal personalities. *Nature*, 447(7144):581–584.

Wolf, M. and Weissing, F. J. 2010. An explanatory framework for adaptive personality differences. *Philosophical Transactions of the Royal Society B*, 365(1560):3959–3968.

Wood, R. I., Kim, J. Y., and Li, G. R. 2016. Cooperation in rats playing the iterated Prisoner's Dilemma game. *Animal Behaviour*, 114:27–35.

Young, L. J. 2003. The neural basis of pair bonding in a monogamous species: a model for understanding the biological basis of human behavior. In Wachter, K. and Bulatao, R. A., editors, *Offspring: human fertility behavior in biodemographic perspective*. National Academies Press (US), Washington (DC).

Zahavi, A. 1975. Mate selection—a selection for a handicap. *Journal of Theoretical Biology*, 53:205–214.

Zahavi, A. 1977. The cost of honesty (further remarks on the handicap principle). *Journal of Theoretical Biology*, 67:603–605.

Zhou, T., Sandi, C., and Hu, H. 2018. Advances in understanding neural mechanisms of social dominance. *Current Opinion in Neurobiology*, 49:99–107.

Index